# SEISMIC
# STRONG MOTION
# SYNTHETICS

This is Volume 4 in

COMPUTATIONAL TECHNIQUES

Edited by BERNI J. ALDER and SIDNEY FERNBACH

# SEISMIC STRONG MOTION SYNTHETICS

*Edited by*   **BRUCE A. BOLT**

Seismographic Station
Department of Geology and Geophysics
University of California
Berkeley, California

1987

ACADEMIC PRESS, INC.
**Harcourt Brace Jovanovich, Publishers**
Orlando   San Diego   New York   Austin
Boston   London   Sydney   Tokyo   Toronto

COPYRIGHT © 1987 BY ACADEMIC PRESS, INC.
ALL RIGHTS RESERVED.
NO PART OF THIS PUBLICATION MAY BE REPRODUCED OR
TRANSMITTED IN ANY FORM OR BY ANY MEANS, ELECTRONIC
OR MECHANICAL, INCLUDING PHOTOCOPY, RECORDING, OR
ANY INFORMATION STORAGE AND RETRIEVAL SYSTEM, WITHOUT
PERMISSION IN WRITING FROM THE PUBLISHER.

ACADEMIC PRESS, INC.
Orlando, Florida 32887

*United Kingdom Edition published by*
ACADEMIC PRESS INC. (LONDON) LTD.
24–28 Oval Road, London NW1 7DX

Library of Congress Cataloging in Publication Data

Seismic strong motion synthetics.

(Computational techniques)
Includes bibliographies and index.
1. Earthquake engineering—Data processing.
2. Earth movements—Mathematical models.  I. Bolt,
Bruce A.  II. Title.  III. Series.
TA654.6.S454  1987     624.1'762      86-20650
ISBN 0–12–112251–4 (alk. paper)

PRINTED IN THE UNITED STATES OF AMERICA

87 88 89 90      9 8 7 6 5 4 3 2 1

# Contents

PREFACE   vii

## Chapter 1   Asymptotic Modeling of Strong Ground Motion Excited by Subsurface Sliding Events
*J. D. Achenbach and John G. Harris*

| | | |
|---|---|---|
| I. | Introduction | 1 |
| II. | Dynamic Fracture Mechanics | 5 |
| III. | Canonical Problems | 12 |
| IV. | Near-Field Ground Motions | 21 |
| V. | Far-Field Ground Motions | 33 |
| VI. | Summary | 48 |
| | Appendix A. The Rayleigh-Wave Components of the Green's Tensor | 49 |
| | Appendix B. Evaluation of the Representation Integral | 50 |
| | References | 52 |

## Chapter 2   Array Analysis and Synthesis Mapping of Strong Seismic Motion
*Norman A. Abrahamson and Bruce A. Bolt*

| | | |
|---|---|---|
| I. | Prediction of Strong Ground Motion | 55 |
| II. | Signal Estimation and Detection | 63 |
| III. | Wave Coherence | 71 |
| IV. | Synthesis Mapping | 76 |
| V. | Computational Results | 80 |
| | References | 88 |

## Chapter 3   Numerical Modeling of Realistic Fault Rupture Processes
*T. Mikumo and T. Miyatake*

| | | |
|---|---|---|
| I. | Introduction | 91 |
| II. | Synthesis from Kinematic Dislocation Models | 94 |
| III. | Three-Dimensional Modeling of Spontaneous Fault Rupture Processes | 108 |

| | | |
|---|---|---|
| IV. | Summary | 146 |
| | References | 148 |

## Chapter 4  Complete Strong Motion Synthetics
*G. F. Panza and P. Suhadolc*

| | | |
|---|---|---|
| I. | Introduction | 153 |
| II. | Automatic Computation of Eigenvalues and Eigenfunctions | 157 |
| III. | Examples of Computation of Synthetic Signals | 170 |
| IV. | Comparison with Real Data | 188 |
| V. | Conclusions | 201 |
| | References | 202 |

## Chapter 5  Techniques for Earthquake Ground-Motion Calculation with Applications to Source Parameterization to Finite Faults
*Paul Spudich and Ralph J. Archuleta*

| | | |
|---|---|---|
| I. | Introduction | 205 |
| II. | The Fault Surface Integral | 212 |
| III. | Examples of Finite Fault Calculations | 247 |
| | References | 263 |

## Chapter 6  Path Effects in Strong Motion Seismology
*John E. Vidale and Donald V. Helmberger*

| | | |
|---|---|---|
| I. | Introduction | 267 |
| II. | Explosion Sources | 269 |
| III. | Earthquake Sources | 289 |
| IV. | Earthquake Applications | 306 |
| V. | Summary | 316 |
| | References | 317 |

INDEX 321

# Preface

Earthquakes continue to present a hazard to man and engineered structures. Because seismically resistant design is critically dependent upon the ground motions which are used in the structural analysis, earthquake engineers need from seismologists realistic ground accelerations that are site specific. The main difficulty is that the sample seismic ground motions available from field recordings are still relatively small. Indeed, in no country has a substantial suite of records been obtained in the near field of an earthquake of magnitude greater than 8. Earthquake sources are also of various types and depths, and occur in various geological structures, so that much interpolation and extrapolation between recorded ground motions is needed. For these reasons, seismologists have endeavored recently to compute ground motions for specified seismic source types and geological conditions. The flourishing of this part of seismology has been much assisted by the great improvement in computing power for large numerical models and display.

The chapters presented in this volume have been chosen to represent modern methods of modeling the production of strong seismic ground motions by realistic seismic sources. The emphasis is on the different ways of numerical treatment and the available computationally rapid and conceptually simple algorithms.

Study of the chapters and references included will provide readers with an understanding of contemporary strong motion prediction methods. The reader should be struck with the rapid progress in ground motion synthetic modeling and how close synthetics now come to explaining observed strong ground motion (at least for earthquakes up to about magnitude 7). It is no longer computationally prohibitive to extend the frequency range of included seismic waves down to about 2 Hz. The material presented indicates that further progress may now be expected in widening the frequency band to 10 Hz, which covers the main interest of structural engineers.

Specifically, it is demonstrated that reasonably accurate synthetic seismograms can be calculated, using either appropriate Green's functions or records of small earthquakes as impulse responses. In this way, large earthquakes can be modeled by assuming an extended source of predetermined dimension and with propagation of rupture in arbitrary directions. The synthetics now may include all types of body and surface waves, including waves converted from crustal structure. The basic modeling techniques include finite-element source construction, wave number integration, and superposition of normal modes. Comparisons between the methods of generating seismic signals in the near field clearly indicate that many basic research problems remain, including the incorporation of fracture dynamics associated with the crack initiation, the realistic treatment of attenuation of seismic energy, and the source of the seismic coda waves. Some of these problems will be made more amenable through analysis of additional observations, using the high resolving power provided by strong motion arrays. These seismic arrays have the promise of disentangling the mix of seismic phases present in strong ground accelerograms and determining the appropriate partition functions for fault ruptures of various geometries and tectonic regimes.

The overall aim of the book, although not covering completely all seismological work on strong ground motion, is to provide an advanced text on numerical modeling for use in graduate and upper division courses in physics, geophysics, and earthquake engineering.

CHAPTER 1

# Asymptotic Modeling of Strong Ground Motion Excited by Subsurface Sliding Events

J. D. ACHENBACH

*Department of Civil Engineering*
*Northwestern University*
*Evanston, Illinois 60201*

JOHN G. HARRIS

*Department of Theoretical and Applied Mechanics*
*University of Illinois*
*Urbana, Illinois 61801*

## I. Introduction

At their low-frequency ends, spectra of seismic waves from earthquakes are primarily controlled by the gross aspects of sliding over the fault plane. Such specific phases of the sliding event as stopping and starting tend to manifest themselves at the high-frequency end of the spectrum. In accordance with these observations, the theories of Aki (1969, 1972) and Brune (1970) predict that the far-field spectral amplitude of the displacement is flat and proportional to the seismic moment at low frequencies and falls off as a negative power of the frequency at high frequencies. The corner frequency $f_0$, which is the transition between these trends, depends on the duration or length of the earthquake rupture. For large earthquakes, frequencies of engineering interest are often above the corner frequency. In the recent literature, the maximum observable frequency $f_{max}$ has been discussed by Hanks and McGuire (1981) and Hanks (1982). It is not yet clear whether the $f_{max}$ is limited by source mechanisms, by attenuation, or by the limitations of strong motion instruments.

In this chapter interest is centered upon the intense bursts of radiation emitted during sudden changes in the rupture-front velocity, which occur when the zone of slip reaches regions of differing stress drop.

The introduction of concepts from dynamic fracture mechanics has given rise to crack models of earthquake faulting. In these models a zone of slip is demarcated by a rupture front, and the stress drop on the slip zone provides the driving force for the slipping mechanisms. These models are equivalent to mode II and mode III crack propagation models of fracture mechanics.

Achenbach (1972, 1977) has discussed dynamic fracture mechanics in some detail. Within the context of linearized elasticity, the fields of stress and particle velocity near a rapidly propagating crack tip are square-root singular, with a coefficient that shows a strong dependence on the crack tip speed. A review of the earlier applications of fracture mechanics to modeling of fault slip has been given by Aki and Richards (1980, pp. 851–911). More recently Madariaga (1983a) has provided a comprehensive summary of seismic source theory, much of which is devoted to the application of dynamic fracture mechanics to fault slip.

Unfortunately there are only a few exact solutions available for radiation from propagating cracks in unbounded media. Notable examples are those by Kostrov (1964) and Richards (1976). The common approach has been numerical, such as the numerical solutions to integral equations by Burridge and Halliday (1971) and Das (1980) and the finite-difference calculations of Madariaga (1976) and Virieux and Madariaga (1982). The singular behavior of the stress fields near the crack tip, which is characteristic of the "brittle" fracture models, can be eliminated by introducing appropriate distributions of cohesive tractions. In a geophysical context, models of this kind for plane strain shear cracks have been discussed by Andrews (1976a,b) and for antiplane shear cracks by Knopoff and Chatterjee (1982) and Chatterjee and Knopoff (1983).

This chapter is primarily concerned with recent work on the application of ray methods to compute ground motions generated by slip over an expanding smoothly curved region in a fault plane. Within the mathematical idealization considered here, the rupture front is a singular curve that forms the transition from continuous stresses and displacements ahead of the rupture front to conditions of sliding in the slip zone. Because the rupture front is only a geometrical definition, it does not have "inertia"; hence its speed can abruptly change when the rupture reaches differing stress or frictional regimes, which are controlled by variations in the local geology.

It may be expected that the strongest motions are radiated at abrupt changes of the rupture-front speed. In this article, such abrupt changes are represented by sudden jumps. Naturally, the idealizations of a discontinuous rupture-front speed and of a singular rupture front produce more severe motions than would be produced by a gradual change of the speed in conjunction with a transition zone for rupture. Hence the results presented here are upper bounds. In terms of the frequency spectrum one can, however, argue that the results apply in a frequency range higher than the corner frequencies but still in a range for which the corresponding wavelengths would be larger than both the distance over which the speed of the rupture front changes and the distance over which the process of complete rupture takes place.

The computations presented here are based mainly on slip displacement distributions near the rupture front that are consistent with mode II and mode III crack models. The distribution of slip just behind the rupture front is taken in the general form $C\eta^{\kappa/2}$, where $\eta$ is the distance from the rupture front. For $\kappa = 1$ this distribution corresponds to brittle fracture. By taking $\kappa \neq 1$, the important case of rupture in the presence of distributions of cohesive tractions at the rupture front is included in the analysis. It is shown that discontinuous changes of the rupture-front speed give rise to the radiation of particle velocities, which at the wavefronts are of the forms $(t - r/c_L)^{(\kappa-1)/2}$ and $(t - r/c_T)^{(\kappa-1)/2}$, while the corresponding frequency spectra are $O(\omega^{-(\kappa+1)/2})$. Thus, the nature of the elastodynamic radiation strongly depends on the value of $\kappa$.

An interesting feature of the results is the focusing of motion caused by the curvature of the propagating rupture front. This kind of focusing should be distinguished from the directional character of the radiation emitted from the propagating rupture front.

The methods discussed here were first introduced by Madariaga (1977) and Achenbach and Harris (1978). Later, by using an exact solution for the radiation emitted by a mode III crack, Madariaga (1983b) described the generation of intense high-frequency emissions by abrupt changes in rupture velocity or stress intensity. Recently, Bernard and Madariaga (1984a) and Spudich and Frazier (1984) have employed the idea of an isochrone to develop asymptotic approximations to near-source wave fields that involve a simple integration along the isochrone (an isochrone is a line on a fault plane from which all the waves reaching an observer at a given instant are emitted). The study of these integrals in combination with the kinematic properties of the isochrones gives rise to approximations to the high-frequency waves emitted by changes in the rupture velocity or the stress intensity at the rupture front. This approach seems to be a promising extension to that discussed in this article.

An early study of the interaction of radiation from a faulting event with a free surface was done by Achenbach and Brock (1973). Other studies, involving both a free and a layered surface have been carried out numerically. Representative examples are the finite-element calculations of Archuleta and Frazier (1978) and the finite-difference calculations of Archuleta and Hartzell (1981) and Bouchon (1980a,b). Exploiting the fact that most high-frequency emissions come from the rupture front when it changes its speed of advance, Harris and Achenbach (1981) approximated the strong motion in the near field (near with respect to the epicenter) by calculating the direct reflections of the ray fields emitted from the curved front as it propagates toward the surface. More recently, Bernard and Madariaga (1984b) have performed a similar calculation. They asymptotically evaluated near-field accelerograms for a circular fault buried in a half-space. Of particular interest in their calculation was the ground motion caused by those rays that had passed through the focal line of the circular fault.

At large distances from the epicenter, the ground motion is primarily caused by surface waves. It is possible to analyze surface waves on the basis of ray theory (Keller and Karal, 1960, 1964). This requires, however, the tracing of rays in complex space. In this article an alternative method is used. The surface-wave field is expressed in terms of a representation integral over a surface that envelops the rupture front and is contained within the half-space. In fact, the surface is taken as a wave front of the wave motion emitted by the rupture front, and the field on this surface is computed by ray methods. The other terms in the representation integral come from the Green's function for the half-space. By the use of asymptotic methods, the representation integral can be evaluated to give a simple explicit form for the surface-wave motion. Using this method, Harris and Achenbach (1983) calculated the Love wave (the lowest mode) excited by radiation emitted from the propagating rupture front of a strike-slip fault, and Harris et al. (1983) calculated the Rayleigh wave excited by radiation emitted from the propagating rupture front of a dip-slip fault.

The results of Harris and Achenbach (1981) show that in the near field the pattern of surface accelerations is mainly determined by the angle at which the ray leaves the fault plane, the orientation of the fault plane, and the nature of the rupture process. Of particular interest is the critical reflection of the incident transverse ray. The point at which it is incident experiences stronger ground acceleration than do neighboring points. In the far field the pattern of surface accelerations is more strongly influenced by the surface conditions than by the rupture processes, as shown by Harris and Achenbach (1981) and Harris et al. (1983). This is to be expected because surface waves dominate here. In the frequency domain, however, the surface waves do carry some information about the rupture processes.

In closing this introduction, we note that strong motion seismology is a large and very active area of research. Its ultimate goal is to predict strong ground motion from a basic understanding of fault mechanics and seismic-wave propagation. Aki (1982, 1983) has recently written two comprehensive, general reviews of strong motion seismology, and Jennings (1983) has provided a review of strong motion studies with an engineering perspective. The goals of the present article are more limited. The faulting models discussed here are deterministic and emphasize the similarity between faulting and crack propagation. Representative examples of work on stochastic models of faulting are Boatwright (1982) and Papageorgiou and Aki (1983a,b). However, the work of this article can be used as an input to stochastic models because it accounts in a realistic way for the underlying stress-release mechanisms and the causal spreading of subevents in the total faulting process.

## II. Dynamic Fracture Mechanics

Dynamic fracture mechanics applies when inertial forces must be included in the governing equations for an accurate analysis of crack-growth initiation, crack propagation, and crack arrest, and hence for an acceptable description of the fields radiated by a crack propagation event.

For essentially brittle fracture, near-tip dynamic effects have been investigated extensively on the basis of linear elastic fracture mechanics. By now, several papers (Achenbach, 1972; Freund, 1975; Kanninen, 1978) have reviewed the computation of elastodynamic stress-intensity factors, and have discussed the influence that inertial effects have on the flux of energy into the crack tip. Analytical methods have yielded three general results for the dynamic fields near a rapidly propagating crack tip. These results are (1) the asymptotic form of the near-tip fields, (2) an expression for the elastodynamic stress-intensity factor in terms of the corresponding quasi-static stress-intensity factor, for the special case of a semi-infinite crack, and (3) a general expression for the flux of energy into a propagating crack tip. In addition, many numerical results have been obtained, both by the finite-element and the finite-difference methods.

The mechanical signals that are generated by dynamic fracture events have also been analyzed in some detail, both in the context of testing methods based on acoustic emission (Achenbach and Harris, 1979) and, as in this chapter, in the context of applications to sliding events on a fault plane.

### A. NEAR-TIP FIELDS

The geometry that will be considered here is shown in Fig. 1. The $z$ axis of a stationary coordinate system is parallel to the crack front, and $x$ points in the direction of crack growth. The position of the crack tip is defined by

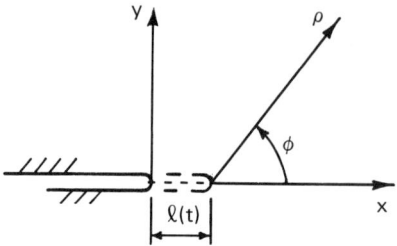

FIG. 1. Propagating crack tip.

$x = l(t)$. A moving polar coordinate system $(\rho, \phi)$ is centered at the moving crack tip. The symbols used here are sometimes used in other contexts in subsequent sections. However, because only the end results of this section are needed for future reference, this should not cause confusion.

Let us first examine the near-tip behavior of the stress components $\tau_{\phi\rho}$ (mode II) and $\tau_{\phi z}$ (mode III) in the polar coordinate system shown in Fig. 1 for the elastodynamic case and for a running crack tip. The stress components are given by

$$\tau_{\phi\rho} = (2\pi\rho)^{-1/2} k_{\mathrm{II}}(t, \dot{l}) T_{\phi\rho}^{\mathrm{II}}(\phi, \dot{l}), \tag{2.1a}$$

and

$$\tau_{\phi z} = (2\pi\rho)^{-1/2} k_{\mathrm{III}}(t, \dot{l}) T_{\phi z}^{\mathrm{III}}(\phi, \dot{l}), \tag{2.1b}$$

where $\dot{l}(t)$ is the speed of the crack tip, $k_{\mathrm{II}}(t, \dot{l})$ and $k_{\mathrm{III}}(t, \dot{l})$ are the elastodynamic stress-intensity factors, and $T_{\phi\rho}^{\mathrm{II}}(0, \dot{l}) = T_{\phi z}^{\mathrm{III}}(0, \dot{l}) = 1$. General expressions for $T_{\phi\rho}^{\mathrm{II}}(\phi, \dot{l})$ and $T_{\phi z}^{\mathrm{III}}(\phi, \dot{l})$ have been derived by Achenbach and Bazant (1975). The functions $T_{\phi\rho}^{\mathrm{II}}(\phi, \dot{l})$ and $T_{\phi z}^{\mathrm{III}}(\phi, \dot{l})$ are independent of the overall geometry and the loading. It is of note that their maximum values move out of the plane $\phi = 0$ (the plane of crack propagation) as $\dot{l}(t)$ increases beyond a certain value. Generally $k_{\mathrm{II}}(t, \dot{l})$ and $k_{\mathrm{III}}(t, \dot{l})$ are much more difficult to compute than the corresponding quasi-static stress-intensity factors.

On the faces of the crack, just behind the crack tip, the relevant displacements corresponding to Eqs. (2.1a) and (2.1b) are

$$u_\rho = \left(\frac{2}{\pi}\right)^{1/2} \frac{\dot{l}}{\mu} \frac{\beta^2(1-\beta^2)^{1/2}}{D(\alpha, \beta)} k_{\mathrm{II}}(t, \dot{l}) \rho^{1/2}, \tag{2.2a}$$

and

$$u_z = \left(\frac{2}{\pi}\right)^{1/2} \frac{\dot{l}}{\mu} \frac{1}{(1-\beta^2)^{1/2}} k_{\mathrm{III}}(t, \dot{l}) \rho^{1/2}, \tag{2.2b}$$

where

$$D(\alpha, \beta) = (\beta^2 - 2)^2 - 4(1 - \alpha^2)^{1/2}(1 - \beta^2)^{1/2}, \quad (2.3)$$

$$\alpha = \dot{l}/c_L, \quad (2.4a)$$

$$\beta = \dot{l}/c_T, \quad (2.4b)$$

and $c_L$ and $c_T$ are the longitudinal and transverse wave speeds, respectively.

For a semi-infinite crack, an interesting relation between $k_I(t, \dot{l})$ and the corresponding quasi-static stress-intensity factor $K_I(t)$ was derived by Freund (1975). The analogous relation for the mode II case is

$$k_{II}(t, \dot{l}) = f(\dot{l}) K_{II}(t), \quad (2.5)$$

where $f(\dot{l})$ depends on the crack-tip speed and the characteristic wave speeds of the material. The function $f(\dot{l})$ is complicated. For the mode-III case, the corresponding relation is, however, of the simple form

$$k_{III}(t, \dot{l}) = (1 - \beta)^{1/2} K_{III}(t), \quad (2.6)$$

where $\beta$ is defined by Eq. (2.4b). Unfortunately, simple relations of the kind given by Eqs. (2.5) and (2.6) hold only for semi-infinite cracks. For cracks of finite lengths, numerical methods are the only practical way to compute elastodynamic stress-intensity factors.

The conditions governing crack-tip motion in mode III can be expressed in terms of $k_{III}(t, \dot{l})$ and an experimentally determined critical value that is assumed to be a property of the material. In conventional (quasi-static) linear elastic-fracture mechanics one has $K_{III} = K_{III}^c$ as the condition for crack instability. In the dynamic generalization, $K_{III}^c$ has two counterparts. First, for the initiation of crack growth

$$k_{III}(t, 0) = K_{III}^d(\dot{\sigma}), \quad (2.7)$$

where $\dot{\sigma}$ represents the loading rate. For perfectly brittle fracture, $K_{III}^d = K_{III}^c$. Similarly, for a propagating crack

$$k_{III}(t, \dot{l}) = K_{III}^D(\dot{l}), \quad (2.8)$$

where $K_{III}^D$ is known as the dynamic fracture toughness.

A more general approach is based on the observation that a propagating crack tip acts as an energy sink. In terms of the energy release rate $G$, a necessary condition for crack-tip motion can be expressed as $G = R(\dot{l})$. Here, $R(\dot{l})$, which is the critical value of $G$, defines the material's resistance to fracture. It represents the energy dissipated in the fracture processes accompanying crack propagation.

The energy release rate $G$ and the flux of energy into the crack tip $F$ are related by $F = G\dot{l}$. For elastodynamic problems where time enters as an independent parameter, it is quite simple to compute the flux of energy into the crack tip. Detailed discussions can be found in Achenbach (1972). In a two-dimensional geometry, and for a combined mode-II and mode-III fracture, $F$ can be expressed in the form

$$F = -\frac{\dot{l}^3}{2\mu c_T^2 D(\alpha, \beta)}(1 - \beta^2)^{1/2}[k_{II}(t, \dot{l})]^2$$
$$+ \frac{\dot{l}}{2\mu(1 - \beta^2)^{1/2}}[k_{III}(t, \dot{l})]^2, \qquad (2.9)$$

where $D(\alpha, \beta)$ is defined by Eq. (2.3). In terms of the flux of energy into the crack tip, the necessary condition for fracture is

$$F = \dot{l}R(\dot{l}). \qquad (2.10)$$

For purely brittle fracture, $\frac{1}{2}R(\dot{l})$ becomes the specific surface energy, i.e., the energy required to create one unit of free surface area.

For the case of mode III fracture, Eq. (2.9) becomes

$$F = \frac{1}{2\mu}\frac{\dot{l}}{(1 - \beta^2)^{1/2}}[k_{III}(t, \dot{l})]^2. \qquad (2.11)$$

For slow propagation of the crack tip (i.e., the limit $\dot{l} \to 0$), Eq. (2.11) reduces to

$$F = \frac{1}{2\mu}[K_{III}(t)]^2 \dot{l}. \qquad (2.12)$$

By combining this result with Eq. (2.10), it follows that $K_{III}$ equals the toughness for fracture initiation, $K_{III}^c$, i.e.,

$$K_{III} = K_{III}^c = [2\mu R(0)]^{1/2}. \qquad (2.13)$$

Thus, for a slowly propagating crack, the usual fracture toughness enters as a material property, and the stress-intensity factor equals the fracture toughness $K_{III}^c$ while the crack propagates. The four relations Eqs. (2.8), (2.10), (2.11), and (2.13) can be used to establish the following theoretical relation between $K_{III}^D$ and $K_{III}^c$:

$$K_{III}^D = \left[(1 - \beta^2)^{1/2} R(\dot{l})/R(0)\right]^{1/2} K_{III}^c, \qquad (2.14)$$

where, as in Eq. (2.13), $R(0)$ corresponds to the quasi-static critical energy release rate.

## B. Radiation from a Crack Edge

We illustrate radiation from a starting rupture front by considering a semi-infinite crack propagating in mode III in an unbounded, homogeneous, isotropic, linearly elastic solid. The geometry is shown in Fig. 1. At time $t = 0$, the crack starts to propagate in the $x$ direction. The position of the crack tip is subsequently defined by $x = vt$; i.e., we consider a constant crack-tip speed.

Just prior to crack propagation $t < 0$ and the antiplane shear stress in the plane $y = 0$ is

$$y = 0, \quad x < 0, \quad \tau_{23} = \tau_f^s \tag{2.15a}$$

$$y = 0, \quad x > 0, \quad \tau_{23} = (2\pi x)^{-1/2} K_{\mathrm{III}}^c + \tau_f^s + O(x^{1/2}), \tag{2.15b}$$

where $\tau_f^s$, which is assumed constant, is the maximum static friction between the crack faces, and $K_{\mathrm{III}}^c$ is the critical value of the mode III stress intensity factor (the static fracture toughness) along the fault plane. When the crack tip propagates and the crack faces slip, the stresses drop to the kinetic friction $\tau_f^k$ along the slip region $-c_T t < x < vt$. The shear stresses in the plane $y = 0$ are then

$$y = 0, \quad -c_T t < x < vt, \quad \tau_{23} = \tau_f^k, \tag{2.16a}$$

$$y = 0, \quad x > vt, \quad \tau_{23} = (2\pi)^{-1/2}(x - vt)^{-1/2} K_{\mathrm{III}}^d + \tau_f^k + O\left[(x - vt)^{1/2}\right]. \tag{2.16b}$$

The term $\tau_f^k$ appears in Eq. (2.16b) because the nonsingular part of the shear stress is continuous at the crack tip. The region defined by $y = 0$, $x < -c_T t$ is as yet unaffected by the crack propagation event.

The radiated fields due to the starting event can be analyzed by solving a superposition problem. Take the situation described by Eqs. (2.15a) and (2.15b) as the initial condition. For the superposition problem, assume that the crack at time $t = 0$ starts to propagate in its own plane at speed $v$ and remove the appropriate tractions from the plane $y = 0$. This gives for the superposition problem the conditions

$$y = 0, \quad -ct < x < 0, \quad \tau_{23} = -\sigma_0 \tag{2.17a}$$

$$y = 0, \quad 0 < x < vt, \quad \tau_{23} = -(2\pi x)^{-1/2} K_{\mathrm{III}}^c - \sigma_0 + O(x^{1/2}), \tag{2.17b}$$

where

$$\sigma_0 = \tau_f^s - \tau_f^k \tag{2.18}$$

is the difference between the maximum static friction and the kinetic friction.

The antiplane displacement $u_3(x, y, t)$ is governed by

$$\frac{\partial^2 u_3}{\partial x^2} + \frac{\partial^2 u_3}{\partial y^2} = \frac{1}{c_T^2}\frac{\partial^2 u_3}{\partial t^2}. \qquad (2.19)$$

Rigorous solutions to this equation, subject to the boundary conditions of Eqs. (2.17a) and (2.17b), can be obtained by considering two kinds of δ-function crack-face loading. The first of these is of the form

$$\begin{aligned} y &= \pm 0, \quad -\infty < x < vt, \\ \mu(\partial u_3/\partial y) &= \delta(x/t + \xi)H(t)H(vt - x). \end{aligned} \qquad (2.20)$$

The particle velocity for this problem is denoted by $v_a^G(x, y, t; \xi)$ or $v_a^G(r, \theta, t; \xi)$. This solution applies to the problem of a solid containing a semi-infinite crack. At time $t = 0$, the crack faces are subjected to delta-function loadings, which propagate with constant velocity $\xi$ in the negative $x$ direction, where $\xi \leq c_T$. In the second problem, we consider crack-face loadings of the form

$$\begin{aligned} y &= \pm 0, \quad -\infty < x < vt, \\ \mu(\partial u_3/\partial y) &= \delta(x - \eta)H(t - \eta/v)H(vt - x). \end{aligned} \qquad (2.21)$$

The solution for the particle velocity to this problem is denoted by $v_b^G(x, y, t; \eta)$ or $v_b^G(r, \theta, t; \eta)$. This solution applies to the same configuration as that for $v_a^G$, except that the crack faces are subjected to time-independent delta-function loadings, which appear at time $t = \eta_d/v$ at position $x = \eta_d$. Now, if the crack faces are subjected to

$$y = \pm 0, \quad -\infty < x < vt, \qquad (2.22)$$
$$\mu(\partial u_3/\partial y) = g(x/t)H(x/t + c_T)H(-x) + f(x)H(x)H(vt - x),$$

where $g(x/t)$ depends on $x/t$ rather than $x$ and $t$ separately, and $f(x)$ depends on $x$ only, superposition can be used to write

$$v_3(r, \theta, t) = \int_0^{c_T} g(-\xi) v_a^G(r, \theta, t; \xi) \, d\xi + \int_0^{vt} f(\eta) v_b^G(r, \theta, t; \eta) \, d\eta, \qquad (2.23)$$

where $v_3$ is the $z$ component of the particle velocity. The first integral defines the particle velocity due to moving loads on the old crack faces.

This integral corresponds to a superposition of loads originating at the old crack tip and moving in the negative $x$ direction with velocities that decrease from $c_T$ to zero. The second integral corresponds to a superposition of stationary loads that are applied at times subsequent to $t = \eta/v$ at points $\eta = vt$.

In this paper we are interested in wave front approximations; i.e., we want to examine the wave fields for times $t$ very near the arrival time $x/c_T$. After some manipulation the wave front approximation to Eq. (2.23) can be expressed as

$$v_3(r,\theta,t) = \int_0^{c_T} g(-\xi) v_a^G(r,\theta,t;\xi)\, d\xi - \frac{v^{3/2}}{\pi\mu} \frac{\sin\frac{1}{2}\theta\, H(t-s_T r)}{r^{1/2}(1-s_T v\cos\theta)^{3/2}}$$

$$\times \int_{s_T r}^t (t-\eta)^{-1/2} f\left[\frac{v(\eta - s_T r)}{1 - s_T v\cos\theta}\right] d\eta \quad (2.24)$$

and

$$v_a^G(r,\theta,t;\xi)$$
$$= -\frac{1}{\pi\mu} \frac{\sin\frac{1}{2}\theta\, (t-s_T r)^{1/2}}{r^{1/2}(1 - s_T v\cos\theta)^{3/2}}$$
$$\times \frac{(v+\xi)^{1/2}[2(1+s_T\xi)(1-s_T v\cos\theta) + (1-s_T v)(1+s_T\xi\cos\theta)]}{(1+s_T\xi)^{3/2}(1+s_T\xi\cos\theta)^2}$$
$$\times H(t-s_T r), \quad (2.25)$$

where $s_T = 1/c_T$. It is of interest to point out that Eq. (2.24) shows that the angular dependence of the radiated particle velocity, even near the wave front, depends on the type of traction acting on the crack faces. It follows from Eqs. (2.24) and (2.25) that for $t$ near $s_T r$ the radiated fields corresponding to the starting event are

$$v_3(r,\theta,t) = (v_3^1 + v_3^2 + v_3^3) H(t - s_T r), \quad (2.26)$$

where $v_3^1$ corresponds to the expanding load $-\sigma_0$ on the old crack faces, $v_3^2$ to the square-root singular tractions $-(2\pi x)^{-1/2} K_{III}^c$ on the new crack faces, and $v_3^3$ to the constant tractions $-\sigma_0$ on the new crack faces. The components $v_3^i$ are given by

$$v_3^1 = -v_3^3 + \frac{\sqrt{2}}{\pi}\frac{\sigma_0}{\mu}\frac{\sin(\theta/2)}{r^{1/2}}\frac{(1+s_T v)^{3/2}(t-s_T r)^{1/2}}{s_T^{3/2}(1-s_T v\cos\theta)^{3/2}(1+\cos\theta)}, \quad (2.27)$$

$$v_3^2 = \frac{K_{\mathrm{III}}^c}{\mu\sqrt{2\pi}}\frac{\sin(\theta/2)}{r^{1/2}}\frac{v}{(1-s_T v\cos\theta)}, \quad (2.28)$$

and

$$v_3^3 = \frac{2\sigma_0}{\pi\mu}\frac{\sin(\theta/2)}{r^{1/2}}\frac{v^{3/2}(t-s_T r)^{1/2}}{(1-s_T v\cos\theta)^{3/2}}. \quad (2.29)$$

As can be seen, the field related to $-\sigma_0$ is of order $(t-s_T r)^{1/2}$ while that corresponding to $-(2\pi x)^{-1/2}K_{\mathrm{III}}^c$ is of order zero.

## III. Canonical Problems

In elastodynamic ray theory, the amplitude of a high-frequency mechanical disturbance or, equivalently, of a propagating discontinuity, is traced as the disturbance propagates along a ray. Among the first contributions to elastodynamic ray theory was the work of Karal and J. B. Keller (1959) and H. Keller (1964). For a recent review and discussion, we refer the reader to Achenbach et al. (1982).

The ray method, which is used here to trace the signal generated by a rupture front, is equivalent to the geometrical theory of diffraction. For slip zones of arbitrary shape, the method is based on solutions for semi-infinite cracks whose surfaces are subjected to analogous disturbances. Solutions for the semi-infinite crack geometry can be obtained analytically by application of Laplace transform techniques in combination with the Wiener–Hopf method (Achenbach, 1973). In particular, wave front motions can be analyzed directly from the Laplace transforms of the solutions by asymptotic considerations. These wave front results can be expressed in terms of emission coefficients that relate the emitted fields to the surface disturbances. Elastodynamic ray theory now provides modifications to the semi-infinite crack results to account for the curvature of the crack edges and for the finite dimensions of the crack. In the usual terminology, the results for the corresponding semi-infinite crack problems are referred to as the canonical solutions.

A geometrical theory of diffraction for scalar waves, which asymptotically approximates diffraction effects, was developed by J. B. Keller (1958). The formulation given by him is, however, not directly applicable to cracks

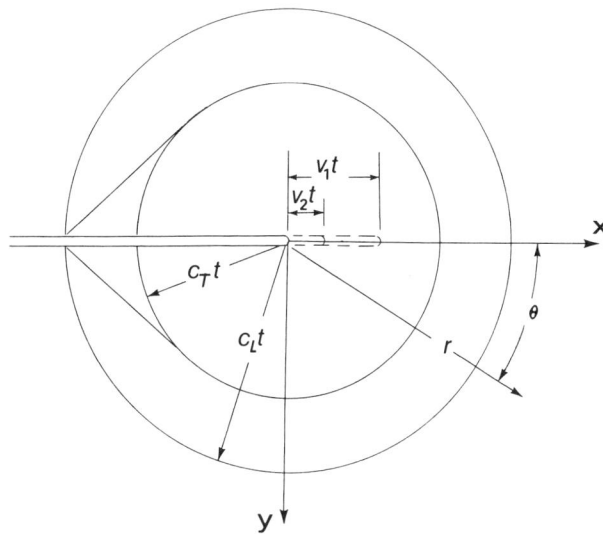

FIG. 2. Two-dimensional geometry for canonical problems.

in solids because wave motions in solids are governed by two wave equations that are coupled by the boundary conditions. An application of geometrical diffraction theory to determine waves excited on the surface of an elastic body by prescribed time-dependent displacements was discussed by Ahluwalia *et al.* (1974). The examples in that paper are, however, for cases of torsional motion, which are governed by a scalar wave equation. More complete groundwork for a three-dimensional geometrical diffraction theory for cracks in an elastic solid can be found in Achenbach *et al.* (1982).

In this section we summarize the solutions to the two relevant canonical problems, the in-plane (mode II) and antiplane (mode III) problems, which we shall need for our subsequent work. The geometry for these problems is shown in Fig. 2.

A. Transient Emissions

We will start with motions generated by in-plane (mode II) slip. At an arbitrary time, which is taken as $t = 0$, the moving edge is located at $x = 0$ in a fixed coordinate system. The instantaneous velocity of the crack tip is $v_1$. At this time the rupture speed instantaneously changes to $v_2$, where $v_2 < v_1$.

Let us consider the slip displacement if the region of slip would propagate a little beyond $x = 0$ with velocity $v_1$. For sufficiently small $t$, the near-edge slip displacements in the $(x, y)$ plane propagate in a self-similar manner, i.e.,

$$u_1(x, 0, t) = f(t - x/v_1). \tag{3.1}$$

Because the edge does not propagate with velocity $v_1$, but rather with velocity $v_2$, the crack opening given by Eq. (3.1) must be closed for $x > v_2 t$. This gives rise to a discontinuity in the wave motion emitted by the edge.

The functional form of $f(t - x/v_1)$ is the controlling factor in the intensity of the motion generated by the change of rupture speed. Because we are only interested in wave front disturbances, it is sufficient to consider small $x/v_1$ and small $t$ and to retain the first term only in an expansion of $f(t - x/v_1)$. In general, we may write

$$u_1(x, 0, t) = C(t - x/v_1)^{\kappa/2} H(t - x/v_1), \tag{3.2}$$

where $C$ is a constant. Here $\kappa = 1$ corresponds to a brittle fracture opening mode, while $\kappa > 1$ corresponds to opening modes that occur in the presence of distributions of cohesive tractions. The form of the displacement shown in Eq. (3.2) is in agreement with our earlier discussion of crack-opening displacements [see Eqs. (2.2a), and (2.2b)].

To analyze the wave front motions caused by the change of rupture speed from $v_1$ to $v_2$, it is convenient to introduce the new variable $q$ defined by

$$q = x - v_2 t. \tag{3.3}$$

As a function of $q$ and $t$, the slip near the rupture front is

$$u_1(x, 0, t) = G(t - s_{12} q), \tag{3.4}$$

where $s_{12} = 1/(v_1 - v_2)$, and

$$G(t) = [(v_1 - v_2)/v_1]^{\kappa/2} F(t), \tag{3.5}$$

$$F(t) = CH(t) t^{\kappa/2}. \tag{3.6}$$

We now superimpose upon the wave field corresponding to Eq. (3.4) the solution to the problem that closes the crack from $v_1 t$ to $v_2 t$. Because of symmetry, only $y \geq 0$ need be considered. This problem has quiescent initial conditions and, at $y = 0$, the boundary conditions

$$y = 0, \quad q < 0, \quad \tau_{21} = 0, \tag{3.7}$$

$$y = 0, \quad -\infty < q < \infty, \quad \tau_{22} = 0, \tag{3.8}$$

$$y = 0, \quad q > (v_1 - v_2)t, \quad u_1 = 0, \tag{3.9}$$

$$y = 0, \quad 0 < q < (v_1 - v_2)t, \quad u_1 = -G(t - s_{12} q). \tag{3.10}$$

The superposition produces the wave front solution for the case in which the rupture front changes speed from $v_1$ to $v_2$.

The case for antiplane (mode III) slip is formulated in an identical way. The displacement, just behind the rupture front propagating with speed $v_1$, is again given by

$$u_3(x, 0, t) = G(t - s_{12}q). \tag{3.11}$$

The solution that we superimpose on the original field to close the crack tip for $x > v_2 t$ satisfies the boundary conditions at $y = 0$

$$y = 0, \quad q < 0, \quad \tau_{23} = 0, \tag{3.12}$$

$$y = 0, \quad q > (v_1 - v_2)t, \quad u_3 = 0, \tag{3.13}$$

$$y = 0, \quad 0 < q < (v_1 - v_2)t, \quad u_3 = -G(t - s_{12}q). \tag{3.14}$$

To consider a sudden increase of the rupture front speed from $v_2$ to $v_1$, we need only reverse the sign of $G(t)$ in Eqs. (3.10) and (3.14).

The analytic methods that can be used to solve the boundary initial value problems stated in Eqs. (3.7)–(3.10) and (3.12)–(3.14) have been discussed in great detail both in Achenbach (1972, pp. 21–27) and in Achenbach (1973, pp. 365–388). Briefly, the principal manipulations will be described. The one-sided Laplace transform is used to eliminate time. The resulting boundary value problems are solved using the Wiener–Hopf technique, and the various Laplace-transformed particle velocities are expressed as integrals where the integrations are to be performed on Cagniard–deHoop contours. These integrals are then evaluated asymptotically to get wave front approximations. This last step, which can cause difficulties because the coordinate $q$ depends on time $t$, is explained in some detail in a note by Harris (1984).

The longitudinal particle velocity $v_r(r, \theta, t)$ near its arrival time $r/c_L$ and the transverse particle velocity $v_\theta(r, \theta, t)$ near its arrival time $r/c_T$ are given by the expressions

$$v_r(r, \theta, t) = \pm \left(\frac{c_L}{2\pi r}\right)^{1/2} E_L^{TV}(\theta) e(t - r/c_L), \tag{3.15}$$

$$v_\theta(r, \theta, t) = \pm \left(\frac{c_T}{2\pi r}\right)^{1/2} \{\text{Re}[E_T^{TV}(\theta)] e(t - r/c_T) \\ - \text{Im}[E_T^{TV}(\theta)] f(t, r/c_T, \theta)\}. \tag{3.16}$$

The + sign is used when the rupture front suddenly speeds up from $v_2$ to $v_1$, and the − sign when it slows down from $v_1$ to $v_2$.

In Eqs. (3.15) and (3.16), the emission coefficients $E_L^{TV}$ and $E_T^{TV}$ govern the angular dependence of the radiation and, to some degree, the signals' strength through the difference ($v_1 - v_2$). The waveform or phase functions $e(t)$ and $f(t, r/c_T, \theta)$ describe the time behavior of the signals near the wave fronts. The $r^{-1/2}$ term indicates the two-dimensional geometric decay. These functions and the other terms in Eqs. (3.15) and (3.16) are defined as

$$E_L^{TV}(\theta) = \sin 2\theta \, F_L(\theta), \tag{3.17}$$

$$E_T^{TV}(\theta) = \cos 2\theta \, F_T(\theta), \tag{3.18}$$

$$F_\alpha(\theta) = \frac{\left(\dfrac{c_T}{c_\alpha}\right)^3 \left(\dfrac{c_R + v_1}{c_T}\right) \left(\dfrac{v_1}{c_T + v_1}\right)^{1/2} \left(\dfrac{v_1 - v_2}{v_1}\right)^{(\kappa+1)/2}}{[1 + (c_R/c_\alpha)\cos\theta][1 - (v_1/c_\alpha)\cos\theta]}$$
$$\times \frac{S_-(-s_{12})[1 + (c_T/c_\alpha)\cos\theta]^{1/2}}{[1 - (v_2/c_\alpha)\cos\theta]^{(\kappa+1)/2} S_-(\zeta_\alpha)}, \tag{3.19}$$

$$\zeta_\alpha = -\cos\theta / \{c_\alpha[1 - (v_2/c_\alpha)\cos\theta]\}, \tag{3.20}$$

$$e(t) = C\left[\frac{\Gamma(\kappa/2 + 1)}{\Gamma(\kappa/2 + 1/2)}\right] t^{(\kappa-1)/2} H(t), \tag{3.21}$$

$$f(t, r/c_T, \theta) = \frac{C}{\pi^{1/2}} \frac{\kappa}{2} \int_{t_h}^m \frac{(t - \eta)^{(\kappa-2)/2}}{(r/c_T - \eta)^{1/2}} d\eta. \tag{3.22}$$

The constants $c_L$, $c_T$, and $c_R$ are the longitudinal, transverse, and Rayleigh wave speeds. The subscript $\alpha$ takes the values L or T for the longitudinal or transverse cases. The function $S_-(\zeta)$ is defined by

$$\ln S_-(\zeta) = -\frac{1}{\pi}\int_a^b \tan^{-1}\left[\frac{4z^2|\gamma_L||\gamma_T|}{(2z^2 - 1/c_T^2 - v_2^2 z^2/c_T^2 + 2v_2 z/c_T^2)^2}\right]\frac{dz}{z - \zeta}, \tag{3.23}$$

where

$$\gamma_\alpha = \left[(1 - v_2 z)^2/c_\alpha^2 - z^2\right]^{1/2}, \quad \alpha = L, T \tag{3.24}$$

$$a = [c_L(1 + v_2/c_L)]^{-1}, \tag{3.25a}$$

$$b = [c_T(1 + v_2/c_T)]^{-1}, \tag{3.25b}$$

and, for $a < \zeta < b$, $\zeta$ should approach the real line through negative imaginary values.

To completely define the emission coefficients $E_L^{TV}$ and $E_T^{TV}$, the radicals need to be specified for complex $\cos\theta$ [note that only $\cos\theta$ appears in Eqs. (3.17)–(3.20)]. The real part of each radical is taken as positive for all $\cos\theta$, and the imaginary part is taken as positive for $\text{Im}(\cos\theta)$ positive and as negative for $\text{Im}(\cos\theta)$ negative. Further, the imaginary part of $E_T^{TV}$ is evaluated by letting $\text{Im}(\cos\theta)$ go to zero through positive values. Checking with Eq. (3.20), we see that this agrees with our definition of $S_-$.

Equations (3.15) and (3.16) represent the longitudinal and transverse cylindrical waves shown in Fig. 2. Also present in the region $\theta_h < |\theta| < \pi$, where $\theta_h = \cos^{-1}(-c_T/c_L)$, is a headwave whose arrival time is $t_h = (r/c_L)\cos(|\theta| - \theta_h)$, and whose termination time is $m = \min(t, r/c_T)$. Achenbach and Harris (1979) and Harris and Pott (1984) studied this headwave when calculating the emission from a propagating mode I crack. The results show that its termination rather than its wave front makes the dominant contribution to the transverse wave emission. The second term in Eq. (3.16) is an approximation to this term, valid for $t$ near $r/c_T$. Near $|\theta| = \theta_h$, Eq. (3.16) is not an accurate description of the transverse wave field because of the confluence of the headwave and cylindrical transverse wave fronts.

Near its arrival time $r/c_T$, the antiplane transverse particle velocity $v_3(r, \theta, t)$ is given by

$$v_3(r, \theta, t) = \pm (c_T/2\pi r)^{1/2} E_T^{TH}(\theta) e(t - r/c_T), \quad (3.26)$$

where

$$E_T^{TH}(\theta) = \frac{\left[2\left(\dfrac{v_1}{c_T}\right)\left(\dfrac{c_T + v_1}{c_T}\right)\right]^{1/2} \left(\dfrac{v_1 - v_2}{v_1}\right)^{(\kappa+1)/2} \sin\left(\dfrac{\theta}{2}\right)}{[1 - (v_1/c_T)\cos\theta][1 - (v_2/c_T)\cos\theta]^{(\kappa+1)/2}}. \quad (3.27)$$

The radicals are defined as before. For $\kappa = 1$, $v_2 = 0$, $v_1 = v$, and a suitable choice of $C$, Eq. (3.26) reduces to Eq. (2.28).

In Figs. 3–6, we have plotted the emission coefficients versus $\theta$ for various parameter values. In all cases Poisson's ratio = 0.25. In Fig. 3 the simplest emission coefficient $E_T^{TH}$ is plotted for various ratios of $v_1/c_T$ and $v_2/v_1$. In general the strength of the emission coefficient decreases as the difference $(v_1 - v_2)$ decreases. In Fig. 4, $E_T^{TH}$ is plotted for different values of $\kappa$, with $v_1/c_T = 0.5$ and $v_2/v_1 = 0.4$. There is a decrease in $E_T^{TH}$ as $\kappa$ increases in value. Thus, as expected, the strongest radiation is emitted for a starting or stopping event when the fracture is brittle. In Fig. 5, $E_L^{TV}$ is plotted for various ratios of $v_1/c_T$ and $v_1/v_2$ and with $\kappa = 1$. Again $E_L^{TV}$

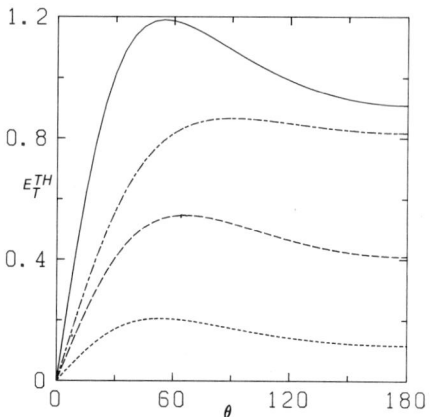

FIG. 3. The emission coefficient for antiplane slip. Solid line, $v_1/c_T = 0.7$, $v_2/v_1 = 0$. Long-short dashed line, $v_1/c_T = 0.5$, $v_2/v_1 = 0$. Long dashed line, $v_1/c_T = 0.5$, $v_2/v_1 = 0.4$. Short dashed line, $v_1/c_T = 0.5$, $v_2/v_1 = 0.8$. The parameter $\kappa = 1$, and Poisson's ratio = 0.25 in all cases.

decreases with $(v_1 - v_2)$ but is never very large. In Fig. 6, both the real and imaginary parts of $E_T^{TV}$ are plotted for various ratios of $v_1/c_T$ and $v_1/v_2$. Again a decrease with $(v_1 - v_2)$ is observed. But, of more interest is the large magnitude of both the real and imaginary parts of $E_T^{TV}$ at the headwave angle. Our expressions for the transverse emission are not very accurate near this angle. Nevertheless, it is reasonable to deduce, from our

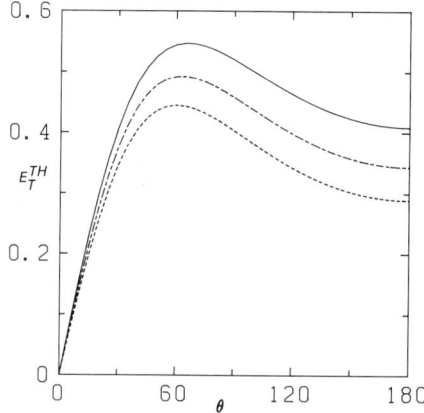

FIG. 4. The emission coefficient for antiplane slip. Solid line, $\kappa = 1$. Long-short dashed line, $\kappa = 1.5$. Short dashed line, $\kappa = 2$. The parameters $v_1/c_T = 0.5$ and $v_2/v_1 = 0.4$, and Poisson's ratio = 0.25 in all cases.

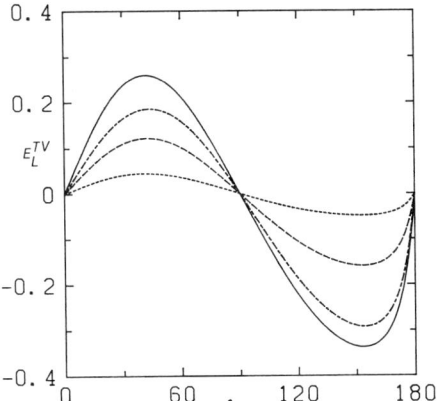

**FIG. 5.** The longitudinal emission coefficient for in-plane slip. Solid line, $v_1/c_T = 0.7$, $v_2/v_1 = 0$. Long-short dashed line, $v_1/c_T = 0.5$, $v_2/v_1 = 0$. Long dashed line, $v_1/c_T = 0.5$, $v_2/v_1 = 0.4$. Short dashed line, $v_1/c_T = 0.5$, $v_2/v_1 = 0.8$. The parameter $\kappa = 1$, and Poisson's ratio $= 0.25$ in all cases.

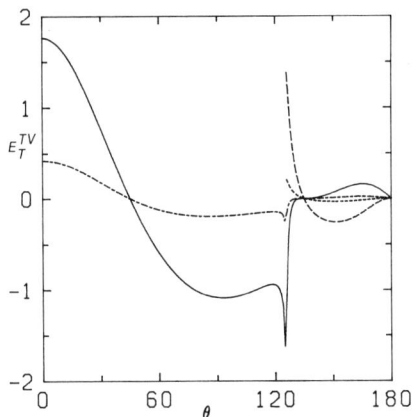

**FIG. 6.** The real and imaginary parts of the transverse emission coefficient for in-plane slip. Solid line, $\text{Re}(E_T^{TV})$, $v_1/c_T = 0.6$, $v_2/v_1 = 0$. Long dashed line, $\text{Im}(E_T^{TV})$, $v_1/c_T = 0.6$, $v_2/v_1 = 0$. Long-short dashed line, $\text{Re}(E_T^{TV})$, $v_1/c_T = 0.5$, $v_2/v_1 = 0.8$. Short dashed line, $\text{Im}(E_T^{TV})$, $v_1/c_T = 0.5$, $v_2/v_1 = 0.8$. The parameter $\kappa = 1$, and Poisson's ratio $= 0.25$ in all cases.

limited calculation, that the transverse emission varies rapidly and is quite strong near the headwave angle. In closing, we note that both $E_L^{TV}$ and $E_T^{TV}$ show a double-couple symmetry that is modified by the multiplicative factors $F_L$ and $F_T$.

## B. Emission Spectra

Wave front approximations, or more precisely approximations near algebraic singularities in the time domain, correspond to high-frequency approximations in the frequency domain (Lighthill, 1970, pp. 51–57). Therefore, when calculating the frequency spectra of the transient radiation from phases of the faulting event, the Fourier transforms can be approximated for large $\omega$, where $\omega$ is the angular frequency. The transform pair used here is

$$f(\omega) = \int_{-\infty}^{\infty} e^{i\omega t} f(t)\, dt, \qquad (3.28a)$$

$$f(t) = \frac{\text{Re}}{\pi} \int_{0}^{\infty} e^{-i\omega t} f(\omega)\, d\omega. \qquad (3.28b)$$

The form of Eq. (3.28b) follows from the standard form of the inverse transform by noting that for $f(t)$ real we have $\bar{f}(\omega) = f(-\omega)$, where the overbar indicates the complex conjugate. By using Eq. (3.28b), we can always assume that $\omega$ is positive. The asymptotic approximation to the Fourier transform of the particle velocity $v_r(r, \theta, \omega)$ corresponding to the longitudinal wave is

$$v_r(r, \theta, \omega) = \pm \left(\frac{c_L}{2\pi r}\right)^{1/2} E_L^{TV}(\theta) e^{ik_L r} F(\omega)(-i\omega)^{1/2}, \qquad (3.29)$$

where $k_L = \omega/c_L$, and $F(\omega)$, the Fourier transform of $F(t)$, is given by

$$F(\omega) = \frac{C\Gamma[(\kappa/2) + 1]}{(-i\omega)^{(\kappa+2)/2}}. \qquad (3.30)$$

The asymptotic approximation to the Fourier transform of the transverse wave velocity $v_\theta(r, \theta, \omega)$ is

$$v_\theta(r, \theta, \omega) = \pm \left(\frac{c_T}{2\pi r}\right)^{1/2}$$
$$\times \{\text{Re}[E_T^{TV}(\theta)] + \text{Im}[E_T^{TV}(\theta)]b(\omega)\} e^{ik_T r} F(\omega)(-i\omega)^{1/2}, \qquad (3.31)$$

where $k_T = \omega/c_T$, and

$$b(\omega) = -\frac{1}{\pi^{1/2}} \int_0^c t^{-1/2} e^{-i\omega t}\, dt, \qquad \omega \geq 0. \qquad (3.32)$$

The function $b(\omega)$ arises from Fourier transforming $f(t, r/c_T, \theta)$, Eq. (3.22), and noting that it is a convolution integral. Also, $c = (r/c_T - t_h)$. The function $b(\omega)$ may be approximated as (Carrier et al., 1966, pp. 255–257)

$$b(\omega) = i\left[(-i\omega)^{-1/2} - e^{-i\omega c}(\pi c)^{-1/2}\omega^{-1}\right]. \tag{3.33}$$

By retaining only the dominant term in Eq. (3.33), Eq. (3.31) becomes

$$v_\theta(r, \theta, \omega) = \pm\left(\frac{c_T}{2\pi r}\right)^{1/2} E_T^{TV}(\theta) e^{ik_T r} F(\omega)(-i\omega)^{1/2}. \tag{3.34}$$

Some care is required when Eq. (3.34) is inverted to get back to the time domain. By approximating $b(\omega)$ by only the first term of Eq. (3.33), we have, in essence, let $t_h \to -\infty$. Thus, inverting Eq. (3.34) in the headwave region, $|\theta| > \theta_h$ leads to a very singular function that must be interpreted as a generalized function (Lighthill, 1970, pp. 42–45). For example, for $\kappa = 1$, the inversion gives

$$f(t, r/c_T, \theta) = -C/2\pi^{1/2}(\ln|r/c_T - t| + C_1), \tag{3.35}$$

where $C_1$ is an arbitrary constant.

Finally the asymptotic approximation to the Fourier transform of the transverse particle velocity $v_3(r, \theta, \omega)$ is

$$v_3(r, \theta, \omega) = \pm(c_T/2\pi r)^{1/2} E_T^{TH}(\theta) e^{ik_T r} F(\omega)(-i\omega)^{1/2}. \tag{3.36}$$

## IV. Near-Field Ground Motions

### A. The Three-Dimensional Geometry

To extend the two-dimensional results given by Eqs. (3.15), (3.16), and (3.26) to a three-dimensional configuration, we consider a curved rupture front of a planar slip zone that advances with a velocity $v$, and we assume that the slip is locally normal and/or tangential to the rupture front. An instantaneous position of $\mathscr{C}$ is shown in Fig. 7. In general the radius of curvature of $\mathscr{C}$, which is defined by $\rho$, may vary along the rupture front. A point on the rupture front is defined by the angle $\psi$. A plane $\mathscr{N}(\psi)$ is perpendicular to the faces of the slip zone, and the tangent to $\mathscr{C}$ is normal to it. It contains a polar coordinate system whose origin is at its intersection with $\mathscr{C}$. The plane $\mathscr{N}(0)$ and its coordinate system $(r, \theta)$ are shown in Fig. 7. At a sudden change in the rupture front speed, the wave front approximations to the waves radiated from an arbitrary point $P$ on the rupture front can be represented by fans of rays in the plane $N(\psi)$ through $P$.

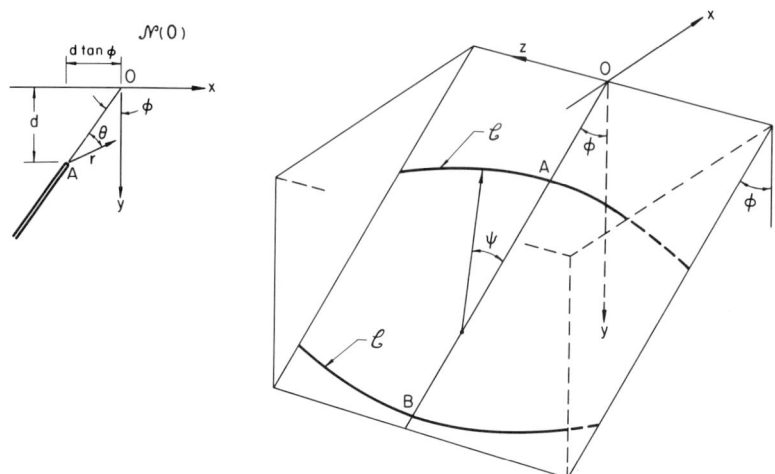

FIG. 7. Rupture front $\mathscr{C}$ of a planar slip zone advancing across an inclined fault plane with a velocity $v$ normal to the front. At the instant shown, point A, which lies in the symmetry plane $\mathcal{N}(0)$, is a distance $d$ below the surface and is moving toward the origin O.

Because we want to approximate the radiated waves at high frequencies, only the local character of the rupture front at $P$ affects them. Therefore, to extend the two-dimensional results to three dimensions, they must be adjusted for the curvature of the rupture front by multiplying by a factor $(1 + r/\bar{\rho})^{-1/2}$, where $\bar{\rho}$ is given by

$$\bar{\rho}(\psi, \theta) = \rho(\psi)/\cos\theta. \tag{4.1}$$

When considering a finite zone of slip, there is an additional complication to be noted. The sudden change in rupture speed also excites Rayleigh waves on the faces of the slip zone, and these waves are subsequently diffracted from another part of the rupture front. For example, in Fig. 7, Rayleigh waves excited at A will be diffracted at B. These diffracted waves can be calculated using techniques described by Achenbach, *et al.* (1982, pp. 133–138, 180–186). However, we have not included these waves here, because we believe that the faces of the slip zone will be too rough to support strong Rayleigh-wave disturbances.

Note that for certain values of $\theta$ the factor $(1 + r/\bar{\rho})$ will vanish. These values of $\theta$, along with $r$ and $\psi$, define a cylindrical surface, called the caustic surface, whose cross section is the evolute of the rupture front. In the case of a circular rupture front, this surface reduces to a straight line through the center of the circle. The three-dimensional approximations are not accurate for points near or on the caustic surface, but the presence of this singularity in our approximations does indicate that the wave fields are particularly strong there. A concise treatment of the wave field near a caustic surface is given by Pierce (1981, pp. 460–469). After a ray has grazed a caustic surface, the factor $(1 + r/\bar{\rho})^{-1/2}$ is replaced by $(-r/\bar{\rho} - 1)^{-1/2}$. In the frequency domain, the ray experiences a $(-\pi/2)$ phase shift, while in the time domain the waveform function $e(t)$ is replaced by its Hilbert transform (Bernard and Madariaga, 1984b).

In Sections IV and V, we shall calculate the particle acceleration of the surface of the half-space when it is struck by radiation emitted from the rupture front. In Fig. 7 we show a fault plane that makes an angle $\phi$ with the normal to the surface. Also shown is a second coordinate system $(x, y, z)$ located with its origin at the intersection of the fault plane, the plane $\mathcal{N}(0)$, and the surface. Thus, point A has coordinates $(-d \tan \theta, d, 0)$. In this chapter, we shall assume that $\mathcal{N}(0)$ is a plane of reflection symmetry. In general, it is difficult to calculate the surface motion out of the plane $\mathcal{N}(0)$. Therefore, except for a perturbation calculation discussed in Appendix B, all the calculations of ground motion will be for points in the plane of symmetry.

There are two special cases of the slip-zone geometry that we shall use in the sequel. In the first case the rupture front is taken as elliptical. The semi-major axis of the ellipse parallels the $z$ axis (Fig. 7) and is of length $a$. The semi-minor axis lies along the fault plane and is of length $b$. The radius of curvature $\rho$ is given by

$$\rho = (a^2 \sin^2 \xi + b^2 \cos^2 \xi)^{3/2}/ab, \tag{4.2}$$

and $\xi$, the eccentricity angle of the ellipse, is related to $\psi$ by

$$\tan \psi = (b/a) \cot \xi. \tag{4.3}$$

Note that for $\psi$ equal to 0 or $\pi$ (points A or B), the radius of curvature for the ellipse is $a^2/b$. In the second case, the rupture front is taken as a circle of radius $a$. In this case the radiation from the rupture front is described by a toroidal coordinate system $(r, \theta, \psi)$. The $(x, y, z)$ coordinate system shown in Fig. 7 and the toroidal system are related as

$$x = -d \tan \phi + (a + r \cos \theta) \cos \psi \sin \phi + r \sin \theta \cos \phi - a \sin \phi, \tag{4.4}$$

$$y = d - (a + r \cos \theta) \cos \psi \cos \phi + r \sin \theta \sin \phi + a \cos \phi, \tag{4.5}$$

$$z = (a + r \cos \theta) \sin \psi. \tag{4.6}$$

## B. In-Plane Reflections

### 1. Longitudinal Emission

At high frequencies or near its wave front, the radiated longitudinal pulse behaves locally as a plane wave, and therefore the pulses reflected from the free surface can be constructed in the same manner as are reflected plane waves. By adding these reflected pulses to the incident pulse, the $x$ and $y$ components of the ground acceleration can be calculated.

Let us consider longitudinal radiation from A. The $x$ and $y$ components $a_1^L$ and $a_2^L$ of the ground acceleration caused by this radiation, evaluated at the surface and in the plane $\mathcal{N}(0)$, are

$$a_i^L(x, t) = R_i^L(\alpha) a_r(R, \theta, t)(1 + \cos\theta R/\rho)^{-1/2}, \qquad (4.7)$$

where $i = 1, 2$. In these expressions, $a_r$ is the two-dimensional, radial component of the particle acceleration from point A, and it can be calculated from Eq. (3.15). It is evaluated on the surface at $r = R$. The radius of curvature of the rupture front $\rho$ is evaluated at $\psi = 0$. Identical expressions can be written for the spectra with $a_r$ calculated using Eq. (3.29). The proportionality factors $R_1^L(\alpha)$ and $R_2^L(\alpha)$ are given by

$$R_1^L(\alpha) = \frac{2(c_L/c_T)\sin 2\alpha \sin\beta}{\sin 2\alpha \sin 2\beta + (c_L/c_T)^2 \cos^2 2\beta}, \qquad (4.8)$$

$$R_2^L(\alpha) = \frac{2(c_L/c_T)^2 \cos 2\beta \sin\alpha}{\sin 2\alpha \sin 2\beta + (c_L/c_T)^2 \cos^2 2\beta}, \qquad (4.9)$$

$$\beta = \cos^{-1}[(c_T/c_L)\cos\alpha]. \qquad (4.10)$$

The distance $R$, the angle $\theta$, and the angle $\alpha$ are related to the position on the surface $x$ by

$$R = \left[d^2 + (d\tan\phi + x)^2\right]^{1/2}, \qquad (4.11)$$

$$\theta = \cos^{-1}(d/R) - \phi, \qquad x > -d\tan\phi, \qquad (4.12a)$$

$$\theta = -\cos^{-1}(d/R) - \phi, \qquad x < -d\tan\phi, \qquad (4.12b)$$

$$\alpha = \pi/2 - (\theta + \phi). \qquad (4.13)$$

In order to investigate how the amplitudes of these accelerations are distributed along the surface, we define the amplitude functions $A_1^L$ and $A_2^L$ as

$$A_i^L(x) = \frac{E_L^{TV}(\theta) R_i^L(\alpha)}{\left[(R/d)|1 + \cos\theta R/\rho|\right]^{1/2}}, \qquad (4.14)$$

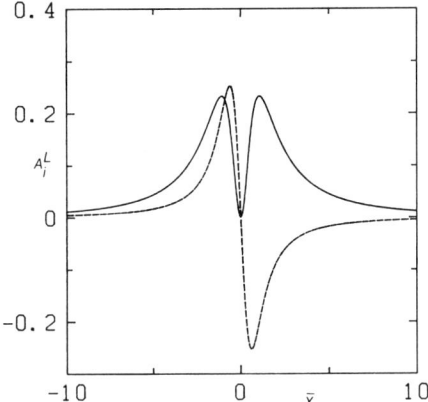

**FIG. 8.** The amplitude of the surface acceleration $A_i^L$ excited by the incident longitudinal emission. Solid line, $i = 1$. Dashed line, $i = 2$. The dip angle $\phi = 0°$ and $a/b = 10$.

where $i = 1, 2$. In Figs. 8 and 9, we have plotted $A_1^L$ and $A_2^L$ against $\bar{x} = x/d$. An elliptical rupture front was assumed, with $a/b = 10$. For the other parameters, the values chosen were $v_1/c_T = 0.5$, $v_2/v_1 = 0$, $\kappa = 1$, $d/a = 0.5$, and Poisson's ratio = 0.25. In Fig. 8 the dip angle $\phi = 0°$ and in Fig. 9 $\phi = 30°$. In both cases, the ground motion is very strong at points directly above A, but dies away rather quickly on each side.

In both Figs. 8 and 9, $a/b = 10$, so that for the range of $x$ considered, the spreading factor $(1 + \cos\theta R/\rho)^{1/2}$ does not vanish. However, had we taken a smaller value of $a/b$, for example, $a/b = 1$, then the spreading

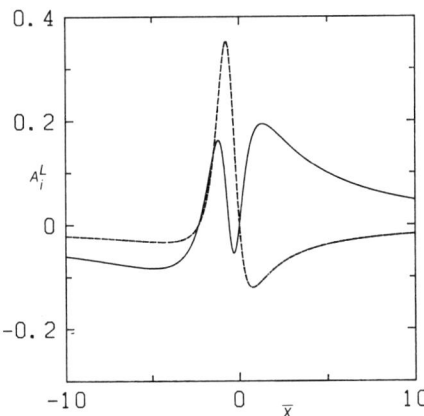

**FIG. 9.** The same as Fig. 8 except $\phi = 30°$.

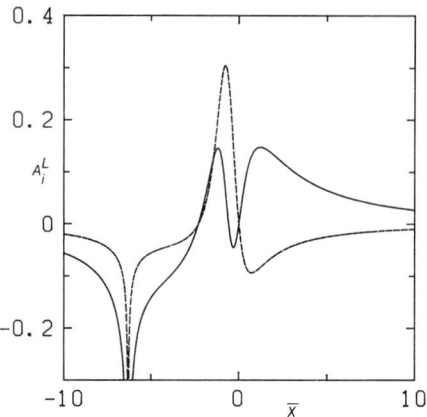

Fig. 10. The same as Fig. 8 except $\phi = 30°$ and $a/b = 1$. Note the effect of the caustic surface.

factor would have vanished at some point where $(x + d \tan \phi) < 0$, causing $A_1^L$ and $A_2^L$ to become unbounded. In Fig. 10 we have plotted $A_1^L$ and $A_2^L$ for $a/b = 1$ and $\phi = 30°$, while leaving the values of all the other parameters unchanged from those in Figs. 8 and 9. Comparison of Figs. 9 and 10 shows that the singularity in the spreading factor violently distorts the graphs of $A_1^L$ and $A_2^L$.

The singularity in the ground motion comes about because the caustic surface associated with the rupture front pierces the surface. For an incident ray from point A to strike the surface to the left of the singularity shown in Fig. 10, it must have grazed the caustic surface. Therefore, in the frequency domain it has been shifted by $(-\pi/2)$. When the caustic surface has been grazed, the $x$ and $y$ components of the ground's acceleration, in the frequency domain, are given by

$$a_i^L(x, \omega) = R_i^L(\alpha) a_r(R, \theta, \omega) e^{-i\pi/2} (-\cos \theta R/\rho - 1)^{-1/2}, \quad (4.15)$$

where $i = 1, 2$. To return to the time domain, we must invert Eq. (4.15) using Eq. (3.28b). This gives

$$a_i^L(x, t) = R_i^L(\alpha) b_r(R, \theta, t)(-\cos \theta R/\rho - 1)^{-1/2}, \quad (4.16)$$

where $i = 1, 2$. The modified radial acceleration $b_r$ is given by

$$b_r(r, \theta, t) = \pm \left(\frac{c_L}{2\pi r}\right)^{1/2} E_L^{TV}(\theta) h(t, r/c_L), \quad (4.17)$$

where

$$h(t, r/c_L) = \frac{C}{\pi} \frac{\Gamma[(\kappa + 2)/2]\Gamma[(3 - \kappa)/2]}{|t - r/c_L|^{(3-\kappa)/2}}$$
$$\times [H(r/c_L - t) - \sin(\kappa\pi/2)H(t - r/c_L)]. \quad (4.18)$$

In Eq. (4.16) $b_r$ is evaluated at $r = R$. In Eq. (4.17) the $+$ and $-$ signs indicate a speeding up or slowing down event, respectively. Note that $h(t, r/c_L)$ does not vanish for $t < r/c_L$. This is because radiation from point B has reached the observation point before that coming from point A. Also note the distortion of the waveform caused by the grazing of the caustic surface. Had we considered point B as the source of our incident radiation, then all the rays emitted from point B that strike the free surface directly above B would have grazed the caustic surface, and in the frequency domain would have experienced a $(-\pi/2)$ phase shift.

2. In-Plane Transverse Emission

We cannot proceed to calculate the ground acceleration when the incident radiation is transverse by working with the transient term Eq. (3.16), because the reflection coefficients become complex when critical reflection occurs. Critical reflection occurs when the angle of incidence $\alpha$, defined by Eq. (4.13), is less than $\alpha_{cr}$ or greater than $(\pi - \alpha_{cr})$, where $\alpha_{cr}$ is given by

$$\alpha_{cr} = \cos^{-1}(c_T/c_L). \quad (4.19)$$

Instead, we use the spectral component of the incident transverse radiation Eq. (3.34). It behaves locally like a plane wave; therefore, the spectral components of the total acceleration of the ground can be calculated exactly as if the incident radiation were a plane wave.

Let us consider in-plane transverse radiation from point A. The $x$ and $y$ spectral components $a_1^T$ and $a_2^T$ of the ground acceleration caused by this radiation, evaluated at the surface and in the plane $\mathcal{N}(0)$, are

$$a_i^T(x, \omega) = R_i^T(\alpha) a_\theta(R, \theta, \omega)(1 + \cos\theta R/\rho)^{-1/2}, \quad (4.20)$$

where $i = 1, 2$. In these expressions $a_\theta$ is the two-dimensional $\theta$ component of the particle acceleration, and it can be calculated from Eq. (3.34). It is evaluated on the surface at $r = R$. The proportionality factors $R_1^T(\alpha)$ and $R_2^T(\alpha)$ are given by

$$R_1^T(\alpha) = \frac{-2(c_L/c_T)^2 \cos 2\alpha \sin \alpha}{\sin 2\alpha \sin 2\gamma + (c_L/c_T)^2 \cos^2 2\alpha}, \quad (4.21)$$

and

$$R_2^T(\alpha) = \frac{2\sin 2\gamma \sin \alpha}{\sin 2\alpha \sin 2\gamma + (c_L/c_T)^2 \cos^2 2\alpha}, \quad (4.22)$$

where

$$\gamma = \cos^{-1}[(c_L/c_T)\cos\alpha]. \quad (4.23)$$

The distance $R$, the angle $\theta$, and the angle $\alpha$ are related to the position $x$ by Eqs. (4.11)–(4.13). In Eqs. (4.21) and (4.22) $\sin\gamma$ is defined as

$$\sin\gamma = [1 - (c_L/c_T)^2 \cos^2\alpha]^{1/2}, \quad \alpha_{cr} < \alpha < (\pi - \alpha_{cr}) \quad (4.24a)$$

$$\sin\gamma = i[\cos^2\alpha(c_L/c_T)^2 - 1]^{1/2}, \quad \alpha < \alpha_{cr} \text{ or } \alpha > (\pi - \alpha_{cr}). \quad (4.24b)$$

In the previous section, we discussed the influence of the caustic surface; therefore, it will be sufficient here to consider points $x$ on the surface that are not reached by rays emitted from A that graze the caustic surface. Thus, we shall assume that $(1 + \cos\theta R/\rho)$ is positive.

In order to investigate how the amplitudes of the accelerations $a_1^T$ and $a_2^T$ are distributed along the surface, we define the (complex) amplitude functions $A_1^T$ and $A_2^T$ as

$$A_i^T(x) = \frac{E_T^{TV}(\theta) R_i^T(\alpha)}{[(R/d)(1 + \cos\theta R/\rho)]^{1/2}}, \quad (4.25)$$

where $i = 1, 2$. In Figs. 11–14, we have plotted the real and imaginary parts of $A_1^T$ and $A_2^T$ against $\bar{x} = x/d$. The rupture front was assumed elliptical,

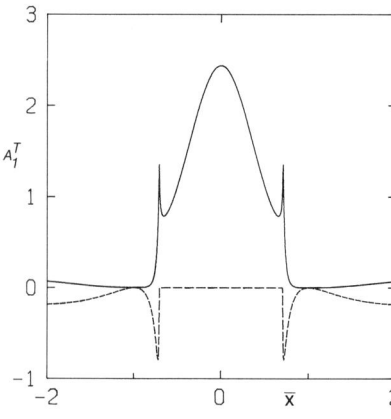

FIG. 11. The real and imaginary parts of the amplitude of the $x$ component of the surface acceleration $A_1^T$ excited by the incident transverse emission. Solid line, Re($A_1^T$). Dashed line, Im($A_1^T$). The dip angle $\phi = 0°$ and $a/b = 10$.

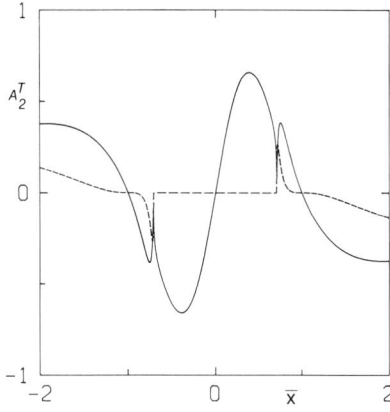

FIG. 12. The same as Fig. 11 for the $y$ component $A_2^T$.

with $a/b = 10$. The other parameters chosen were $v_1/c_T = 0.5$, $v_2/v_1 = 0$, $\kappa = 1$, $d/a = 0.5$, and Poisson's ratio = 0.25. In Figs. 11 and 12 the dip angle $\phi = 0°$, and in Figs. 13 and 14, $\phi = 30°$. In general the $x$ component of the ground motion is strong at points directly above A, the $y$ component less so. The most distinctive feature of all the plots is, however, the extreme behavior at points $x$ on the surface corresponding to the critical angles $\alpha_{cr}$ and $(\pi - \alpha_{cr})$. We have seen behavior like this earlier. A glance at Fig. 6 shows that a similar phenomenon takes place in the transverse emission near the headwave angle $\theta_h$. The phenomena in both cases are similar. At the two points on the surface struck by a transverse ray incident at $\alpha_{cr}$ or

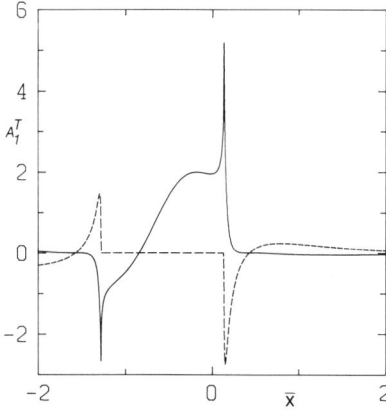

FIG. 13. The same as Fig. 11 except $\phi = 30°$.

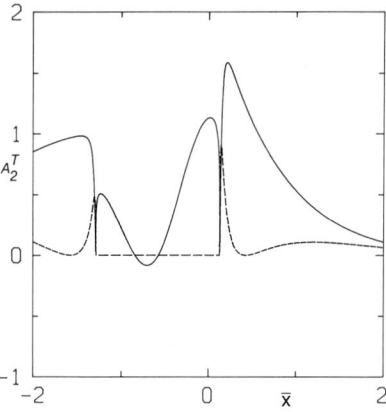

FIG. 14. The same as Fig. 12 except $\phi = 30°$.

$(\pi - \alpha_{cr})$, both reflected rays and a headwave are being excited. The confluence of the headwave with the reflected waves causes the strong spikes or sudden zeros. Our calculations are not as accurate as would be desirable near these two points, but they do indicate that particularly strong ground motion does take place near these points. Note that when $\alpha < \alpha_{cr}$ or $\alpha > (\pi - \alpha_{cr})$, $A_1^T$ and $A_2^T$ both acquire imaginary parts.

To learn more about the ground's acceleration we need to invert Eq. (4.20) using Eq. (3.28b). To make our calculations simpler, we shall assume that $\phi$ is such that the rays emitted from A do not leave the slip zone at an angle $|\theta| > \theta_h$. In other words $E_T^{TV}$ is real. Carrying out the inversion, we get for the $x$ and $y$ components of acceleration, the expression

$$a_i^T(x, t) = \{\text{Re}[R_i^T(\alpha)]a_\theta(R, \theta, t) - \text{Im}[R_i^T(\alpha)]b_\theta(R, \theta, t)\}$$
$$\times (1 + \cos\theta R/\rho)^{-1/2}, \qquad (4.26)$$

where $i = 1, 2$ and $a_\theta$ is calculated using Eq. (3.16) for $E_T^{TV}$ real and setting $r = R$. The term $b_\theta$ is given by

$$b_\theta(r, \theta, t) = \pm(c_T/2\pi r)^{1/2} E_T^{TV}(\theta) h(t, r/c_T), \qquad (4.27)$$

where $h(t, r/c_T)$ is given by Eq. (4.18) with $c_L$ replaced by $c_T$. In Eq. (4.27), $r$ is evaluated at $R$ for points on the surface. Note that $h(t, R/c_T)$ does not vanish for $t < R/c_T$. This is because a critically reflected longitudinal ray skimmed the surface and arrived at the observation point before the incident shear ray arrived. Note further the distortion of the waveform caused by the critical reflection.

Had we considered rays emitted from point B then the calculation would have been more complicated. If we had taken modest values for the dip angle $\phi$ and points $x$ on the surface at or near a point directly above B, then the rays would have been emited at angles $|\theta| > \theta_h$ so that $E_T^{TV}$ would be complex. Moreover, the rays would have grazed the caustic surface picking up a $(-\pi/2)$ phase shift in the frequency domain. Finally, the rays might have been critically reflected at the surface.

## C. Antiplane Reflections

### 1. Thin Layer Approximation

A surface layer can have a very considerable effect upon ground motion (see, e.g., Bouchon, 1980a,b). We shall attempt to account for an overlaying layer by considering a relatively thin one, whose presence can be modeled as a boundary condition on the substrate. Though the two assumptions, that the frequency be high and the layer be thin, limit the applicability of our calculations, the results do illustrate the effect of the layer. When considering an overlaying layer, we shall confine ourselves to discussing antiplane radiation from the rupture front. The layer will have a definite frequency response; consequently, we shall do much of our work in the frequency domain and only transform our final results to the time domain.

The boundary condition can be derived by considering a balance of linear momentum for an element of the layer shown in Fig. 15 (Simons, 1975). The result is

$$\frac{\partial u_3}{\partial y} = -Ghk_T\left[\frac{1}{k_T}\frac{\partial^2 u_3}{\partial x^2} + v_T^2 k_T u_3\right], \quad y = 0, \qquad (4.28)$$

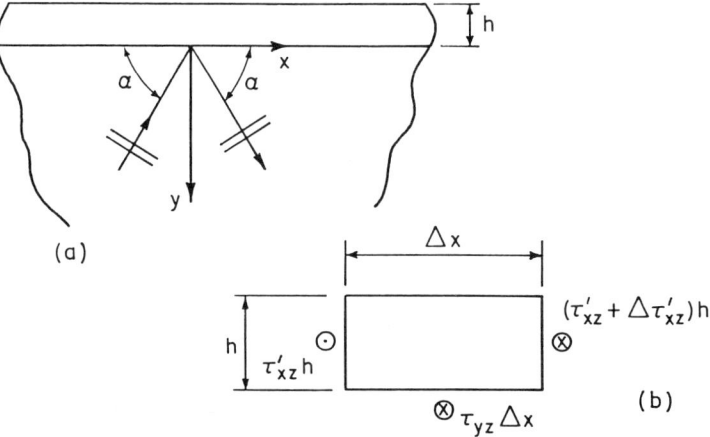

**Fig. 15.** A layer of thickness $h$, which overlays an elastic half-space. A free-body diagram of an element of the layer.

where $u_3$ is the $z$ component of displacement. Here $h$ is the thickness of the layer, $G = \mu'/\mu$ and $\nu_T = c_T/c_T'$, where the prime identifies the parameters of the layer.

The simplest wave motion that satisfies the equation for antiplane wave motion

$$\nabla^2 u_3 + k_T^2 u_3 = 0 \quad (4.29)$$

and the boundary condition Eq. (4.28) is a superposition of incident and reflected plane waves; namely,

$$u_3(x, y, \omega) = \exp(ik_T x \cos\alpha)$$
$$\times [\exp(-ik_T y \sin\alpha) + R^{TH}(\alpha) \exp(ik_T y \sin\alpha)], \quad (4.30)$$

where $\alpha$ is the angle of incidence shown in Fig. 15, and $R^{TH}(\alpha)$ is the reflection coefficient

$$R^{TH}(\alpha) = \frac{i\sin\alpha - Ghk_T(\nu_T^2 - \cos^2\alpha)}{i\sin\alpha + Ghk_T(\nu_T^2 - \cos^2\alpha)}. \quad (4.31)$$

Note that $R^{TH}(\alpha)$ when viewed as a function of the complex variable $\alpha$, has a pole at $\alpha = \alpha^*$, where

$$i\sin\alpha^* + Ghk_T(\nu_T^2 - \cos^2\alpha^*) = 0. \quad (4.32)$$

The existence of a pole in Eq. (4.31) indicates that another solution of Eqs. (4.28) and (4.29) is the surface wave

$$u_3^L(x, y, \omega) = \exp[ik_T(x\cos\alpha^* + y\sin\alpha^*)], \quad (4.33)$$

where $\alpha^*$ follows from Eq. (4.32) as

$$\sin\alpha^* = i\left\{\left[1 + 4(Ghk_T)^2(\nu_T^2 - 1)\right]^{1/2} - 1\right\} \Big/ 2Ghk_T, \quad (4.34)$$

or for $hk_T < 1$

$$\alpha^* \cong iGhk_T(\nu_T^2 - 1). \quad (4.35)$$

Note that we have assumed that $\nu_T^2 > 1$; i.e., the transverse wave speed in the layer is smaller than that in the substrate. It can be shown that when $hk_T < 1$, Eq. (4.33) approximates the lowest Love-wave mode in the layer–substrate system.

2. Antiplane Transverse Emission

By using the spectral component of the incident, antiplane, transverse radiation, Eq. (3.36), and the reflection coefficient $R^{TH}(\alpha)$, Eq. (4.31), we can derive an approximation to the acceleration of the layer. We consider radiation from point A, evaluate our results at the surface, confine them to the plane of symmetry $\mathcal{N}(0)$, and assume that the rays from A have not grazed the caustic surface. The result is

$$a_3^L(x, \omega) = \frac{2i \sin \alpha}{i \sin \alpha + Ghk_T(\nu_T^2 - \cos^2 \alpha)} a_3(R, \theta, \omega)(1 + \cos \theta R/\rho)^{-1/2}, \quad (4.36)$$

where $a_3$ is calculated using Eq. (3.36).

To see what happens in the time domain, we need to invert Eq. (4.36) using Eq. (3.28b). There are, however, some difficulties in carrying out this inversion. The spectral component $a_3$ was calculated on the assumption that $\omega > \omega_{min}$, where $\omega_{min}$ depends on how fast the rupture front changes its speed. But the boundary condition Eq. (4.28) was derived under the assumption that $(\omega h/c_T) < 1$ or $\omega < \omega_{max}$, where $\omega_{max} = c_T/h$. Therefore the expression for $a_3^L$, Eq. (4.36), is accurate for $\omega_{min} < \omega < \omega_{max}$ and, though we can invert Eq. (4.36) mathematically, its meaning is limited by $\omega_{min}$ and $\omega_{max}$.

Taking the case $\kappa = 1$, we can invert Eq. (4.36) to get

$$a_3^L(x, t) = \pm \left(\frac{c_T}{2\pi R}\right)^{1/2} \frac{E_T^{TH}(\theta)}{(1 + \cos \theta R/\rho)^{1/2}} \frac{2c_T \sin \alpha}{Gh(\nu_T^2 - \cos^2 \alpha)}$$

$$\times C\left(\frac{\pi^{1/2}}{2}\right) \exp\left[\frac{-c_T \sin \alpha}{Gh(\nu_T^2 - \cos^2 \alpha)}\left(t - \frac{R}{c_T}\right)\right] H\left(t - \frac{R}{c_T}\right). \quad (4.37)$$

We can see the conflict in our approximations by examining the exponential term. Both $(t - R/c_T)$ and $h$ are small, but their ratio need not be. Equation (4.37) should be compared with Eqs. (4.7) and (4.26). The frequency-dependent reflection coefficient causes considerable modifications to the waveform function.

### V. Far-Field Ground Motions

In the far field, the body waves have decayed, and the ground motion is dominated by Rayleigh and Love waves. This section deals with the surface waves excited by the radiation emitted from the rupture front when it changes its speed of advance. For the three-dimensional case, the calculations are complicated. To explain them as clearly as possible, we shall display in some detail a simpler two-dimensional case and then discuss the three-dimensional case by analogy. The two-dimensional example is that of a strike-slip fault in a half-space with a thin overlaying layer. The change in rupture front speed generates antiplane radiation that excites Love-wave motion.

## A. A Simple Two-Dimensional Example

Up until this point, we have been able to interpret our results in terms of rays. However, when we come to consider surface waves, the ray approach runs into some complications. This can be illustrated by considering the case of antiplane radiation in a half-space covered by a thin layer. Our earlier discussion has revealed that Love-wave motion (Section IV.C.1) can be excited by a ray incident at the imaginary angle $\alpha^*$, that is, by a complex ray. Moreover, the Love wave itself, because its phase function depends upon $\alpha^*$, can be thought of as a complex ray. Therefore, to calculate the excitation and propagation of a Love wave, we have to track complex rays from the source to the observation point through complex space. This can be done, as we indicated in the introduction, but it is an awkward procedure to implement. Instead we shall construct a representation integral for Love waves and evaluate it asymptotically. In this way the problem becomes one in complex analysis rather than one in complex differential geometry.

### 1. Integral Representation

Consider a two-dimensional half-space containing a semi-infinite crack. The geometry is shown in Fig. 16. Further assume that there is a thin layer on the surface that can be modeled by the boundary condition given by Eq. (4.28). We seek a representation in the frequency domain of the solution to the reduced wave equation, Eq. (4.29), within this half-space. Figure 16 shows a curve $\mathscr{S} = \mathscr{S}_r + \mathscr{S}_s + \mathscr{S}_i$ with $\mathscr{S}_i$ surrounding the tip of the crack. If the Green's function for the half-space with the boundary condition Eq. (4.28) is $u_3^G(\mathbf{x}, \mathbf{x}')$ (the $\omega$ dependence has been suppressed until it is needed) and if the displacement $u_3(\mathbf{x})$ is an outgoing wave at infinity, then the displacement at an arbitrary position $\mathbf{x}$ can be expressed in the form

$$u_3(\mathbf{x}) = \int_{\mathscr{S}_i} \left[ u_3^G(\mathbf{x}', \mathbf{x}) \frac{\partial u_3}{\partial n}(\mathbf{x}') - u_3(\mathbf{x}') \frac{\partial u_3^G}{\partial n}(\mathbf{x}', \mathbf{x}) \right] d\mathscr{S}(\mathbf{x}'), \quad (5.1)$$

where the normal derivatives are calculated by differentiating with respect to the argument $\mathbf{x}'$. The wave field $u_3$ can be decomposed into three parts, namely,

$$u_3(\mathbf{x}) = u_3^I(\mathbf{x}) + u_3^S(\mathbf{x}) + u_3^{MS}(\mathbf{x}). \quad (5.2)$$

Here $u_3^I$ represents the displacement in an unbounded medium, $u_3^S$ repre-

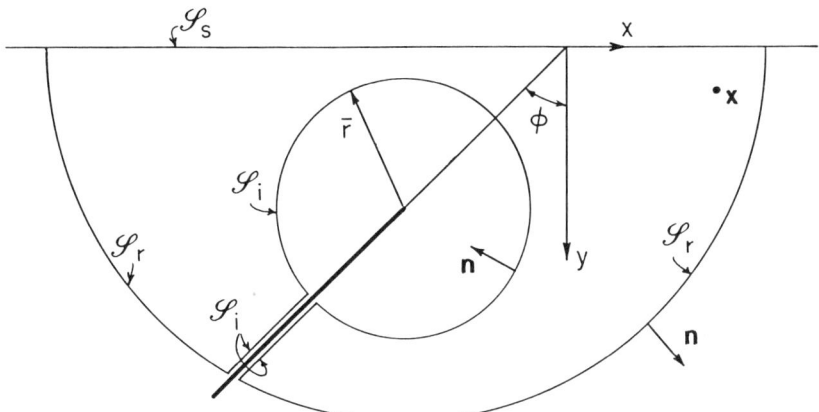

FIG. 16. The contour or surface $\mathscr{S} = \mathscr{S}_r + \mathscr{S}_s + \mathscr{S}_i$ for deriving the integral representations for the surface wave.

sents the scattering of $u_3^I$ from the surface of the half-space and $u_3^{MS}$ represents multiple wave interactions between the source and the surface of the half-space. Because $u_3^S$ satisfies the source-free equation within $\mathscr{S}_i$, the term with $u_3^S$ vanishes. Next, the terms with $u_3^{MS}$ in Eq. (5.1) are neglected, because, in the time domain, these disturbances arrive at the surface much later than $u_3^I$. Finally, because we are only interested in the Love-wave contribution to the field $u_3$, only the Love-wave component of the Green's function, which is denoted by $u_3^{GL}(x, x')$, need be used in Eq. (5.1). After these simplifications Eq. (5.1) reduces to

$$u_3^L(x) = \int_{\mathscr{S}_i} \left[ u_3^{GL}(x', x) \frac{\partial u_3^I}{\partial n}(x') - u_3^I(x') \frac{\partial u_3^{GL}}{\partial n}(x', x) \right] d\mathscr{S}(x'), \quad (5.3)$$

where $u_3^L$ is the Love-wave displacement. To complete Eq. (5.3), it only remains to compute $u_3^{GL}$.

The antiplane Green's function $u_3^G$ for the half-space is the solution to

$$\nabla^2 u_3^G + k_T^2 u_3^G = -\delta(x - x')\delta(y - y'), \quad (5.4)$$

which satisfies the boundary condition Eq. (4.28). The Green's function is obtained by first representing the solution to Eq. (5.4) in an unbounded medium as an integral over a spectrum of plane waves. Each member of this spectrum can be considered as a wave incident upon the plane $y = 0$, and hence each corresponding reflected wave can be obtained by multiplication by the reflection coefficient. By summing over all the reflected waves,

we obtain
$$u_3^{GR}(x, x') = -\frac{i}{4\pi}\int_{\mathscr{B}} R^{TH}(\alpha)\exp\{ik_T[|x - x'|\cos\alpha + (y + y')\sin\alpha]\}\,d\alpha. \quad (5.5)$$

The contour $\mathscr{B}$ starts at $(\pi - i\infty)$ and ends at $i\infty$ in the complex $\alpha$ plane. By deforming this contour to one of steepest descents, we pick up a contribution from the pole of $R^{TH}(\alpha)$ when

$$(y + y')/|x - x'| < Ghk_T(\nu_T^2 - 1). \quad (5.6)$$

The approximation to $\alpha^*$ Eq. (4.35) has been used to obtain Eq. (5.6). The pole contribution is the Love-wave contribution to the Green's function, namely,

$$u_3^{GL}(\mathbf{x}, \mathbf{x}') = \alpha^* \exp\{ik_T[|x - x'|\cos\alpha^* + (y + y')\sin\alpha^*]\}, \quad (5.7)$$

where again we have used the approximation to $\alpha^*$ and terms of $O[(hk_T)^2]$ have been neglected in the amplitude. This result is almost identical to the one given by J. B. Keller and Karal (1960).

2. Calculation of the Love Wave in the Frequency Domain

To evaluate Eq. (5.3) we take the contour $\mathscr{S}_i$ to be a wave front of the radiation emitted from the rupture front plus some portions along either surface of the fault plane. Because we are interested in high-frequency radiation, we may approximate $u_3^I$ by using the particle velocity Eq. (3.36), and because we want $k_T r$ to be large we may evaluate Eq. (5.3) by the method of stationary phase. The $(x, y)$ coordinate system shown in Fig. 16 and the $(r, \theta)$ system used to calculate the antiplane emission, Eq. (3.36), are related by the equations

$$x = -d\tan\phi + r\sin(\theta + \phi) \quad (5.8)$$
$$y = d - r\cos(\theta + \phi). \quad (5.9)$$

Note that the coordinate systems discussed here agree with those in Fig. 7. The radius of $\mathscr{S}_i$ is $\bar{r}$ (Fig. 16). We shall assume that $-\pi/2 < \phi < \pi/2$, but that $(x + d\tan\phi) \geq 0$. This convention differs from the one used for the near-field results.

Upon substitution of

$$u_3^I(\mathbf{x}) = v_3(r, \theta, \omega)/(-i\omega) \quad (5.10)$$

and $u_3^{GL}$ from Eq. (5.7) into Eq. (5.3), we find that the Love-wave particle displacement is given by

$$u_3^L(x, y, \omega) = \mp i[ik_T\bar{r}/(2\pi)]^{1/2}\alpha^* F(\omega) W^L(x, y)$$
$$\times \exp(ik_T d\sin\alpha^*)\exp(ik_T\bar{r})I(\alpha^*, \phi), \quad (5.11)$$

where
$$W^L(x, y) = \exp\{ik_T[(x + d\tan\phi)\cos\alpha^* + y\sin\alpha^*]\}, \quad (5.12)$$
and
$$I(\alpha^*, \phi) = \int_{-\pi}^{\pi} E_T^{TH}(\theta)[\sin(\phi + \theta + \alpha^*) + 1]$$
$$\times \exp[-ik_T\bar{r}\sin(\phi + \theta + \alpha^*)] d\theta. \quad (5.13)$$

The complex angle $\alpha^*$ is given by Eq. (4.34). The stationary phase point is located at $\bar{\theta} = \pi/2 - \phi - \alpha^*$. Asymptotically evaluating Eq. (5.10) we get for the particle acceleration

$$a_3^L(x, y, \omega) = A^L(-i\omega)^2 F(\omega) W^L(x, y), \quad (5.14)$$

where
$$A^L = \pm 2\alpha^* E_T^{TH}(\pi/2 - \phi)\exp(ik_T d\alpha^*). \quad (5.15)$$

The + sign is to be used when the rupture front suddenly increases its speed of propagation from $v_2$ to $v_1$, and the − sign when it suddenly decreases its speed from $v_1$ to $v_2$. In calculating the final expressions, Eqs. (5.14) and (5.15), we retained terms of $O(k_T h)$ only and, accordingly, used the approximation Eq. (4.35) for $\alpha^*$. Note that Eqs. (5.14) and (5.15) are independent of $\bar{r}$ as they should be.

Next, consider the physical conclusions that can be drawn from the result. First, we note that the emission coefficient $E_T^{TH}$ is a maximum for $|\phi|$ near 30° though it is still quite large when $|\phi|$ is near 0°. Next, we note that the strength of the Love wave depends on the boundary condition at $y = 0$ and the depth $d$ of the rupture front. Equation (5.15) shows the dependence on the boundary condition through the multiplying term $\alpha^*$ and the dependence on $d$ through the exponential term. Because $k_T h$ must be some number smaller than one, the frequency dependence of the Love-wave acceleration at the surface $y = 0$ is mainly determined by the exponential dependence on $d$ and the function $F(\omega)$. On the other hand, the body-wave acceleration contains the term $(k_T r)^{-1/2} F(\omega)$, as can be seen in Eq. (3.36). Hence, the body wave decays faster with increasing frequency and increasing distance from the source than does the Love wave, and in the far field the Love wave will dominate.

3. Transient Love Wave

When we want to invert Eq. (5.14) to get the time-domain results, we encounter the same limitation as encountered in Section IV.C.2 when we inverted the near-field results, namely, that the expression Eq. (5.14) is accurate only in the frequency window $\omega_{\min} < \omega < \omega_{\max}$. With this cautionary note in mind, we can express the transient Love-wave acceleration as

$$a_3^L(x, y, t) = \pm 2 E_T^{TH}(\pi/2 - \phi)[Gh(\nu_T^2 - 1)/c_T] a^L(x, y, t), \quad (5.16)$$

where

$a^L(x, y, t)$

$$= C\Gamma\left(\frac{\kappa + 2}{2}\right) \frac{\text{Re}}{\pi} \left[ \exp\left(\frac{i\kappa\pi}{4}\right) \int_0^\infty \exp\left\{-i\omega\left[t - \frac{(x + d\tan\phi)}{c_T}\right]\right\} \right.$$

$$\left. \times \omega^{2-\kappa/2} \exp\left\{-\omega^2 \left[\frac{Gh}{c_T^2}(\nu_T^2 - 1)(y + d)\right]\right\} d\omega \right]. \quad (5.17)$$

In general, Eq. (5.17) can be expressed in terms of Kummer's confluent hypergeometric series. Rather than do this, however, we shall consider the limiting case $\kappa = 2$. This case corresponds to yielding at the rupture front instead of brittle rupture. Then Eq. (5.17) becomes

$$a^L(x, y, t) = \frac{C[t - (x + d\tan\phi)/c_T]}{4\pi^{1/2}[(Gh/c_T^2)(\nu_T^2 - 1)(y + d)]^{3/2}}$$

$$\times \exp\left\{\frac{-[t - (x + d\tan\phi)/c_T]^2}{4(Gh/c_T^2)(\nu_T^2 - 1)(y + d)}\right\}. \quad (5.18)$$

By examining Eqs. (5.17) and (5.18), we can infer that the dynamical parameter $\kappa$ can have a considerable effect on both the shape of the waveform and the decay with depth $d$.

B. RAYLEIGH WAVES

1. Integral Representation

Rayleigh waves excited by radiation from a propagating rupture front that suddenly changes its speed of advance will be investigated, first in the frequency domain and then in the time domain. The overall geometry is shown in Fig. 7. The calculation is analogous to the one for two-dimensional Love waves. Using the Green's displacement and stress tensors, $u_{i;k}^G$ and $\tau_{ij;k}^G$, for the elastic half-space (Achenbach et al., 1982, pp. 63–75) and the reciprocity identity, the particle displacement $u_k$ can be expressed as a surface integral

$$u_k(\mathbf{x}) = \int_{\mathscr{S}_i} \left[ u^G_{i;k}(\mathbf{x}',\mathbf{x})\tau_{ij}(\mathbf{x}') - u_i(\mathbf{x}')\tau^G_{ij;k}(\mathbf{x}',\mathbf{x}) \right] n_j \, d\mathscr{S}(\mathbf{x}'), \quad (5.19)$$

where $\mathscr{S}_i$ is a surface enclosing the rupture front (Fig. 16) and $n_j$ is a unit normal pointing inward. The displacement $u_i$ within the integral is approximated by $u_i^I$, the wave field radiated by rupture in an unbounded medium. The Green's tensors within the integral are broken up into a sum of a Rayleigh-wave term, indicated by $u^{GR}_{i;k}$ and $\tau^{GR}_{ij;k}$, and body-wave terms. The integral containing the former gives the Rayleigh wave excited by the rupture. Thus, the Rayleigh-wave displacement $u_k^R$ is given by

$$u_k^R(\mathbf{x}) = \int_{\mathscr{S}_i} \left[ u^{GR}_{i;k}(\mathbf{x}',\mathbf{x})\tau^I_{ij}(\mathbf{x}') - u^I_i(\mathbf{x}')\tau^{GR}_{ij;k}(\mathbf{x}',\mathbf{x}) \right] n_j \, d\mathscr{S}(x'). \quad (5.20)$$

We are interested in calculating $u_k^R$ at high frequencies or, equivalently, in the far field; therefore, we need only the far-field approximations to $u^{GR}_{i;k}$ and $\tau^{GR}_{ij;k}$. Equations describing $u^{GR}_{i;k}$ in the far field are summarized in Appendix A. Similarly, we use Eqs. (3.29) and (3.34) to calculate far-field expressions for $u^I_i$ and $\tau^I_{ij}$. Because these terms consist of a longitudinal and transverse part, Eq. (5.20) breaks into two integrals. One is evaluated by taking $\mathscr{S}_i$ to be the wave front of the longitudinal part of $u_i^I$ and the other by taking $\mathscr{S}_i$ to be the wave front of the transverse part. Each surface integral is evaluated by the method of stationary phase. This is consistent with our previous far-field or high-frequency approximations.

2. Calculating the Rayleigh Waves in the Frequency Domain

Let us consider the special case of a circular rupture front of radius $a$. In this case the coordinate system $(r, \theta, \psi)$, shown in Fig. 7, is a toroidal one that is related to the $(x, y, z)$ system by Eqs. (4.4)–(4.6). We take the surface $\mathscr{S}_i$ to be

$$r = \bar{r}, \quad -\psi_1 < \psi < \psi_2, \quad -\pi < \theta < \pi. \quad (5.21)$$

We shall take the angles $\psi_1$, $\psi_2 < \pi$, because only the upper part of the rupture front is assumed significant to the excitation of Rayleigh waves. To further simplify the analysis, we consider the point of observation to lie in the plane of reflection symmetry $\mathscr{N}(0)$. Observation points not quite in the symmetry plane are considered in the closing part of Appendix B. Finally, we adopt the convention that the distance $(x/d + \tan \phi)$ is always positive; thus, the observation point lies ahead of the rupture front when $\phi$ is positive and behind when $\phi$ is negative. This is the same convention that we adopted earlier in the two-dimensional Love-wave calculation.

For this case the stationary-phase points of Eq. (5.20) are

$$\psi = 0, \tag{5.22a}$$

$$\theta = \pi/2 - \phi - \theta_\alpha, \quad \alpha = L, T, \tag{5.22b}$$

where

$$\theta_\alpha = i\cosh^{-1}(c_\alpha/c_R), \quad \alpha = L, T. \tag{5.23}$$

The stationary-phase calculations are straightforward but lengthy. The most important details are given in Appendix B. The end result of these calculations is a high-frequency approximation to the Rayleigh-wave acceleration, $\mathbf{a}^R(x, y, 0, \omega)$ given by the expressions

$$\begin{aligned}\mathbf{a}^R(x, y, 0, \omega) = \pm\, 4\mathbf{U}(\mathbf{x})[g_L E_L^{TV}(\pi/2 - \phi - \theta_L)\exp(-ik_L \bar{\mathbf{p}}^L \cdot \mathbf{x}_A) \\ - \nu^{-1}\eta g_T E_T^{TV}(\pi/2 - \phi - \theta_T)\exp(-ik_T \bar{\mathbf{p}}^T \cdot \mathbf{x}_A)] \\ \times D^{-1} F(\omega)(-i\omega)^2 e^{i\pi/2}, \end{aligned} \tag{5.24}$$

where

$$\mathbf{U}(\mathbf{x}) = (2\nu_R/\nu^2)[\mathbf{d}^L \exp(ik_L \mathbf{p}^L \cdot \mathbf{x}) + \nu\eta\, \mathbf{d}^T \exp(ik_T \mathbf{p}^T \cdot \mathbf{x})], \tag{5.25}$$

$$g_\alpha = \left[1 + \frac{d}{a}\left(\frac{x}{d} + \tan\phi\right)\frac{\sin(\phi + \theta_\alpha)}{\cos\theta_\alpha}\right]^{-1/2}, \quad \alpha = L, T, \tag{5.26}$$

$$\mathbf{p}^\alpha = \cos\theta_\alpha \mathbf{i} + \sin\theta_\alpha \mathbf{j}, \quad \alpha = L, T, \tag{5.27}$$

$$\mathbf{d}^L = \mathbf{p}^L, \tag{5.28a}$$

$$\mathbf{d}^T = \mathbf{k} \times \mathbf{p}^T, \tag{5.28b}$$

$$\mathbf{x} = x\mathbf{i} + y\mathbf{j}, \tag{5.29a}$$

$$\mathbf{x}_A = -d\tan\phi\, \mathbf{i} + d\mathbf{j}. \tag{5.29b}$$

The $+$ sign is used for a speeding up event and the $-$ sign for a slowing down event. The term $D$ is given by Eq. (A7). The overbars in Eq. (5.24) indicate the complex conjugate. The constant $\nu = c_L/c_T$, and the constant $\eta = \cot 2\theta_T$. The terms $g_\alpha$ ($\alpha = L, T$), which are the spreading factors for the Rayleigh waves, describe how the Rayleigh wave decays as it spreads across the surface of the half-space. Because $\theta_\alpha$ is complex, $g_\alpha$ is complex; accordingly, the branch of the square root in Eq. (5.26) is selected so that $\mathrm{Re}(g_\alpha)$ is positive. Finally, note that, though these results were calculated for the special case of a circular slip zone, they can be generalized to the case of a slip zone whose rupture front at $\psi = 0$ has an arbitrary radius of curvature $\rho$, by replacing $a$ in Eq. (5.26) by $\rho$.

## 3. Transient Rayleigh Waves

To calculate the transient waveforms of the particle accelerations, we must invert Eq. (5.24) using Eq. (3.28b). These operations give the expressions for the Rayleigh-wave particle acceleration $\mathbf{a}^R(x, y, z, t)$ in the symmetry plane $\mathcal{N}(0)$ as

$$\mathbf{a}^R(x, y, 0, t) = (2\nu_R C/\nu^2 D)(c_R/d)^{(4-\kappa)/2}[a_1(x, y, t)\mathbf{i} + a_2(x, y, t)\mathbf{j}], \quad (5.30)$$

where

$$a_i(x, y, t) = \pm \frac{4}{\pi} \Gamma\left(\frac{\kappa+2}{2}\right) \Gamma\left(\frac{4-\kappa}{2}\right) \sum_{\alpha,\beta=L,T} a_i^{\alpha\beta}(x, y, t), \quad (5.31)$$

$$a_i^{\alpha\beta}(x, y, t) = \frac{A_i^{\alpha\beta} \cos\{[(4-\kappa)/2]\tan^{-1}(B^{\alpha\beta}) - \theta_i^{\alpha\beta} - (\kappa\pi/4)\}}{(C^{\alpha\beta})^{(4-\kappa)/2}\left[1 + (B^{\alpha\beta})^2\right]^{(4-\kappa)/4}}, \quad (5.32)$$

$$A_i^{\alpha\beta} \exp(i\theta_i^{\alpha\beta}) = S^{\alpha\beta} d_i^\alpha E_\beta^{TV}(\pi/2 - \phi - \theta_\beta) g_\beta, \quad (5.33)$$

$$B^{\alpha\beta} = \frac{[(c_R t/d) - (x/d + \tan\phi)]}{C^{\alpha\beta}}, \quad (5.34)$$

$$C^{\alpha\beta} = \left[1 - (c_R/c_\alpha)^2\right]^{1/2}(y/d) + \left[1 - (c_R/c_\beta)^2\right]^{1/2}. \quad (5.35)$$

The $+$ and $-$ signs in Eq. (5.31) have their usual meanings. In Eqs. (5.31)–(5.33) $i = 1, 2$ and in Eqs. (5.32)–(5.35) $\alpha, \beta = L, T$. The terms $S^{\alpha\beta}$ are given by

$$S^{LL} = 1, \quad (5.36a)$$

$$S^{TL} = \nu\eta, \quad (5.36b)$$

$$S^{LT} = \eta/\nu, \quad (5.36c)$$

$$S^{TT} = -\eta^2. \quad (5.36d)$$

The parameter $\kappa$ must satisfy the inequalities $1 \leq \kappa < 4$, where $\kappa \geq 1$ for physical reasons (Section III.A), while $\kappa < 4$ in order that the particular transform inversion used to calculate Eqs. (5.30)–(5.35) remains valid. This range of $\kappa$ encompasses all physically reasonable values.

We next consider the physical conclusions that can be drawn from our calculations. First we note that the arrival time, which follows from $B^{\alpha\beta} = 0$, suggests that the Rayleigh wave emanates from a point directly above A (Fig. 7). However, no Rayleigh wave can exist at the surface unless the incident waves properly couple to the surface (Ewing *et al.*, 1957, pp.

63–66). As a consequence, a position at the surface where a Rayleigh wave is observed must satisfy one of the inequalities

$$(x/d + \tan \phi) > \left[(c_\alpha/c_R)^2 - 1\right]^{-1/2}, \quad \alpha = L, T. \quad (5.37)$$

Taking the more stringent of the two, and setting Poisson's ratio = 0.25, gives $(x/d + \tan \phi) > 2.3$. Because our calculation is a high-frequency or far-field one, this inequality is always satisfied.

The depth dependence of the Rayleigh wave manifests itself through the normalizing term $(c_R/d)^{(4-\kappa)/2}$ and through the term $(C^{\alpha\beta})^{(\kappa-4)/2}$. Note that for $\kappa = 1$, the decay with depth is greatest. But the depth dependence also manifests itself in another way: $d$ is the natural normalizing distance. The normalized time is $\bar{t} = (c_R t)/d$, and the normalized lengths are $\bar{x} = x/d$ and $\bar{y} = y/d$. This leaves only the ratio $\bar{d} = d/a$ in $g_\beta$ ($\beta = L, T$).

In the plots of the results, two conventions regarding $\phi$ need to be noted. First, recall that the angle $\phi$ can take positive and negative values but $(x + d \tan \phi)$ remains positive. This, however, means that $\theta$ is positive in a downward sense for $\phi$ positive, but is positive in an upward sense for $\phi$ negative. Second, the horizontal distance from the rupture front in the plane of symmetry is $(x + d \tan \phi)$ so that when $\phi$ is varied this distance changes even though $x$ is fixed. In all the plots the parameters are assigned fixed values: $v_1/c_T = 0.5$, $v_2/c_T = 0$, $\bar{y} = 0$, $\bar{d} = 0.5$, and Poisson's ratio = 0.25. Moreover, all the plots are for stopping events.

Figures 17 and 18 show the normalized components of particle acceleration $a_i$ ($i = 1, 2$) for a stopping event plotted versus $\tau = (\bar{t} - \bar{x})$. In Fig. 17,

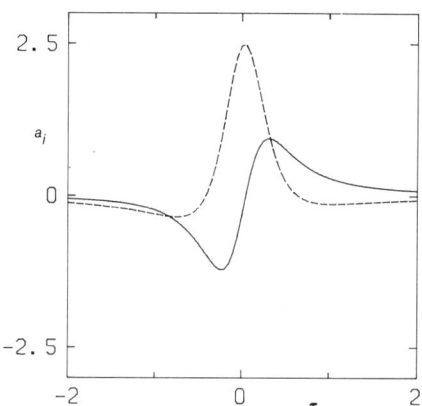

FIG. 17. The Rayleigh-wave acceleration components $a_i$ evaluated at the surface for a stopping event. Solid line, $i = 1$. Dashed line, $i = 2$. The parameter $\kappa = 1$, $\phi = 0°$, and $\bar{x} = 10$.

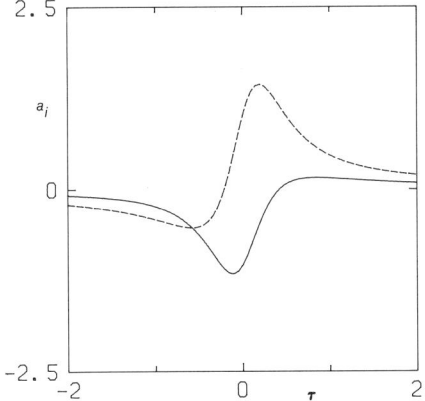

**FIG. 18.** The same as Fig. 17 except $\kappa = 2$.

$\kappa = 1$ and in Fig. 18, $\kappa = 2$. The angle $\phi = 0°$ and $\bar{x} = 10$ in both cases. The values $\kappa = 1$ and $\kappa = 2$ correspond to brittle fracture and to some yielding at the rupture front, respectively. Comparison of Fig. 17 with Fig. 18 shows that increasing $\kappa$ reduces the sharpness of the signal though the basic waveform remains unchanged. This is an expected result, and it agrees with our earlier observations of the behavior of $E_T^{TH}$ as $\kappa$ is varied.

For a stopping event, Figs. 19–22 show the normalized components of particle acceleration $a_i$ for a stopping event plotted versus $\tau = (\bar{t} - \bar{x})$. In Fig. 19, $\phi = 30°$ and $\bar{x} = 5$; in Fig. 20, $\phi = 0°$ and $\bar{x} = 5$; in Fig. 21, $\phi = -30°$ and $\bar{x} = 5$; and in Fig. 22, $\phi = -30°$ and $\bar{x} = 20$. The parame-

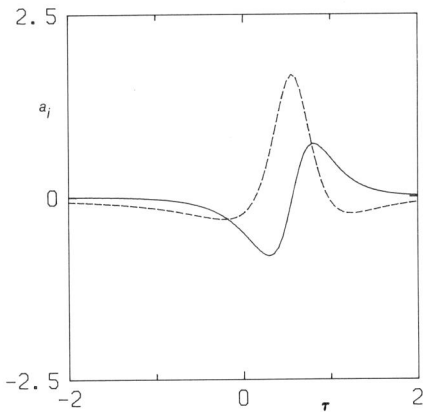

**FIG. 19.** The same as Fig. 17 except $\phi = 30°$ and $\bar{x} = 5$.

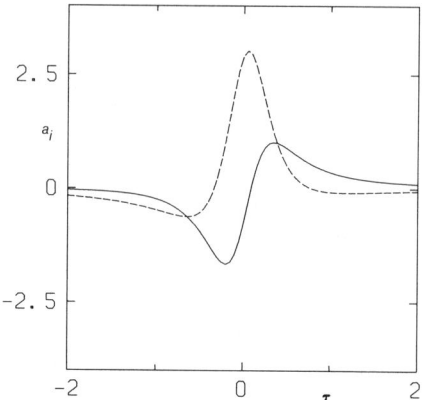

**FIG. 20.** The same as Fig. 17 except $\bar{x} = 5$.

ter $\kappa = 1$ throughout. Comparison of Figs. 19–21 shows that $\phi = 0°$ and $\phi = -30°$ both produce stronger signals than $\phi = 30°$. This is caused, in part, by the normalized distance $(\bar{x} + \tan \phi)$ getting smaller. In addition, however, the $\phi = 0$ case is strong because $|E_T^{TV}|$ is large near $\pi/2 - \theta_T$ (recall $\theta_T$ is complex), and the $\phi = -30°$ case is strong because, as shall be shown, $|g_T|$ is large near $\bar{x} = 5$. Comparison of either Fig. 17 with Fig. 20, or Fig. 21 with Fig. 22 shows the geometrical decay of the signals as $\bar{x}$ becomes larger.

Figure 23 shows the magnitudes of the spreading factors $g_L$ and $g_T$ plotted against $\bar{x}$ for $\phi = -30°$. Each plot begins at that value of $\bar{x}$ at

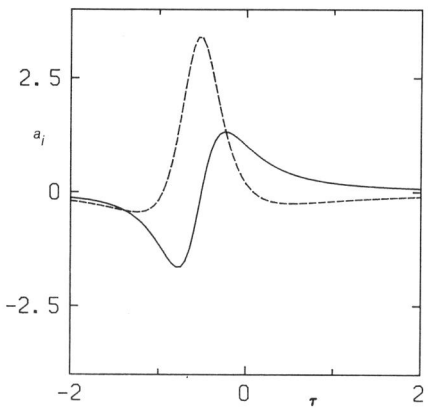

**FIG. 21.** The same as Fig. 17 except $\phi = -30°$ and $\bar{x} = 5$.

1. MODELING OF STRONG GROUND MOTION                    45

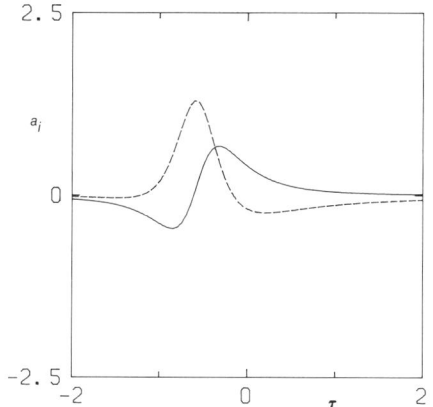

**FIG. 22.** The same as Fig. 17 except $\phi = -30°$ and $\bar{x} = 20$.

which the Rayleigh wave first appears. For example, according to Eq. (5.37), $\bar{x} > 2.9$ for the case of $|g_T|$. Note that both $|g_L|$ and $|g_T|$ go through a maximum and then start to fall off approximately as $(\bar{x})^{-1/2}$ (the geometrical decay for a Rayleigh wave excited by a buried point source). These maxima are the result of a focusing effect caused by the curved rupture front. Recall that it was earlier observed that the points at which the spreading factor $(1 + r/\bar{\rho})^{-1/2}$ becomes unbounded define a caustic surface. Note, however, that in the case of Rayleigh waves the spreading

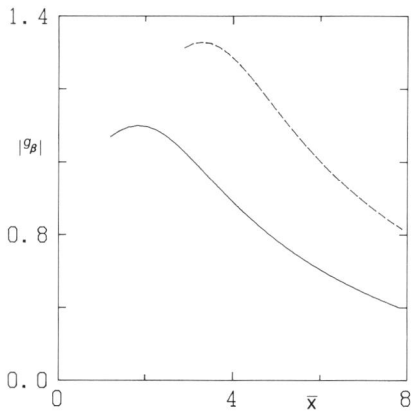

**FIG. 23.** The magnitude of the spreading factor $|g_\beta|$. Solid line, $\beta = L$. Dashed line, $\beta = T$. The dip angle $\phi = -30°$.

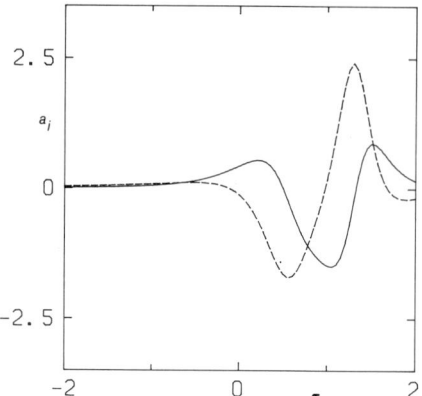

FIG. 24. The Rayleigh-wave acceleration components $a_i$ evaluated at the surface for a combined starting and stopping event. Solid line, $i = 1$. Dashed line, $i = 2$. The parameter $\bar{\delta} = 0.1$, $\phi = 30°$, $\bar{x} = 5$, and $\kappa = 1$.

factors $|g_L|$ and $|g_T|$ become unbounded only in complex space; i.e., the caustic surface for Rayleigh waves lies entirely in complex space.

In Figs. 17–22 we considered a single stopping event. A starting and stopping event has been combined in Figs. 24 and 25. The Rayleigh-wave particle acceleration for this combined event is still given by Eq. (5.30). However, Eq. (5.31) needs to be rewritten as

$$a_i(x, y, t) = \frac{4}{\pi} \Gamma\left(\frac{\kappa + 2}{2}\right) \Gamma\left(\frac{4 - \kappa}{2}\right) \sum_{\alpha, \beta = L, T} \left[ \left(\frac{d}{d_1}\right)^{(4-\kappa)/2} a_{i1}^{\alpha\beta}(x, y, t) - \left(\frac{d}{d_2}\right)^{(4-\kappa)/2} a_{i2}^{\alpha\beta}\left(x, y, t - \frac{2\delta}{v}\right)\right], \quad i = 1, 2. \quad (5.38)$$

The components $a_{i1}^{\alpha\beta}$ are given by Eqs. (5.32)–(5.35) with $d$ replaced by $d_1 = (d + \delta \cos \phi)$, and the components $a_{i2}^{\alpha\beta}$ are given by the same equations with $d$ replaced by $d_2 = d - \delta \cos \phi$ and $t$ replaced by $t - (2\delta/v)$. The distance $2\delta$ is the distance the rupture front travels at a constant speed $v$ after starting at $t = 0$. The normalized $\delta$ is $\bar{\delta} = \delta/a$. The distance $d$ is the depth of the midpoint and is used as the normalizing distance. Figures 24 and 25 show the particle acceleration components $a_i$ ($i = 1, 2$) plotted against $\tau = (\bar{t} - \bar{x})$. In Fig. 24, $\bar{\delta} = 0.1$, and in Fig. 25 $\bar{\delta} = 0.05$. The angle $\phi = 30°$, and $\bar{x} = 5$ in both cases. Comparison of Fig. 19 with Figs. 24 and 25 shows that the signals of the combined event are essentially the derivatives of the single event.

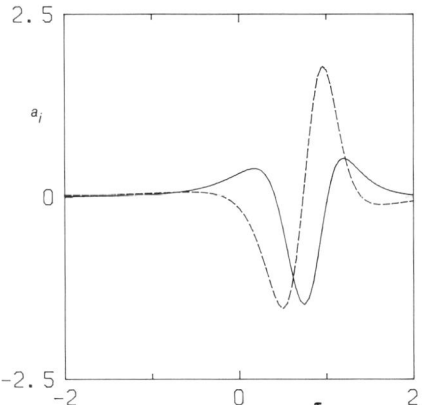

FIG. 25. The same as Fig. 24 except $\bar{\delta} = 0.05$.

## C. Love Waves

We close this article with a brief look at the excitation of Love waves in three dimensions. The starting point of our calculation is an integral representation completely analogous to Eq. (5.20). This expression is

$$u_k^L(\mathbf{x}) = \int_{\mathscr{S}_i} \left[ u_{i;k}^{GL}(\mathbf{x}', \mathbf{x}) \tau_{ij}^I(\mathbf{x}') - u_i^I(\mathbf{x}') \tau_{ij;k}^{GL}(\mathbf{x}', \mathbf{x}) \right] n_j \, d\mathscr{S}(\mathbf{x}'), \quad (5.39)$$

where $u_k^L$ is the Love-wave displacement, and $u_{i;k}^{GL}$ and $\tau_{ij;k}^{GL}$ are the Love-wave components of the Green's displacement tensor and stress tensor, respectively. They satisfy the boundary condition Eq. (4.28). The derivation of $u_{i;k}^{GL}$ parallels that of the Rayleigh wave, which we describe in Appendix A. The one component of $u_{i;k}^{GL}$ that we need to know is the displacement in the $z$ direction generated by a point load in the $z$ direction. It is given by

$$u_{3;3}^{GL}(\mathbf{x}, \mathbf{x}') = \left( \frac{k_T}{2\pi R} \right)^{1/2} \frac{\alpha^* \cos^2 \chi}{\mu}$$

$$\times \exp\{ ik_T [R \cos \alpha^* + (y + y') \sin \alpha^*] \} \exp\left( \frac{-i\pi}{4} \right), \quad (5.40)$$

where

$$R = [(x - x') + (z - z')]^{1/2}, \quad (5.41)$$

$$\cos \chi = (x - x')/R, \quad (5.42)$$

and the angle $\alpha^*$ is given by Eq. (4.35). In Eq. (5.40) we have retained only the terms of $O(hk_T)$ in the amplitude. Moreover, we have assumed that $k_T R \gg 1$.

Consider again the circular rupture zone described by Eqs. (4.4)–(4.6). We allow that $-\pi/2 < \phi < \pi/2$, but assume, as we have done previously, that $(x + d \tan \phi)$ is always positive. Moreover, we take the observation point to lie in the plane of symmetry $\mathcal{N}(0)$. The terms $u_k^I$ and $\tau_{ij}^I$ in Eq. (5.39) are constructed from Eq. (3.36), and $\mathcal{S}_i$ is taken as the wave front described by Eq. (5.21). With this information we can construct the integral Eq. (5.39) and evaluate it by the method of stationary phase. The stationary point is

$$\psi = 0, \qquad \theta = \pi/2 - \phi - \alpha^*. \tag{5.43}$$

Upon completion of the calculation, the Love-wave particle acceleration in three dimensions is obtained as

$$a_3^L(x, y, 0, \omega) = g a_3^L(x, y, \omega), \tag{5.44}$$

where $a_3^L(x, y, \omega)$ is given by Eq. (5.14). The spreading factor $g$ is given by

$$g = \left[1 + \frac{d}{a}\left(\frac{x}{d} + \tan \phi\right) \sin(\phi + \alpha^*)\right]^{-1/2}. \tag{5.45}$$

This is the same as $g_\alpha$, Eq. (5.26), where $\alpha^*$ replaces $\theta_\alpha$, and the approximation $\cos \alpha^* \cong 1$ has been used. Note that $g$ is complex. The square root is determined so that $\text{Re}(g) \geq 0$. The transient case follows from Eqs. (5.16) and (5.17), provided the phase angle associated with $g$ is taken into account.

## VI. Summary

Elastic waves that are emitted when slip occurs on a fault plane and the strong ground motions that these waves excite when they interact with the surface of the earth have been analyzed by the use of asymptotic methods.

When a zone of sliding suddenly starts to grow or suddenly changes its rate of growth, each point on the rupture front produces rays of longitudinal and transverse wave motion. The nature of these wave motions is determined by local conditions on the rupture front, such as the rupture process, the radius of curvature of the rupture front, and the angle that each ray makes with the fault plane at the points of emission. The waves propagating along these rays have been approximated by calculating the waves produced by propagation of a semi-infinite crack in fracture modes II and III and subsequently correcting them for the local curvature of the rupture front. Both the discontinuities at the wave fronts and the high-frequency spectral components of the emitted disturbances have been calculated.

In the near field, the ground motion is dominated by the reflection of emitted disturbances, while in the far field, it is dominated by surface waves. Green's tensors for an elastic half-space, or for an elastic half-space supporting a very thin surface layer, have been used to construct representation integrals for the near- and far-field ground motions. These integrals have been evaluated over wave fronts of the waves excited at the rupture front, and they are therefore mathematical expressions of Huyghen's principle. The integrations have been carried out asymptotically in keeping with the earlier approximations to the emissions from the rupture front. The formulation of the problem is quite general, but explicit, closed-form results have been given only for events along the surface that lie in a plane of reflection symmetry of the slip zone.

In the near field the pattern of surface accelerations is mainly determined by the angle at which the ray leaves the fault plane, the orientation of the fault plane, the nature of the rupture process, and the angle of incidence. Of particular interest is the critical reflection of the incident transverse ray. The point at which it is incident experiences stronger ground acceleration than do neighboring points. In the far field the pattern of surface acceleration is more strongly influenced by the surface conditions than by the rupture processes, which is to be expected because surface waves dominate there. In the frequency domain, the surface waves carry, however, some information about the rupture processes. Synthetic accelerograms have been computed for both the near and far fields.

**Appendix A. The Rayleigh-Wave Components of the Green's Tensor**

Consider an elastic half-space with a traction-free boundary. The Green's tensor $u_{i;k}^G$ represents the displacement components excited by a time-harmonic point load, applied at $\mathbf{x} = \mathbf{x}'$ in the direction $\mathbf{i}_k$. The tensor $\tau_{ij;k}^G$ is the associated stress. Both tensors are derived by Achenbach *et al.* (1982, pp. 63–75). The Rayleigh-wave contribution to $u_{i;k}^G$ is given by

$$u_{i;k}^{GR}(\mathbf{x}, \mathbf{x}') = A[k_R \mathbf{p} \cdot (\mathbf{x} - \mathbf{x}')]^{-1/2} U_i(\mathbf{x}) \overline{U}_k(\mathbf{x}'), \qquad (A1)$$

where

$$\mathbf{U}(\mathbf{x}) = (2\nu_R/\nu^2)[\mathbf{d}^L \exp(ik_L \mathbf{p}^L \cdot \mathbf{x}) + \nu\eta \, \mathbf{d}^T \exp(ik_T \mathbf{p}^T \cdot \mathbf{x})]. \quad (A2)$$

The overbar indicates the complex conjugate. This is an asymptotic result that is only accurate if $k_R|\mathbf{x} - \mathbf{x}'| \gg 1$. The various ancillary formulas are

$$A = (e^{i\pi/4}k_L)/[\mu(2\pi)^{1/2}D], \tag{A3}$$

$$\mathbf{p} = [(x - x')\mathbf{i} + (z - z')\mathbf{k}]/[(x - x')^2 + (z - z')^2]^{1/2}, \tag{A4}$$

$$\mathbf{p}^\alpha = \cos\theta_\alpha \mathbf{p} + \sin\theta_\alpha \mathbf{j}, \quad \alpha = L, T, \tag{A5}$$

$$\mathbf{d}^L = \mathbf{p}^L, \tag{A6a}$$

$$\mathbf{d}^T = (\mathbf{p} \times \mathbf{j}) \times \mathbf{p}^T \tag{A6b}$$

$$D = [\nu_R \nu^2(\nu_R^2 - \nu^2)^{1/2}]^{-1} \frac{d}{dt}\left[4t^2(t^2 - \nu^2)^{1/2}(t^2 - 1)^{1/2} - (\nu^2 - 2t^2)^2\right], \tag{A7}$$

$$\theta_\alpha = i\cosh^{-1}(c_\alpha/c_R), \quad \alpha = L, T, \tag{A8}$$

$$\eta = \cot 2\theta_T. \tag{A9}$$

The derivative in Eq. (A7) is evaluated at $t = \nu_R$. Also $\nu = c_L/c_T$, and $\nu_R = c_L/c_R$. Wave numbers are defined by $k_\alpha = \omega/c_\alpha$ ($\alpha = L, T, R$), where $\omega$ is the angular frequency, and $c_\alpha$ are the L, T, and Rayleigh wave speeds.

## Appendix B.  Evaluation of the Representation Integral

The surface of integration $\mathcal{S}_i$ of Eq. (5.20) is the toroidal surface Eq. (5.21). Using the Rayleigh-wave Green's tensors $u_{i;k}^{GR}$ and $\tau_{ij;k}^{GR}$, where the former is given by Eq. (A1), in combination with the displacement $u_i^I$, which can be calculated from Eqs. (3.29) and (3.34), we write the representation integral Eq. (5.20) as a sum of eight integrals each of which has in its integrand the term $\exp[i(k_\alpha \mathbf{p}^\alpha \cdot \mathbf{x}' - k_\beta \bar{\mathbf{p}}^\beta \cdot \mathbf{x}]$ ($\alpha, \beta = L, T$), where the overbar indicates the complex conjugate. The vectors $\mathbf{p}^\alpha$ are given by Eq. (A5) with the coordinates $(x, z)$ and $(x', z')$ interchanged. It can be shown that the physically relevant, stationary-phase point of each integrand occurs at that point $\mathbf{x}'$ for which $\mathbf{n} = \mathbf{p}^\alpha$, where $-\mathbf{n}$ is a unit vector in the $r$ direction (Fig. 7). It can also be shown that as a consequence only four integrals are nonzero at their stationary-phase points. Each of these integrals has the general form

$$I^{\alpha\beta} = \frac{(k_R k_\alpha)^{1/2}}{2\pi i} \int_{-\pi}^{\pi} \int_{\psi_1}^{\psi_2} \left[\frac{(a\bar{r})(a + \bar{r}\cos\theta)}{f(\theta, \psi)}\right]^{1/2} H^\alpha(\mathbf{x}', \mathbf{x}) E_\alpha^{TV}(\theta)$$
$$\times \exp\{ik_\alpha[\bar{r} + F^{\alpha\beta}(\theta, \psi)]\}\, d\psi\, d\theta, \quad \alpha, \beta = L, T. \tag{B1}$$

The functions $H^\alpha$ assume the simple forms
$$H^L = 1, \tag{B2a}$$
$$H^T = \eta \tag{B2b}$$
at the stationary-phase points $\mathbf{n} = \mathbf{p}^L$ and $\mathbf{n} = \mathbf{p}^T$, respectively. The toroidal coordinates $(r, \theta, \psi)$ are related to the Cartesian coordinates $(x, y, z)$ by Eqs. (4.4)–(4.6). However, it is necessary to introduce yet another coordinate system to locate the observation point $\mathbf{x}$, namely,
$$x = \chi \cos \varepsilon, \tag{B3a}$$
$$z = \chi \sin \varepsilon, \tag{B3b}$$
with $y$ unchanged. With these new coordinates, we can express $f(\theta, \psi)$ and $F^{\alpha\beta}(\theta, \psi)$ as
$$\begin{aligned} f(\theta, \psi) = \{ & [(a + \bar{r}\cos\theta)\sin\psi - \chi \sin\varepsilon]^2 \\ & + [(a + \bar{r}\cos\theta)\cos\psi \sin\phi + \bar{r}\sin\theta\cos\phi \\ & - a\sin\phi - d\tan\phi - \chi\cos\varepsilon]^2\}^{1/2} \end{aligned} \tag{B4}$$
$$\begin{aligned} F^{\alpha\beta}(\theta, \psi) = & f(\theta, \psi)\cos\theta_\alpha \\ & + [\bar{r}\sin\theta\sin\phi - (a + \bar{r}\cos\theta)\cos\psi\cos\phi]\sin\theta_\alpha \\ & + (d + a\cos\phi)\sin\theta_\alpha + y\cos\theta_\alpha\tan\theta_\beta, \quad \alpha, \beta = L, T. \end{aligned} \tag{B5}$$

The stationary-phase points of the integral Eq. (B1) are found by solving the two simultaneous equations $F_\theta^{\alpha\beta} = 0$ and $F_\psi^{\alpha\beta} = 0$ for $\theta$ and $\psi$. In general these are complicated transcendental equations that can only be solved explicitly when $\varepsilon = 0$ (i.e., when the observation point lies in the plane of symmetry). For $\varepsilon = 0$, the stationary-phase point of interest is given by $\psi = 0$ and $\theta = \pi/2 - \phi - \theta_\alpha$. Performing the stationary-phase calculation we obtain
$$I^{\alpha\beta} = g_\alpha E_\alpha^{TV}(\pi/2 - \phi - \theta_\alpha) H^\alpha \exp\left[i\left(k_\alpha \mathbf{p}^\alpha \cdot \mathbf{x}_A - k_\beta \bar{\mathbf{p}}^\beta \cdot \mathbf{x}\right)\right],$$
$$\alpha, \beta = L, T, \tag{B6}$$
where $g_\alpha$, $E_\alpha^{TV}$, and $H^\alpha$ are given by Eqs. (5.26), (3.17)–(3.19), (B2a), and (B2b), respectively. The vectors $\mathbf{x}$ and $\mathbf{x}_A$ are given by Eq. (5.29a) and (5.29b). Note that $\mathbf{p} = -\mathbf{i}$ in this appendix whereas $\mathbf{p} = \mathbf{i}$ in Eq. (5.27). This difference arises because we moved the complex conjugate from the $\mathbf{x}$ term to the $\mathbf{x}_A$ term in Eq. (B6) to arrive at Eqs. (5.24)–(5.29).

The complex factors $g_\alpha$ in Eq. (B6) express the focusing effect that the slip-zone geometry has upon the Rayleigh wave. It can be shown that
$$\frac{-[(\pi/2) + \tan^{-1}(\tan\phi|\cot\theta_\alpha|)]}{2} < \arg(g_\alpha) < 0, \quad \alpha = L, T. \tag{B7}$$
The upper limit occurs as $a \to \infty$, while the lower limit occurs as $\chi \to \infty$.

Let us now assume that the observation point **x** has a small $z$ component. Thus, the angle $\varepsilon$ is small and the equations $F_\psi^{\alpha\beta} = 0$ and $F_\theta^{\alpha\beta} = 0$ can be solved for values of $\psi$ and $\theta$, correct to $O(\varepsilon^2)$. The critical value of $\theta$ remains unchanged, while $\psi = \varepsilon g_\alpha^2(x/a)$. The phase of Eq. (B6) is thus changed. However, its amplitude remains unchanged to $O(\varepsilon^2)$. Note that $\psi$ is complex because the $g_\alpha$ are complex; that is, the "flash point" moves into complex space.

## Acknowledgments

The work of the first author was supported by the National Science Foundation Grant EAR 82-07090. That of the second author was supported by the National Science Foundation Grant MEA 83-11385. Both authors wish to thank James Loewenherz for his help with the computer work and Isabel Von Driska for her excellent typing of the manuscript.

## References

Achenbach, J. D. (1972). Dynamic effects in brittle fracture. *In* "Mechanics Today" (S. Nemat-Nasser, ed.), Vol. 1, pp. 1–57. Pergamon, New York.

Achenbach, J. D. (1973). "Wave Propagation in Elastic Solids." North-Holland Publ., Amsterdam.

Achenbach, J. D. (1977). Wave propagation, elastodynamic stress singularities, and fracture. *In* "Proceedings of the 14th IUTAM Congress" (W. T. Koiter, ed.), pp. 71–87. North-Holland Publ., Amsterdam.

Achenbach, J. D., and Bazant, Z. P. (1975). Elastodynamic near-tip stress and displacement fields for rapidly propagating cracks in orthotropic materials. *ASME J. Appl. Mech.* **42**, 183–189.

Achenbach, J. D., and Brock, L. M. (1973). Surface motions due to sub-surface sliding. *Bull. Seismol. Soc. Am.* **63**, 1473–1486.

Achenbach, J. D., and Harris, J. G. (1978). Ray method for elastodynamic radiation from a slip zone of arbitrary shape. *J. Geophys. Res.* **83**, 2283–2291.

Achenbach, J. D., and Harris, J. G. (1979). Acoustic emission from a brief crack propagation event. *ASME J. Appl. Mech.* **46**, 104–112.

Achenbach, J. D., Gautesen, A. K., and McMaken, H. (1982). "Ray Methods for Waves in Elastic Solids." Pitman, Boston, Massachusetts.

Ahluwalia, D. S., Keller, J. B., and Jarvis, R. (1974). Elastic waves produced by surface displacements. *SIAM J. Appl. Math.* **26**, 108–119.

Aki, K. (1969). Scaling law of seismic spectrum. *J. Geophys. Res.* **72**, 1217–1231.

Aki, K. (1972). Scaling law of earthquake-source time function. *Geophys. J. R. Astron. Soc.* **31**, 3–25.

Aki, K. (1982). Strong motion prediction using mathematical modelling techniques. *Bull. Seismol. Soc. Am.* **72**, S29–S41.

Aki, K. (1983). Strong-motion seismology. *In* "Earthquakes: Observation, Theory and Interpretation" (H. Kanamori and E. Boschi, eds.), Proc. Int. Sch. Phys. "Enrico Fermi," Course LXXXV, pp. 223–250. North-Holland Publ., Amsterdam.

Aki, K., and Richards, P. G. (1980). "Quantitative Seismology," Vol. 2, pp. 851–911. Freeman, San Francisco, California.

Andrews, D. J. (1976a). Rupture propagation with finite stress in antiplane strain. *J. Geophys. Res.* **81**, 3575–3582.

Andrews, D. J. (1976b). Rupture velocity of plane strain shear cracks. *J. Geophys. Res.* **81**, 5679–5687.

Archuleta, R. J., and Frazier, G. A. (1978). Three-dimensional numerical simulations of dynamic faulting in a half-space. *Bull. Seismol. Soc. Am.* **68**, 541–572.

Archuleta, R. J., and Hartzell, S. H. (1981). Effects of fault finiteness and near-source ground motion. *Bull. Seismol. Soc. Am.* **71**, 939–957.

Bernard, P., and Madariaga, R. (1984a). A new asymptotic method for the modeling of near-field accelerograms. *Bull. Seismol. Soc. Am.* **74**, 539–557.

Bernard, P., and Madariaga, R. (1984b). High-frequency seismic radiation from a buried circular fault. *Geophys. J. R. Astron. Soc.* **78**, 1–17.

Boatwright, J. (1982). A dynamic model for far-field accelerations. *Bull. Seismol. Soc. Am.* **72**, 1049–1068.

Bouchon, M. (1980a). The motion of the ground during an earthquake 1. The case of a strike-slip fault. *J. Geophys. Res.* **85**, 356–366.

Bouchon, M. (1980b). The motion of the ground during an earthquake 2. The case of a dip-slip fault. *J. Geophys. Res.* **85**, 367–375.

Brune, J. (1970). Tectonic stress and the spectra of seismic shear waves from earthquakes. *J. Geophys. Res.* **75**, 4997–5009.

Burridge, R., and Halliday, G. S. (1971). Dynamic shear cracks with friction as models for shallow focus earthquakes. *Geophys. J. R. Astron. Soc.* **25**, 261–283.

Carrier, G. F., Krook, M., and Pearson, C. E. (1966). "Functions of a Complex Variable." McGraw-Hill, New York.

Chatterjee, A. K., and Knopoff, L. (1983). Bilateral propagation of a spontaneous two-dimensional anti-plane shear crack under the influence of cohesion. *Geophys. J. R. Astron. Soc.* **73**, 449–473.

Das, S. (1980). A numerical method for determination of source time functions for general three-dimensional rupture propagation. *Geophys. J. R. Astron. Soc.* **62**, 591–604.

Ewing, W. H., Jardetzky, W. S., and Press, F. (1957). "Elastic Waves in Layered Media." McGraw-Hill, New York.

Freund, L. B. (1975). Dynamic crack propagation. *In* "The Mechanics of Fracture" (S. Nemat-Nasser, ed.), Vol. AMD-19, pp. 105–134. Am. Soc. Mech. Eng., New York.

Hanks, T. C. (1982). $f_{max}$. *Bull. Seismol. Soc. Am.* **72**, 1867–1880.

Hanks, T. C., and McGuire, R. K. (1981). The character of high-frequency strong ground motion. *Bull. Seismol. Soc. Am.* **71**, 2071–2095.

Harris, J. G. (1984). Wave-front approximations in a moving coordinate system. *ASME J. Appl. Mech.* **51**, 934–935.

Harris, J. G., and Achenbach, J. D. (1981). Near-field surface motions excited by radiation from a slip zone of arbitrary shape. *J. Geophys. Res.* **86**, 9352–9356.

Harris, J. G., and Achenbach, J. D. (1983). Love waves excited by discontinuous propagation of a rupture front. *Geophys. J. R. Astron. Soc.* **72**, 337–351.

Harris, J. G., and Pott, J. (1984). Surface motion excited by acoustic emission from a buried crack. *ASME J. Appl. Mech.* **51**, 77–83.

Harris, J. G., Achenbach, J. D., and Norris, A. N. (1983). Rayleigh waves excited by the discontinuous advance of a rupture front. *J. Geophys. Res.* **88**, 2233–2239.

Jennings, P. C. (1983). Engineering seismology. *In* "Earthquakes: Observation, Theory and Interpretation" (H. Kanamori and E. Boschi, eds.), Proc. Int. Sch. Phys. "Enrico Fermi," Course LXXXV, pp. 138–173. North-Holland Publ., Amsterdam.

Kanninen, M. F. (1978). A critical appraisal of solution techniques in dynamic fracture mechanics. *In* "Numerical Methods in Fracture Mechanics" (A. R. Luxmoore and D. R. J. Owen, ed.). Univ. College of Swansea, Swansea, Wales.

Karal, F. C., and Keller, J. B. (1959). Elastic wave propagation in homogeneous and inhomogeneous media. *J. Acoust. Soc. Am.* **31**, 694–705.

Keller, H. (1964). Propagation of stress discontinuities in inhomogeneous elastic media. *SIAM Rev.* **6**, 356–382.

Keller, J. B. (1958). A geometrical theory of diffraction. *In* "Calculus of Variations and its Applications" (L. M. Graves, ed.), Vol. 8, pp. 27–52. Am. Math. Soc., Providence, Rhode Island.

Keller, J. B., and Karal, F. C. (1960). Surface wave excitation and propagation. *J. Appl. Phys.* **31**, 1039–1046.

Keller, J. B., and Karal, F. C. (1964). Geometrical theory of elastic surface wave excitation and propagation. *J. Acoust. Soc. Am.* **36**, 32–40.

Knopoff, L., and Chatterjee, A. K. (1982). Unilateral extension of a two-dimensional shear crack under the influence of cohesive forces. *Geophys. J. R. Astron. Soc.* **68**, 7–25.

Kostrov, B. V. (1964). Self-similar problems of propagation of shear cracks. *J. Appl. Math. Mech. (Engl. Transl.)* **28**, 1077–1087.

Lighthill, M. J. (1970). "Fourier Analysis and Generalised Functions." Cambridge Univ. Press, London and New York.

Madariaga, R. (1976). Dynamics of an expanding circular fault. *Bull. Seismol. Soc. Am.* **66**, 639–666.

Madariaga, R. (1977). High frequency radiation from crack (stress drop) models of earthquake faulting. *Geophys. J. R. Astron. Soc.* **51**, 625–651.

Madariaga, R. (1983a). Earthquake source theory: A review. *In* "Earthquakes: Observation, Theory and Interpretation" (H. Kanamori and E. Boschi, eds.), Proc. Int. Sch. Phys. "Enrico Fermi," Course LXXXV, pp. 1–44. North-Holland Publ., Amsterdam.

Madariaga, R. (1983b). High frequency radiation from dynamic earthquake fault models. *Ann. Geophys.* **1**, 17–23.

Papageorgiou, A. S., and Aki, K. (1983a). A specific barrier model for the quantitative description of inhomogeneous faulting and the prediction of strong ground motion I. Description of the model. *Bull. Seismol. Soc. Am.* **73**, 693–722.

Papageorgiou, A. S., and Aki, K. (1983b). A specific barrier model for the quantitative description of inhomogeneous faulting and the prediction of strong ground motion II. Applications of the model. *Bull. Seismol. Soc. Am.* **73**, 953–978.

Pierce, A. D. (1981). "Acoustics." McGraw-Hill, New York.

Richards, P. G. (1976). Dynamic motions near an earthquake fault: A three dimensional solution. *Bull. Seismol. Soc. Am.* **66**, 1–31.

Simons, D. A. (1975). Scattering of a Love wave by the edge of a thin surface layer. *ASME J. Appl. Mech.* **42**, 842–846.

Spudich, P., and Frazier, L. N. (1984). Use of ray theory to calculate high-frequency radiation from earthquake sources having spatially variable rupture velocity and stress drop. *Bull. Seismol. Soc. Am.* **74**, 2061–2082.

Virieux, J., and Madariaga, R. (1982). Dynamic faulting by a finite difference method. *Bull. Seismol. Soc. Am.* **72**, 345–369.

CHAPTER 2

# Array Analysis and Synthesis Mapping of Strong Seismic Motion

Norman A. Abrahamson

*Bechtel Civil and Minerals, Inc.*
*San Francisco, California 94119*

Bruce A. Bolt

*Seismographic Station*
*Department of Geology and Geophysics*
*University of California, Berkeley*
*Berkeley, California 94720*

## I. Prediction of Strong Ground Motion

### A. Seismic Intensity

In general wave physics, energy signifies the total energy associated with the presence of a wave, while intensity generally denotes the rate of transport of wave energy. It is unfortunate that, in seismology, seismic intensity has come to mean a measure of the strong ground shaking defined by field observations. In this chapter, seismic intensity is used in the wave theoretic sense.

The total seismic energy density (kinetic and potential) at time $t$ is

$$W(t) = \rho \dot{u}^2(t) \tag{1}$$

for traveling plane waves, where $u(t)$ is the displacement of a particle of the ground and $\rho$ is the density. Also, the wave intensity is

$$I(t) = \rho c \dot{u}^2(t). \tag{2}$$

In other words, $I$, the rate of transport of seismic energy per unit area, is the wave velocity $c$ times the energy per unit volume $W$.

In the three-dimensional case, intensity **I** is a vector whose component **I · n**, is the rate in which energy is transported in the direction of unit vector **n** across a small plane element at right angles to **n**. We then have the equation of conservation of seismic wave energy as

$$\partial W/\partial t = -\nabla \cdot \mathbf{I}. \tag{3}$$

It may be shown (Bullen and Bolt, 1985), that for seismic shear waves at the surface, the equation for seismic energy at distance $r$ from the (point) source is

$$E = (2\pi \rho r^2 \beta / F_s^2) \int \dot{u}^2 \, dt, \tag{4}$$

where $F_s$ is the free surface effect and $\beta$ is the shear wave velocity.

Seismic energy is usually measured in joules (or ergs). A convenient logarithmic scale defines the intensity level in decibels [given as $120 + 10 \log_{10}(I/\mathrm{W\ m}^{-2})$], where the magnitude of the intensity vector is measured in watts per square meter. (The characteristic intensity level at the threshold of hearing, is 40 dB at 100 Hz.) The estimation of the power output in watts for seismic sources is a fundamental question, calculated approximately (neglecting the source's directional nature and media damping) by dividing the intensity at distance $r$ in meters from the source by $4\pi r^2$. The power output of a rupturing fault in a shallow dislocation is of the order $10^{14}$ W (compared to about $10^5$ W for rocket motors in space craft). The corresponding seismic intensity is $10^5$ W m$^{-2}$ or 170 dB. More precise intensity estimates require rather dense measurements of seismic waves around the fault source, for example, by seismic strong motion arrays.

The above formulas are generally applicable to linear theory and simple plane or spherical waves with appropriate geometrical spreading terms. Although we are dealing in this chapter with observations of seismic waves close to the fault, it is often necessary to distinguish between near-field and far-field terms. Here it is sufficient to define far field as meaning that part of the earth's surface whose distance $r$ from the source is large compared to $c/\omega$, where $\omega$ is the radiating frequency. In general, it is found that in the far field, the plane wave relations become a good approximation for estimating the power of the seismic source, even though there are additional terms in the full solution of the wave field near the source.

In most of the work on synthetic modeling of strong seismic motion, the field variable mapped is ground acceleration, velocity, or displacement. Conversion to energy and intensity is given by Eqs. (2) and (4).

B. The Earthquake Source

The basic physical model for the construction of synthetic seismograms is that which H. F. Reid developed from studies of the 1906 San Francisco earthquake. The elastic rebound theory states that strains build up in the faulted rocks until a failure point is reached. Rupture then takes place and the strained rock rebounds on each side of the fault under its own elastic stresses until the strain is largely or wholly relieved. Reid (1911) stated, "It is probable the whole movement at any one point does not take place at once, but proceeds in irregular steps. The more or less sudden stopping of the movement, and the friction, gives rise to the vibrations that are propagated to a distance. The sudden starting of the motion would produce vibrations, just as would its sudden stopping." The recognition is clear that regular motions occur along the fault and are caused by intermittent locking, stress variations, or "roughness." Alternatively, it may be said that the fault surface contains "asperities." The slip over an area of the fault reduces the stress suddenly, producing a local "stress drop" $\Delta\sigma$ given by

$$\Delta\sigma = \mu d/w, \tag{5}$$

where $\mu$ is the rock rigidity, $d$ the fault offset, and $w$ the width of the rupture plane (assumed rectangular).

Haskell (1966) developed a model, "in which the fault displacement is represented by coherent waves only over segments of the fault and the radiations from adjacent sections are assumed to be statistically independent or incoherent." In this model the idea of statistical randomness of fault slip along the fault plane is introduced. More recently, Das and Aki (1977a,b) considered a fault plane having various barriers distributed over it. They conceived that rupture would start near one of the barriers and then propagate over the fault plane until it was brought to rest or slowed at the next barrier. Sometimes the barriers are broken by the dislocation; sometimes the barriers remain unbroken, but the dislocation reinitiates on the far side and continues; sometimes the barriers are not broken initially but, because of local repartitioning of the stresses and possibly nonlinear effects, it eventually breaks, usually with the occurrence of aftershocks. Mikumo and Miyatake (Chapter 3) also discuss irregular rupture models.

The starting point for the interpretation of observed near-field seismograms and the computation of synthetic seismograms for points near to the source is therefore an elastic rebound model that involves the dislocation moving along a fault plane over which roughnesses of various types are distributed stochastically. There are different kinds of geological fault rupture involving purely horizontal slip (strike-slip) or purely vertical slip (dip-slip). The different source radiation patterns affect the wave patterns generated by the different fault mechanisms. (See Vidale and Helmburger,

Chapter 6, Figs. 23–25, for radiation pattern pictures.) The theory must also incorporate the effects of the moving source, producing Doppler-like effects, depending on the speed of fault rupture in the direction of the faulting (Boore and Joyner, 1978).

## C. Numerical Prediction Models

The main lines of approach to modeling mathematically the earthquake strong motion seismic waves are threefold. The first model uses a kinematic approach in which the time history of the slip on the generating fault is known *a priori*. Several defining parameters may be specified, such as shape, rise time, and amplitude of the source function (or source time function and slip); the velocity of the slip over the fault surface; and the final area over which the slip occurred. A Green's function representation is usually adopted to calculate the resulting displacements of the medium. Green's functions for the various classifications of faulting have been constructed, and numerous theoretical papers using this approach have been published. Mikumo and Miyatake discuss this in detail in Section II, Chapter 3. The process is a kind of inverse curve fitting whereby the parameters of the source are varied in order to estimate by inspection the closeness of fit of the synthetic ground motion with observed distant radiated seismic waves. Once the seismic source is defined by this process, using distant recordings, then the near-field parameters can be used to estimate the ground motions near to the source.

The second approach is to use the differential equations involving the forces that produce the rupture. This dynamic procedure has received considerable emphasis recently. The basic model is a shear crack, which is initiated in the preexisting stress field and which causes stress concentrations around the tip of the crack. These concentrations, in turn, cause the crack to grow (Kostrov, 1966). Studies more concerned with the seismic waves that are recorded in the near field need a numerical technique, such as finite elements or finite differences, to incorporate realistic structural conditions. Mikumo and Miyatake in Section III, Chapter 3, use this approach.

The third approach is more general, takes a less specific view of the earthquake source, and proceeds without a specific kinematic or dynamic model of slip on the fault. For example, recent work by Backus (1977) takes up the fundamental idea of the uniqueness of various source descriptions. The representation of an arbitrary source of seismic waves may be given in terms of moment tensors. The representation of the seismic source is an expansion of spatial moments.

The observed complexity of strong motion records is a function of the structural complications of the path along which the waves propagate to the site as well as the complexity of the earthquake source. Three types of records are available. The primary seismogram in most cases is the accelerogram, because instruments are designed to record ground accelerations in the frequency range normally of interest to engineers. There are also complementary records of ground velocity and displacement. The availability of these three time functions is of great assistance in the interpretation of strong motion records. Generally, the accelerograms contain many high-frequency pulses and considerable variability in amplitudes. Integration to wave velocity considerably smooths these records and emphasizes frequencies in the middle range of interest. A second integration produces relatively smooth ground displacements with a simpler pattern of dominant waves, usually with periods beyond 1 sec. Sometimes, however, due to problems with baseline corrections and instrumental drift, the integrations produce large amplitude, long-period variations in displacement records, which may or may not be physically related to the actual seismic waves.

Although complex in general, there is an underlying pattern to strong motion records that follows seismic wave theory. There is an initial portion of ground motion consisting mainly of longitudinal P waves. Depending on the distance between the site and the source, there is then an onset of S waves, which are superimposed on the P waves still arriving from other parts of the moving dislocation. The greatly enhanced shaking continues, consisting of a mixture of S and P waves, but with S motions becoming richer as the duration increases. In the final portion of horizontal component records, surface waves of both Rayleigh and Love-wave types often predominate and in general are mixed with S body waves. Again, depending on the distance of the site from the causative fault and also on the structure of the intervening rocks and soils, the surface waves are dispersed into trains with certain frequency characteristics as a function of time (Hanks, 1975). This record coda is likely to be significantly affected by the focal depth of the faulted surface; the greater the depth, the smaller the amplitudes of surface waves contained in the strong motions. The record may contain pulses that can be explained in terms of the finiteness of the extended source. Thus, there may be wave pulses that may be identified as "break-out phases" and "stopping phases." Because source dimensions are not likely to be known *a priori*, such details are difficult to include in synthetic ground motions.

A further modeling difficulty is the treatment of waves arising primarily from the (unknown) distribution of roughness along the fault and, consequently, the roughness distribution density $\Phi(x)$. This stochastic component of strong ground motion has motivated an approach to modeling artificial time histories by using random number generators (Penzien, 1970). As yet, no roughness distribution densities have been proposed for the different classes of earthquakes.

## D. Use of Strong Motion Arrays

The reliable synthesis of recorded ground motions requires detailed and reliable interpretation of the ground motions in terms of the seismic source parameters. A significant advance in the resolution of the latter can be obtained by the use of arrays of strong motion accelerometers (Iwan, 1978). A common time base allows the phases of the seismic waves to be correlated between recording elements. It allows direct measurement of the rupture velocity of the source and the spatial coherency of the ground motion. Both of these are key parameters in modeling strong motion. The role of rupture velocity was mentioned in the previous section. The spatial coherence becomes especially important for the coda waves in order to determine if they are caused by rupture of multiple "asperities" or scattering (Aki and Chouet, 1975). There will be a fundamental difference in modeling if the coda are modeled by multiple subevents or by scattering of the main event.

A seismic wave front propagates across the array and triggers each instrument in a sequence that depends upon the azimuth of arrival and wave velocity. (See Section IV for an analogy with radio telescope arrays.) Thus, both the direction of the approach of each wave type and apparent velocity can be determined by cross-correlating the recorded ground motion at the array elements. From a different perspective, an array can be "steered" toward a particular section of the nearby rupturing fault to determine what type and amount of seismic energy is, at that time, radiating from that section of the fault. Indeed, the array response can be locked onto the rupture front in order to follow the dislocation as it moves from one end of the fault break to the other. Design of hybrid arrays can be arranged to deal with both source mechanism estimation and wave propagation problems.

The first large hybrid array of strong motion recorders to commence operation is located in northeast Taiwan. This array, called SMART 1 (Strong Motion Array in Taiwan number 1), initially had 37 triaxial digital accelerometers located in a concentric plan. There are now 3 concentric rings, each with 12 instruments, with radii of 200 m (I ring), 1000 m (M ring) and 2000 m (O ring), respectively; a central station (C00) at the center of the rings; and two additional stations on an extended arm. These dimensions and configuration minimize spatial aliasing for strong ground motions with frequencies in the range between 5 Hz and 2 sec. The omnidirectional nature of a circular array best suits the widely varied azimuthal distribution of likely sources of strong earthquakes in northern Taiwan. A detailed description of the SMART 1 array is given by Bolt *et al.* (1982).

## 2. ARRAY ANALYSIS AND SYNTHESIS MAPPING

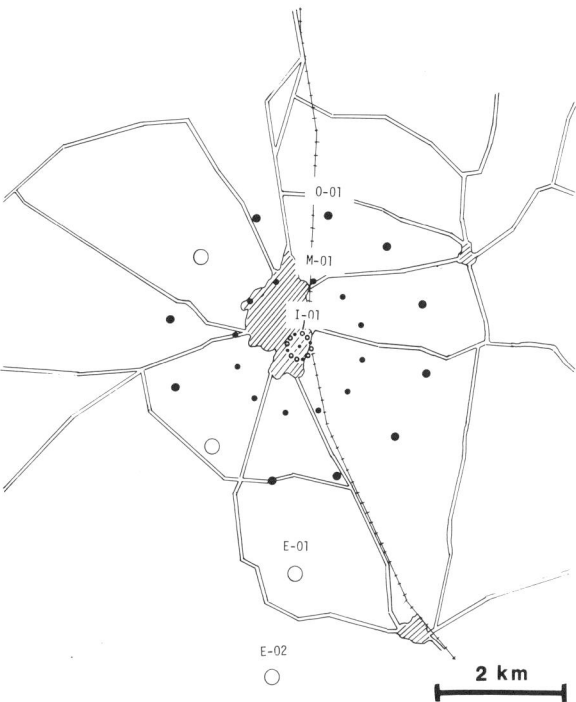

**FIG. 1.** Geometry of the SMART 1 array in Taiwan. The solid circles indicate stations that were operating during the January 29, 1981, event. The open circles indicate stations that were operational later.

On January 29, 1981, at 04:51 (UTC) a large, shallow-focus earthquake occurred off the northeastern coast of Taiwan and triggered all 27 strong motion recorders installed at that time in SMART 1 (see Fig. 1), which is located 30 km NNW of the epicenter. A peak horizontal acceleration of $0.24g$ was recorded by the array. Analysis of this earthquake and its recordings by SMART 1 will be used as illustrations in Section V of this chapter.

A 1979 earthquake in Imperial Valley, California, also provided valuable strong ground motion recordings in the near-source region. The rupture that propagated through the accelerometer network has been modeled in a variety of ways. One approach used was to discretize the fault into a finite sum of point sources. The time of the generation of each source and the amplitude and direction of slip were fit by waveform modeling. Hartzell and Helmburger (1982) used forward modeling of displacements recorded by a linear array of accelerometers normal to the Imperial Valley fault at El Centro. They estimated the average rupture velocity at 2.5–2.7 km/sec.

Olson and Apsel (1982) used a least-squares inversion of long-period ($T > 1$ sec) accelerations from the El Centro array to estimate the temporal and spatial distribution of slip. They obtained a horizontal rupture velocity between 4.0 and 5.0 km/sec. Hartzell and Heaton (1983) modeled both teleseismic body waves and strong motion recordings simultaneously. They inferred an average rupture velocity of 2.5 km/sec. Their analysis suggested an acceleration of the rupture front from 2.2 km/sec near the hypocenter to 2.8 km/sec approximately 12 km northward along the fault. Waveform modeling has been successful at matching observed long-period strong ground motion, but it is computationally expensive and to date has been unable to match recordings at frequencies much greater than 1 Hz. This is because, at high frequencies, the waveforms are sensitive to the fine details of the seismic source and the structure between the fault and receiver.

The rupture process of the 1979 earthquake has also been specially studied using the recordings from an independent 213-m linear differential array just southwest of El Centro, California. Niazi (1982) measured the time-dependent polarization of the relatively long-period ($T > 1$ sec) P waves recorded on the horizontal components. The results indicated a rupture velocity between 2–3 km/sec during the first 5 sec after which the P waves on the horizontal components were contaminated by S-wave arrivals. Chu (1984) examined the Fourier amplitude spectra for a Doppler shift from a moving source with a constant velocity. His estimate of the mean rupture velocity was $2.09 \pm 0.05$ km/sec. Spudich and Cranswick (1984) used a cross-correlation method to measure the slowness of the P and S waves at the differential array. With detailed knowledge of the velocity structure, the slowness observed at the differential array was correlated to a source position on the fault surface. In this manner, the fault surface was contoured to show the progress of the rupture surface. They estimated the average rupture velocity to be between 2.6–3.2 km/sec and suggested that the rupture velocity briefly accelerated to near the P-wave velocity. Although slowness mapping provided additional insights into the rupture process, a major difficulty in the application of the method was the restriction to a scalar slowness rather than a vector slowness caused by the unidimensional array. The absence of a second dimension to the array resulted in a surface of possible source positions for any given slowness measured at the array making it impossible to determine uniquely the rupture velocity.

In addition to source properties, seismological goals of array recordings include the measurement of coherency of the seismic waves and the determination of the dependence of coherency on frequency, station separation, and wave type (Bolt et al., 1984). A qualitative description of wave coherency observed during the 1971 San Fernando earthquake is given in Liu and Heaton (1984). Although absolute time is not available for the San

Fernando strong motion recordings, these authors were able to observe coherent phase arrivals in the velocity and acceleration time histories across distances of tens of kilometers.

In the following sections, we outline the theory of array processing applicable to the analysis and ultimately to the computational synthesis of strong seismic wave motion. The theory is also needed, along with theoretical source modeling, in the design of optimal strong motion arrays for specific purposes.

## II. Signal Estimation and Detection

### A. Array Correlation

Linear array analysis can be expressed as filtering followed by a summation as shown in Fig. 2. A standard assumption in array analysis is that the recorded signal consists of a deterministic signal plus noise. For example,

$$u_j(t) = u(\mathbf{x}_j, t) = s(t) + \varepsilon(\mathbf{x}_j, t), \tag{6}$$

where $u_j(t)$ is the output (acceleration, velocity, or displacement) of the $j$th seismometer at position $\mathbf{x}_j$, $s(t)$ is the deterministic signal, and $\varepsilon(\mathbf{x}_j, t)$ is the noise. In matrix notation Eq. (6) becomes $\mathbf{u}(\mathbf{x}, t) = s(t)\zeta + \varepsilon(\mathbf{x}, t)$, where $\zeta = \text{col}[1, 1, \ldots, 1]$. The form of Eq. (6) assumes that the signal arrives simultaneously at each station. This condition can be satisfied by introducing a delay $\tau_j$ to the output of the $j$th seismometer. For a plane wave, the delay is

$$\tau_j = \mathbf{k} \cdot \mathbf{x}_j / \omega, \tag{7}$$

where $\mathbf{k}$ is the wave number and $\omega$ is the frequency of the plane wave.

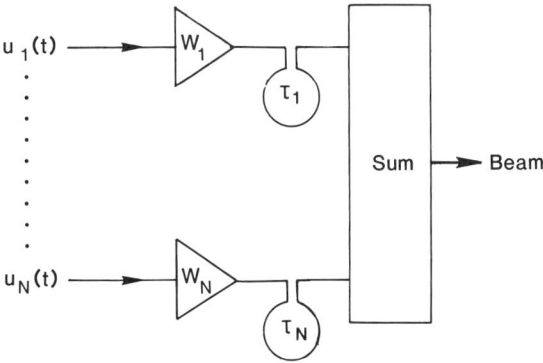

FIG. 2. General linear array processing.

In some analyses it is sufficient to consider just a single harmonic component of the signal,

$$s(\mathbf{x}, t) = A \cos(\mathbf{k} \cdot \mathbf{x} - \omega t + \delta), \tag{8}$$

where $A$ and $\delta$ are constants. The wave (phase) velocity is $c = \omega/|\mathbf{k}|$. In Eq. (8), there are two basic problems to be addressed: (a) the *detection problem*, which tests the presence of a harmonic (i.e., $A \neq 0$), and (b) the *fitting problem*, which estimates the wave number $\mathbf{k}$, the amplitude $A$, and the phase shift $\delta$, with statistical uncertainties (Brillinger, 1985).

### B. Beamforming

In array analysis, the standard method for improving the signal-to-noise ratio (SNR) is beamforming. In beamforming, a delay $\tau_j$ is inserted in $u_j(t)$. The time shifted signals are then averaged to increase the SNR. The estimate of the signal is called the *beam* and is given by

$$\hat{s}(t) = \frac{1}{N} \sum_{j=1}^{N} u_j(t + \tau_j) = s(t) + \frac{1}{N} \sum_{j=1}^{N} \varepsilon_j(t), \tag{9}$$

where $N$ is the number of stations. For any zero-mean noise, this estimate for $s(t)$ is unbiased. Assuming the $\varepsilon_j(t)$ are uncorrelated with variance $\sigma_\varepsilon^2$, beamforming reduces the variance of $\hat{s}(t)$ from $\sigma_\varepsilon^2$ to $\sigma_\varepsilon^2/N$. Capon *et al.* (1967) showed that this estimate also gives both the linear, minimum-variance estimate that is unbiased and the maximum likelihood estimate for independent, Gaussian, zero-mean noise.

The output of the array processor shown in Fig. 2 is

$$\hat{s}(t) = \mathbf{u}^T(t) \cdot \mathbf{W}, \tag{10}$$

where $\mathbf{W}$ is the vector of weights (filters) and T denotes the transpose. If the filters are normalized so that they sum to unity (e.g., $\mathbf{W}^T \cdot \boldsymbol{\zeta} = 1$), then the estimate is unbiased. The variance is given by

$$\mathrm{Var}[\hat{s}] = \mathbf{W}^T \mathbf{C}_{\varepsilon\varepsilon} \mathbf{W}, \tag{11}$$

where $\mathbf{C}_{\varepsilon\varepsilon}$ is the covariance matrix of the noise. Minimization of the variance with respect to $\mathbf{W}$ can be solved using Lagrange multipliers. The solution is

$$\mathbf{W} = \mathbf{C}_{\varepsilon\varepsilon}^{-1} \boldsymbol{\zeta} / \boldsymbol{\zeta}^T \mathbf{C}_{\varepsilon\varepsilon}^{-1} \boldsymbol{\zeta}. \tag{12}$$

If the noise is spacially white ($\mathbf{C}_{\varepsilon\varepsilon} = \mathbf{I}\sigma_\varepsilon^2$), then $\mathbf{W} = \boldsymbol{\zeta}/N$, which is the beamforming estimate [see Eq. (9)].

For the maximum likelihood estimate, $s$ is chosen such that the realized probability density function of $\mathbf{u} = s\boldsymbol{\zeta} + \boldsymbol{\varepsilon}$ is a maximum. Assuming

Gaussian, zero-mean noise, the likelihood function is given by

$$L = (2\pi)^{T/2}|\mathbf{C}_{\varepsilon\varepsilon}|^{-1/2}\exp\{-(\varepsilon^T\mathbf{C}_{\varepsilon\varepsilon}^{-1}\varepsilon)/2\},$$
$$= (2\pi)^{T/2}|\mathbf{C}_{\varepsilon\varepsilon}|^{-1/2}\exp\{-(\mathbf{u} - s\zeta)^T\mathbf{C}_{\varepsilon\varepsilon}^{-1}(\mathbf{u} - s\zeta)/2\}, \quad (13)$$

where $T$ is the number of time samples. Setting $d(\log L)/ds = 0$ yields

$$\hat{s} = \mathbf{u}^T\mathbf{C}_{\varepsilon\varepsilon}^{-1}\zeta / \zeta^T\mathbf{C}_{\varepsilon\varepsilon}^{-1}\zeta, \quad (14)$$

which uses weights identical to the linear minimum-variance weights given in Eq. (12) and corresponds to beamforming if $\mathbf{C}_{\varepsilon\varepsilon} = \mathbf{I}\sigma_\varepsilon^2$.

These estimates were derived in the time domain, but as Capon *et al.* pointed out, there are three main problems with actual computations in the time domain:

1. A large amount of computer time is required.
2. There is sensitivity to the assumption that the noise is stationary.
3. There is sensitivity to the assumption that the signal is identical across the array.

Both problems 2 and 3 commonly occur with seismic array data. To correct these difficulties, analysis may be profitably carried out in the frequency wave number domain.

## C. Frequency–Wave-Number Analysis

The recorded signals $u(x, t)$ represent a time-dependent, three-dimensional wave field. The wave field may be written as a three-dimensional (generalized) Fourier transform

$$\mathbf{u}(\mathbf{x}, t) = \int_{-\infty}^{\infty}\int_{-\infty}^{\infty}\mathbf{u}(\mathbf{k}, \omega)\exp\{-i(\mathbf{k}\cdot\mathbf{x} - \omega t)\}\,d\mathbf{k}\,d\omega, \quad (15)$$

where $\mathbf{k}$ is the wave number vector and $\omega$ the frequency. The inverse relation is

$$\mathbf{u}(\mathbf{k}, \omega) = \frac{1}{(2\pi)^3}\int_{-\infty}^{\infty}\int_{-\infty}^{\infty}\mathbf{u}(\mathbf{x}, t)\exp\{i(\mathbf{k}\cdot\mathbf{x} - \omega t)\}\,d\mathbf{x}\,dt. \quad (16)$$

The amplitude-squared spectrum is given by

$$f_{uu}(\mathbf{k}, \omega) = |\mathbf{u}(\mathbf{k}, \omega)|^2. \quad (17)$$

To estimate $\mathbf{u}(\mathbf{k}, \omega)$, the spatial integral in Eq. (16) is replaced by a weighted sum over the sampled station distribution. The weights correspond to the filters shown in Fig. 2. The spectral estimate in Eq. (17) is

$$f_{uu}(\mathbf{k}, \omega) = \left| \sum_{j=1}^{N} W_j u(\mathbf{x}_j, \omega) \exp\{i\mathbf{k} \cdot \mathbf{x}\} \right|^2 \quad (18)$$

$$= \sum_{j=1}^{N} \sum_{l=1}^{N} W_j W_l u(\mathbf{x}_j, \omega) \bar{u}(\mathbf{x}_l, \omega) \exp\{i\mathbf{k} \cdot (\mathbf{x}_j - \mathbf{x}_l)\}$$

$$= \mathbf{W}\overline{\mathbf{W}}^T \overline{\mathbf{U}}^T(\mathbf{k}) \mathbf{S}(\omega) \mathbf{U}(\mathbf{k}), \quad (19)$$

where $\mathbf{S}(\omega)$ is the *cross-spectral matrix* and the overbar indicates the complex conjugate. The elements of the *beamsteering vector* are

$$U_j(\mathbf{k}) = \exp\{i\mathbf{k} \cdot \mathbf{x}_j\}, \quad (20)$$

and have the effect of advancing the phase of the sinusoid observed at station $j$ by an amount corresponding to the time delay with respect to the array origin of a plane wave propagating with wave number $\mathbf{k}$. The estimation of $\mathbf{S}(\omega)$ for stationary signals is typically performed by averaging over several time windows (Capon, 1969). This is unsatisfactory for nonstationary seismic data, such as strong motion records, where we are often restricted to a short time window. As an alternative, smoothing may be performed over near-neighbor frequencies rather than over multiple time windows. The elements of the sample cross-spectral matrix are then

$$S_{jl} = \sum_{m=-M}^{M} a_m u\left(\mathbf{x}_j, \frac{\omega + 2\pi m}{T}\right) \bar{u}\left(\mathbf{x}_l, \frac{\omega + 2\pi m}{T}\right), \quad (21)$$

where $2M + 1$ is the number of discrete frequencies and $a_m$ are the weights used in the frequency smoothing and are normalized so that they sum to unity. A Hamming window function is a common choice for the $a_m$. The two methods of estimating $\mathbf{S}(\omega)$ have similar statistical properties. In both cases, the values to be smoothed may be approximated by independent, identically distributed, complex normal variates. The smoothing over of the near-neighbor frequencies is applied to the individual cross-spectral estimates and not to the Fourier spectral estimates, because the complex Fourier transform values have mean zero but the product inside the sum in Eq. (21) has a nonzero mean over narrow frequency bands.

The cross-spectral estimates are often normalized to unit amplitude. This normalization reduces site amplification effects. With this normalization, the cross-spectral matrix becomes simply the matrix of exponential phase differences

$$S_{jl}(\omega) = \sum_{m=-M}^{M} a_m \exp\left\{i\left[\varphi\left(\mathbf{x}_j, \frac{\omega + 2\pi m}{T}\right) - \varphi\left(\mathbf{x}_l, \frac{\omega + 2\pi m}{T}\right)\right]\right\}, \quad (22)$$

where $\varphi$ is the Fourier phase.

The frequency smoothing in Eqs. (21) and (22) uses a weighted arithmetic mean and is not appropriate for wideband analysis. Over a wide frequency band, the arithmetic mean of the product inside the sum in Eq. (21) will be zero. Therefore, smoothing a slow deterministic signal over a wide frequency band results in $S_{jl} = 0$. As an alternative, a weighted geometric mean can be used in Eq. (21). The geometric mean is valid over an arbitrarily wide frequency band.

The choice of the spatial weighting scheme used in Eq. (19) is the only difference between two commonly used techniques called the *conventional* and *high-resolution* frequency–wave-number methods. As shown earlier, the beamforming estimate uses a weighting scheme $W_j = 1/N$. These weights yield the beamforming or *conventional* (CV) estimate of the power spectrum

$$P^{CV}(\mathbf{k}, \omega) = (1/N^2)\bar{\mathbf{U}}^T(\mathbf{k})\mathbf{S}(\omega)\mathbf{U}(\mathbf{k}). \tag{23}$$

Using singular value decomposition (SVD) on the square matrix $\mathbf{S}(\omega)$, McLaughlin (1983) shows that

$$P^{CV}(\mathbf{k}, \omega) = (1/N^2)\bar{\mathbf{U}}^T(\mathbf{k})\mathbf{V}(\omega)\Lambda(\omega)\bar{\mathbf{V}}^T(\omega)\mathbf{U}(\mathbf{k}), \tag{24}$$

where the columns of $\mathbf{V}$ are the eigenvectors of $\mathbf{S}$, and $\Lambda$ is the diagonal matrix of eigenvalues of $\mathbf{S}$. Equation (24) can be written as

$$P^{CV}(\mathbf{k}, \omega) = \frac{1}{N^2} \sum_{j=1}^{N} \lambda_j(\omega)\bar{\mathbf{U}}^T(\mathbf{k})\mathbf{V}_j(\omega)\bar{\mathbf{V}}_j^T(\omega)\mathbf{U}(\mathbf{k}), \tag{25}$$

where $\lambda_j$ are the diagonal elements of $\Lambda$, and $\mathbf{V}_j$ are the columns of $\mathbf{V}$. This decomposition provides a convenient comparison with the high-resolution frequency–wave-number method discussed next.

The *high resolution* (HR) method proposed by Capon (1969, 1973) uses a more complex weighting scheme than beamforming. The weights $W_j$ are chosen such that they satisfy two conditions:

1. A pure plane wave with wave number $\mathbf{k}_0$ should pass undistorted.

2. The variance of the noise (signals with wave numbers $\mathbf{k} \neq \mathbf{k}_0$) should be minimized in a least-squares sense.

The weights that satisfy these conditions are generally different for each frequency and wave number. Capon shows that these weights are equivalent to estimating the power spectrum by

$$P^{HR}(\mathbf{k}, \omega) = [\bar{\mathbf{U}}^T(\mathbf{k})\mathbf{S}^{-1}(\omega)\mathbf{U}(\mathbf{k})]^{-1}. \tag{26}$$

The HR statistic involves the inverse of $\mathbf{S}(\omega)$; however, $\mathbf{S}(\omega)$ is often numerically singular and usually requires prewhitening before it can be inverted numerically. A method of treatment is to add a small constant (damping value) to the diagonal elements of $\mathbf{S}(\omega)$ to shift the eigenvalues away from zero. If $\mathbf{S}(\omega)$ is normalized, then Capon (1969) recommends a damping value of 0.05.

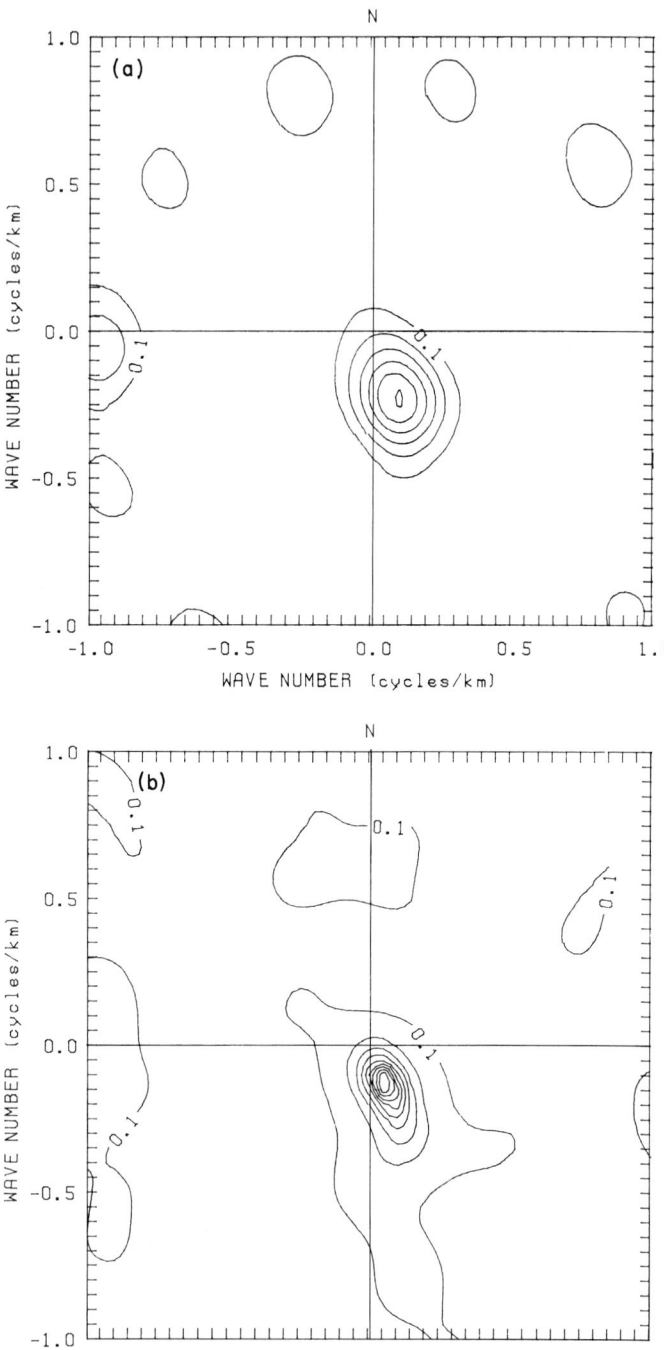

Using SVD, Eq. (26) becomes

$$P^{HR}(\mathbf{k}, \omega) = \left[ \sum_{j=1}^{N} \lambda_j^{-1}(\omega) \overline{\mathbf{U}}^T(\mathbf{k}) \mathbf{V}_j(\omega) \overline{\mathbf{V}}_j^T(\omega) \mathbf{U}(\mathbf{k}) \right]^{-1}. \quad (27)$$

By comparison of Eqs. (25) and (27), it can be seen that a difference between the CV and HR methods is the weight given to the eigenvectors of $\mathbf{S}(\omega)$. The CV method applies a weight that is linearly proportional to the associated eigenvalue, while in the HR method the weight is inversely proportional to the associated eigenvalue. The large eigenvalues are associated with signals and the small eigenvalues are associated with noise. Therefore, the CV statistic gives large weights to the eigenvectors of the signals, while the HR statistic gives the largest weights to the eigenvectors of the noise. The weighting scheme used in the HR method is a numerical technique that takes advantage of the statistical properties of the noise to increase the resolution.

A third frequency–wave-number estimate has been developed recently by Brillinger (1985). This method is the maximum likelihood estimate and differs from the HR statistic in the assumptions of the signal and noise. The signal is assumed to consist of a single harmonic at frequency $\omega_0$. The method uses a more general noise model in that the noise may be correlated between stations.

The Brillinger (BR) method uses the same spatial weighting scheme as the HR method, but it involves a modified form of the cross-spectral matrix. For simplicity, assume that the signal frequency can be expressed as $2\pi m/T$ with $m$ an integer. The center frequency $\omega_0$ is removed from the smoothed cross-spectral matrix. In Eq. (22), this corresponds to removing the case $m = 0$ from the summation. Define the modified cross-spectral matrix

$$S'_{jl}(\omega) = \sum_{m=-M, m \neq 0}^{M} a_m u\left(\mathbf{x}_j, \frac{\omega + 2\pi m}{T}\right) \bar{u}\left(\mathbf{x}_l, \frac{\omega + 2\pi m}{T}\right), \quad (28)$$

which leads to

$$S'_{jl}(\omega) = S_{jl}(\omega) - u(\mathbf{x}_j, \omega) \bar{u}(\mathbf{x}_l, \omega). \quad (29)$$

Brillinger (1985) shows that the maximum likelihood statistic is

$$P^{BR}(\mathbf{k}, \omega) = [\overline{\mathbf{U}}^T(\mathbf{k}) \mathbf{S}^{-1}(\omega) \mathbf{U}(\mathbf{k}) / \overline{\mathbf{U}}^T(\mathbf{k}) \mathbf{S}'^{-1}(\omega) \mathbf{U}(\mathbf{k})] - 1, \quad (30)$$

---

FIG. 3. Spectrum of the S waves at 1 Hz from SMART 1, event 5. (a) Conventional wave number. (b) High resolution wave number.

which can be compared directly with Eqs. (23) and (26) for the CV and HR methods. The HR statistic $[\overline{U}^T(k)S^{-1}(\omega)U(k)]^{-1}$ may be viewed as a measure of the power present at wave number **k** in the frequency band $[\omega - 2\pi M/T, \omega + 2\pi M/T]$. Brillinger points out that if the HR statistic is computed excluding the center frequency $\omega$, then the statistic becomes $[\overline{U}^T(k)S'(\omega)^{-1}U(k)]^{-1}$. Therefore, a measure of the power at the discrete frequency $\omega$ is given by

$$[\overline{U}^T(k)S^{-1}(\omega)U(k)]^{-1} - [\overline{U}^T(k)S'^{-1}(\omega)U(k)]^{-1}. \quad (31)$$

Dividing Eq. (31) by $[\overline{U}^T(k)S'^{-1}(\omega)U(k)]^{-1}$ yields the BR statistic. In this sense, the BR statistic compares the power at the discrete center frequency $\omega$ with the weighted sum of the power at neighboring frequencies. For this reason, the estimate is sensitive to the assumption that the signal is a single harmonic.

A comparison of the three frequency–wave-number methods is shown in Brillinger (1985) for a harmonic signal in the presence of correlated noise. For the test case, Brillinger shows that the HR and BR statistics have better resolution than the CV statistic. Figure 3 gives a comparison of the wave-number statistics for the S-wave portion of SMART 1, event 5 (see Section I).

D. SIGNAL DETECTION

Developing specific tests for signal detection requires a known null distribution for the case $A = 0$ [see Eq. (8)]. The statistical properties of the CV and HR distributions are discussed in Capon (1970) and Capon and Goodman (1970). Assuming a boxcar frequency window $[a_m = 1/(2M + 1)]$, the null distribution of the CV statistic is given in terms of the chi-squared distribution by

$$(1/N^2)2(2M + 1)\overline{U}^T(k)C_{\varepsilon\varepsilon}U(k)\chi^2_{2(2M+1)}, \quad (32)$$

and the null distribution of the HR statistic is given by

$$[\overline{U}^T(k)C_{\varepsilon\varepsilon}^{-1}U(k)]^{-1}\chi^2_{2(2M+2-N)}, \quad (33)$$

where

$$C_{\varepsilon\varepsilon} = E[\mathbf{u}(\omega + 2\pi m/T)\overline{\mathbf{u}}^T(\omega + 2\pi m/T)], \quad m \neq 0. \quad (34)$$

Both the CV and HR distributions contain the unknown parameters $C_{\varepsilon\varepsilon}$ and therefore may not be used to construct tests of prespecified size. In contrast, the null distribution of the BR statistic may be approximated in terms of Fisher's distribution by

$$(2M + 1 - N)^{-1} F_{[2, 2(2M+1-N)]}. \tag{35}$$

For this reason, the BR statistic is well suited for the detection problem. The hypothesis that no signal is present can be rejected with a specified "false alarm" probability for large values of Eq. (31).

### III. Wave Coherence

The coherence of strong ground motions over short distances has important implications for waveform modeling and the generation of synthetic seismograms. As previous studies show (Olson and Apsel, 1982; Hartzell and Heaton, 1983; Mikumo and Miyatake, Chapter 3, this volume; Vidale and Helmburger, Chapter 6, this volume) waveform modeling of strong ground motion does not closely match observed high-frequency waves. This mismatch can be explained in part by a lack of spatial coherence at high frequencies. Possible causes of the lack of spatial coherence are wave scattering and multiple diffraction and refraction of waves as they pass through complex velocity structures. (See Spudich and Archuleta, Chapter 5, this volume, for further discussion on the numerical problems of modeling high-frequency waves.) We can develop empirical guidelines for the maximum frequency that can accurately be modeled by measuring the spatial coherence of strong ground motion over short distances using recordings from dense, strong motion arrays. In other words, attempts at deterministic waveform modeling at frequencies above the cutoff for coherent energy are impractical.

A. TWO-STATION COHERENCE ESTIMATES

Coherence measures express in quantitative form the likeness of data content among data channels. The simplest coherence measures are for two-data channels. A time-domain measure of the likeness of two time series is the sample *cross covariance* given by

$$C_{12}(\tau) = \sum_{i=1}^{T} u_1(t_i) u_2(t_i + \tau), \tag{36}$$

where $T$ is the number of time samples. A normalized correlation is found by dividing by the autocovariances and is given by

$$R_{12}(\tau) = C_{12}(\tau)[C_{11}(0) C_{22}(0)]^{1/2}. \tag{37}$$

This statistic is in the range $[-1, 1]$, where a value of 1 indicates perfect correlation. The *coherency* is defined in the frequency domain and is given by

$$\gamma_{12}(\omega) = S_{12}(\omega)/[S_{11}(\omega)S_{22}(\omega)]^{1/2}. \tag{38}$$

The *coherence* is defined as the square of the modulus of the coherency and is given by

$$|\gamma_{12}(\omega)|^2 = S_{12}(\omega)S_{21}(\omega)/S_{11}(\omega)S_{22}(\omega). \tag{39}$$

The coherence is a normalized measure, $0 \leq |\gamma_{12}(\omega)|^2 \leq 1$, where a value of 1 indicates complete coherence. Note that the cross-spectral matrix **S** must be smoothed over a frequency band [e.g., $M \neq 0$ in Eq. (21)] or the coherence is always unity.

## B. Multistation Coherence Estimates

Three multichannel coherence measures commonly used in exploration seismology are *unnormalized correlation*, *normalized correlation*, and *semblance*. These coherence measures are described in detail by Neidell and Tanner (1971).

The unnormalized correlation is simply the sum of all the possible two-station cross covariances and is given by

$$\sum_{l=1}^{N-1} \sum_{j=1}^{N-l} C_{jl}(\tau_l - \tau_j), \tag{40}$$

where $N$ is the number of channels, $\tau_j$ is the lag applied to channel $j$ (see Fig. 2), and the cross covariance is

$$C_{jl}(\tau_l - \tau_j) = \sum_{i=1}^{T} u_j(t_i + \tau_j)u_l(t_i + \tau_l). \tag{41}$$

The normalized correlation is

$$R_N(\tau) = \frac{1}{(N-1)N} \sum_{l=1}^{N} \sum_{j=1, j\neq l}^{N} \frac{C_{jl}(\tau_l - \tau_j)}{[C_{jj}(0)C_{ll}(0)]^{1/2}}. \tag{42}$$

In the frequency domain, the corresponding coherence statistic is

$$|\gamma_c(\mathbf{k}, \omega)|^2 = \left| \frac{1}{(N-1)N} \sum_{l=1}^{N} \sum_{j=1, j\neq l}^{N} \frac{S_{jl}(\omega)\overline{U}_j(\mathbf{k})U_l(\mathbf{k})}{[S_{jj}(\omega)S_{ll}(\omega)]^{1/2}} \right|^2. \tag{43}$$

In matrix notation Eq. (43) becomes

$$|\gamma_c(\mathbf{k}, \omega)|^2 = \left[ \frac{1}{(N-1)N} \overline{\mathbf{U}}^T(k)\hat{\mathbf{S}}(\omega)\mathbf{U}(k) - \frac{1}{N-1} \right]^2, \tag{44}$$

## 2. ARRAY ANALYSIS AND SYNTHESIS MAPPING

where the hat indicates which elements of the cross-spectral matrix are normalized by the geometric mean auto power. This coherence measure is in the range $0 \leq |\gamma_c(\mathbf{k}, \omega)|^2 \leq 1$.

An alternate time-domain correlation measure is the semblance coefficient $S_c$, which measures the power in the seismic beam divided by $N$ times the sum of the power in the individual recordings. It is defined algebraically as

$$S_c(\tau) = \frac{\sum_{i=1}^{T}\left[\sum_{j=1}^{N} u_j(t_i + \tau_j)\right]^2}{N \sum_{i=1}^{T} \sum_{j=1}^{N} u_j^2(t_i)}. \tag{45}$$

Equation (45) can be written as

$$S_c(\tau) = \frac{\sum_{l=1}^{N} \sum_{j=1}^{N} C_{jl}(\tau_l - \tau_j)}{N \sum_{j=1}^{N} C_{jj}(0)}. \tag{46}$$

The semblance coefficient is a normalized correlation measure and varies in the range of $1/(N) \leq S_c \leq 1$, where unity corresponds to complete wave coherence. The semblance is a biased measure of correlation because it includes terms that do not contain cross-channel information [e.g., $C_{jj}(0)$]. The semblance penalizes both amplitude and phase variations in the signal across the array.

In the frequency–wave-number domain, the semblance coherence, $|\gamma_s(\mathbf{k}, \omega)|^2$, is given by

$$|\gamma_s(\mathbf{k}, \omega)|^2 = \frac{\left|\sum_{l=1}^{N} \sum_{j=1}^{N} S_{jl}(\omega) \overline{U}_j(\mathbf{k}) U_l(\mathbf{k})\right|^2}{\left[N \sum_{j=1}^{N} S_{jj}(\omega)\right]^2}. \tag{47}$$

The semblance coherence is in the range $1/N^2 \leq |\gamma_s|^2 \leq 1$. In matrix notation

$$|\gamma_s(\mathbf{k}, \omega)|^2 = \frac{[\overline{\mathbf{U}}^T(\mathbf{k}) \mathbf{S}(\omega) \mathbf{U}(\mathbf{k})]^2}{[N \operatorname{tr} \mathbf{S}(\omega)]^2}, \tag{48}$$

where tr indicates the trace. In this form, $|\gamma_s|^2$ can be compared to the

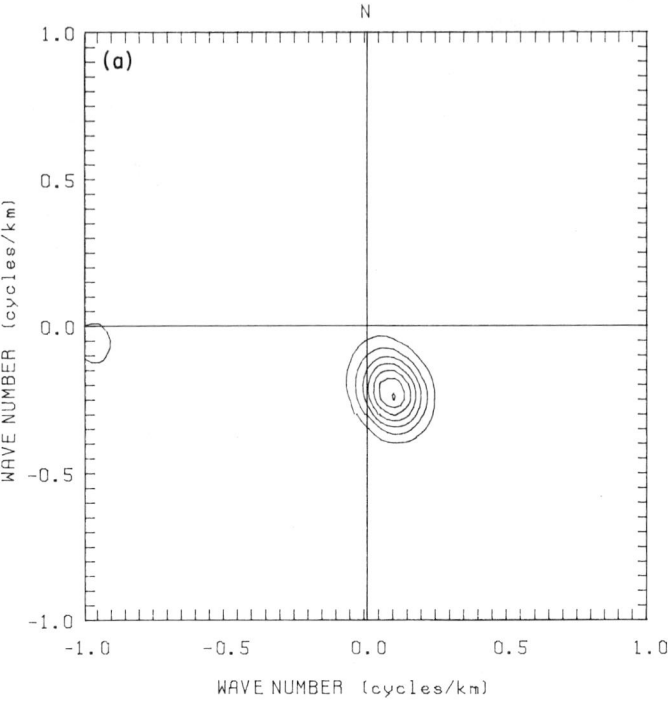

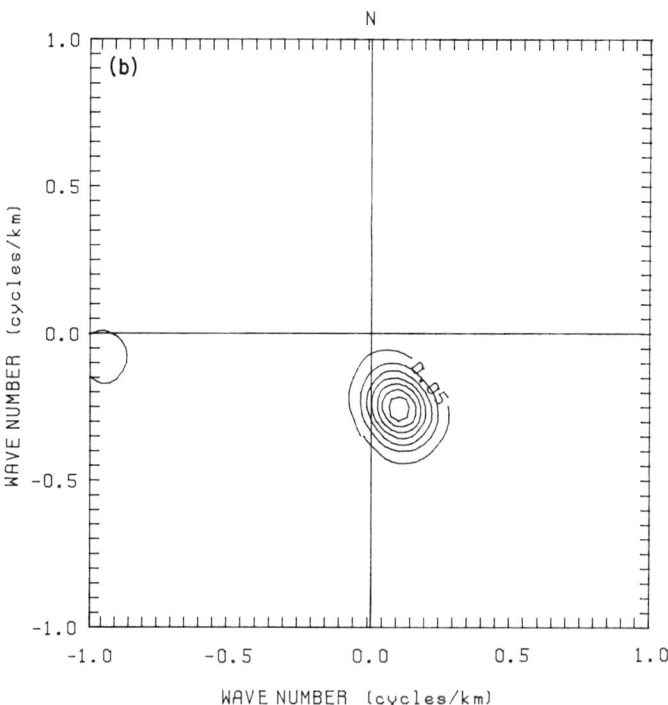

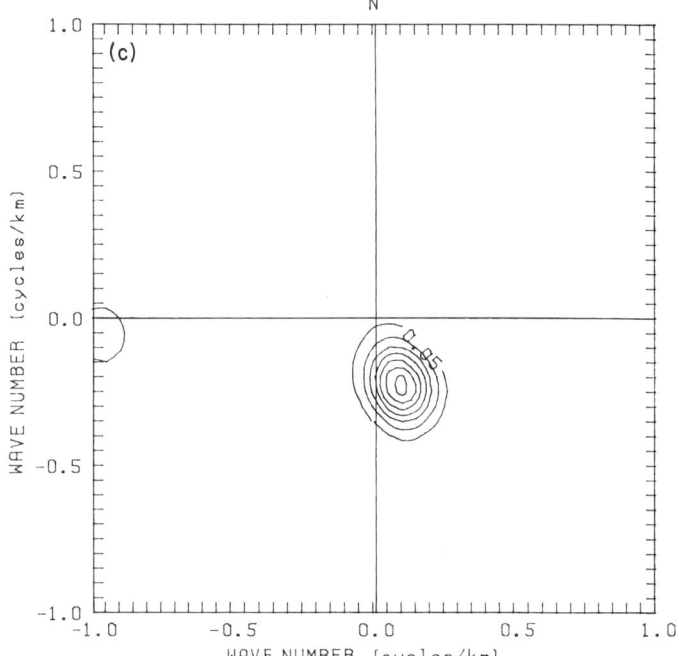

**FIG. 4.** Multistation coherence measures of the S waves at 1 Hz from SMART 1, event 5. (a) Normalized correlation coherence, Eq. (44). (b) Semblance coherence, Eq. (48). (c) Squared CV statistic.

frequency–wave-number statistics given in Section II. In particular, if the cross-spectral matrix $S(\omega)$, defined in Eq. (21), is normalized, then $\operatorname{tr} S(\omega) = N$, and the $|\gamma_s(\mathbf{k}, \omega)|^2$ is the square of the CV statistic given in Eq. (23). Therefore, the square of the CV statistic can be thought of as the semblance without the penalty for amplitude variation. In this way, the squared CV statistic is a measure of the coherence of the phasing of the ground motion across the array.

## C. Statistical Aspects

The statistical properties of the two-station coherence, $|\gamma_{12}(\omega)|^2$, are not simple. For a box car frequency window $[a_m = 1/(2M + 1)]$, Enochson and Goodman (1965) suggest that a reasonable approximation of the distribution of $\tanh^{-1}|\gamma_{12}(\omega)|^2$ is a normal distribution with mean $\tanh^{-1}|\gamma_{12}(\omega)|^2 + 1/4M$ and variance $1/2(2M - 1)$, where $M$ is defined in Eq. (21). For the null distribution, $|\gamma_{12}(\omega)|^2 = 0$, and the approximate

$100\alpha$-percent point of the estimated coherence is given by $1 - (1 - \alpha)^{1/2M}$ (Brillinger, 1981).

The semblance is an energy ratio and can be related to an $F$ statistic. When the noise is assumed to be Gaussian and uncorrelated between stations, and the signal and noise are both band limited with a constant amplitude spectrum, then the semblance can be approximated by an $F$ statistic with $\nu_1$ and $\nu_2$ degrees of freedom and the noncentrality parameter $\lambda$ (Douze and Laster, 1979). The approximation is

$$S_c = \frac{F(\nu_1, \nu_2, \lambda)}{F(\nu_1, \nu_2, \lambda) + N - 1}, \tag{49}$$

where the degrees of freedom $\nu_1$ and $\nu_2$ and the noncentrality parameter $\lambda$ are defined as $\nu_1 = 2\Delta fT$, $\nu_2 = \nu_1(N - 1)$, and $\lambda = N\nu_1 SNR^2$, where $\Delta f$ is the frequency band, $T$ is the time window in seconds, and $SNR^2$ is the mean signal power divided by the mean noise power over the frequency band. For comparison, the CV statistic is a chi-squared variate as given in Eq. (32).

D. EXAMPLE

As an example, the coherence measures given by Eqs. (44) and (48) were estimated using the SMART 1 recordings from the January 29, 1981, Taiwan earthquake (see Section I). The coherence statistics $|\gamma_c(\mathbf{k}, \omega)|^2$ and $|\gamma_s(\mathbf{k}, \omega)|^2$ measured at 1 Hz using the N–S recordings of the S waves are shown in Figs. 4a and b. Recall that $|\gamma_c(\mathbf{k}, \omega)|^2$ employs the normalized form of the cross-spectral matrix and therefore only measures the coherence of the phase. The semblance coherence $|\gamma_s(\mathbf{k}, \omega)|^2$ measures the coherence of both the phase and the amplitude. As a comparison, the squared CV statistic estimated using the normalized cross-spectral matrix is shown in Fig. 4c. In this case, the wave number spectra from these three coherence measures are very similar.

## IV. Synthesis Mapping

In this section, we describe the basic model of array processing used in radio-astronomy, with emphasis on aspects that have applications in high-resolution analysis of seismic sources.

2. ARRAY ANALYSIS AND SYNTHESIS MAPPING 77

There are two basic designs of arrays: *fixed* redundancy and *variable* redundancy. The most widely used of the latter has equally spaced antennas at fixed locations combined with one or more movable antennas. Shifts of the movable antennas allow changes in incremental spacing to optimize the array for the particular source under study. One design principle in both types of configurations is to seek minimum redundancy (Moffet, 1968; Bracewell, 1973). In this way, the number of different spacings that can be obtained simultaneously with a given number of antennas is maximized.

In both astronomy and seismology, the design of two- and three-dimensional arrays is more of an empirical matter than the design of one-dimensional arrays, because there are few general principles to apply, such as minimum redundancy, if sources can be widely separated or in unknown azimuths. If a source configuration is known or assumed, optimum design can again be attempted.

A. RESPONSE OF AN INTERFEROMETER

Radio telescope arrays are used to produce maps of the radio intensity (brightness) of the sky. The mapping is accomplished by regarding a radio telescope array as an ensemble of two-element interferometers and the overall array response is analyzed in terms of this simpler instrument (Thompson *et al.*, 1980).

Consider an interferometer whose two antennas are located at $x_1$ and $x_2$. The signals $u_1(t)$ and $u_2(t)$ from the two antennas are combined in a *correlator* (basically a multiplier and a time averager) as shown in Fig. 5. The output of the correlator is the cross covariance of the input waveforms and is given by

$$C_{12}(\tau) = \frac{1}{T} \int u_1(t) u_2(t + \tau) \, dt, \qquad (50)$$

where $\tau$ is the lag time applied to $u_2(t)$ and $T$ is the time length of the data. In radio astronomy, the cross covariance is called the *visibility*. For complex $u_1$ and $u_2$, $u_1$ in Eq. (50) should be replaced by its complex conjugate, and the correlator output is given by the real part of $C_{12}$. In the frequency domain, the visibility is the cross spectrum and is given by

$$V_{12}(\omega) = S_{12}(\omega) = \int_{-\infty}^{\infty} C_{12}(\tau) \exp\{-i\tau\omega\} \, d\tau. \qquad (51)$$

Since radio sources are in the far field, the waves are usually assumed to be plane waves. Consider a radio point source that generates plane waves

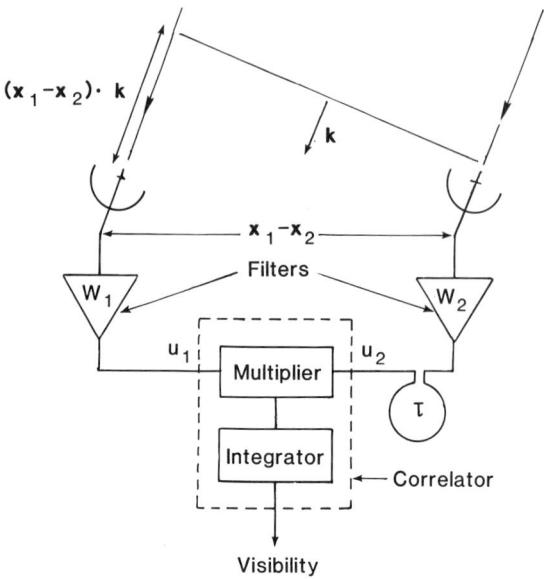

FIG. 5. Geometry of an interferometer.

with wave number **k** at frequency $\omega$. The geometric delay between the two signals that is due to the alignment of the interferometer with respect to the source is given by $\tau = (\mathbf{x}_1 - \mathbf{x}_2) \cdot \mathbf{k}/\omega$ [see Eq. (7)]. In radio astronomy, a *baseline vector* **B** is defined as $\mathbf{B} = \mathbf{x}_1 - \mathbf{x}_2$. The geometric delay is applied by multiplying $V_{12}(\omega)$ by $\exp\{-i\mathbf{B} \cdot \mathbf{k}\}$. Let $I(\mathbf{k}, \omega)$ represent the intensity measured in W/m² [see Eq. (2)] over a frequency band centered at $\omega$, then the received intensity in the beam is $\rho c |u_1(t) u_2(t + \tau)|$ [see Eq. (2)]. In the frequency domain, the received intensity is $(\rho c/T) V_{12}(\omega) \exp\{-i\mathbf{B} \cdot \mathbf{k}\}$, where $V_{12}(\omega)$ is smoothed over the desired frequency band, and we have

$$(\rho c/T) V_{12}(\omega) \exp\{-i\mathbf{B} \cdot \mathbf{k}\} = I(\mathbf{k}, \omega). \tag{52}$$

The intensity estimates in Eq. (52) consider just a single interferometer (station pair). We consider all interferometers by integrating over the baseline vector **B**. Prior to integrating, a two-dimensional spatial taper function $\Gamma(\mathbf{B})$ is introduced due to the finite array aperture. We then have

$$\int \Gamma(\mathbf{B}) I(\mathbf{k}, \omega) \, d\mathbf{B} = \frac{2\rho c}{T} \int \Gamma(\mathbf{B}) S(\mathbf{B}, \omega) \exp\{i\mathbf{B} \cdot \mathbf{k}\} \, d\mathbf{B}. \tag{53}$$

The intensity is independent of the baseline vector **B** and can be taken out of the integral on the left-hand side of Eq. (11). The intensity is therefore given by

$$I(\mathbf{k}, \omega) = \frac{2\rho c}{T} \int \Gamma(\mathbf{B}) S(\mathbf{B}, \omega) \exp\{-i\mathbf{B} \cdot \mathbf{k}\} \, d\mathbf{B} \Big/ \int \Gamma(\mathbf{B}) \, d\mathbf{B}. \quad (54)$$

Equation (54) is a key equation for synthesis mapping. An assumption made while inverting Eq. (53) is that visibilities from different points of the source can be added independently, that is, the source is spatially incoherent, so that signal components radiating from different points on the source are uncorrelated. (This assumption may not be always appropriate for seismic sources.)

B. Fourier Transform Mapping

The Fourier transform in Eq. (54) is evaluated numerically. For a small number of station pairs, the direct Fourier transform can be used to evaluate Eq. (54) at the individual sample points. For a large number of station pairs, it is worthwhile to interpolate the measured visibilities onto a uniform grid so that a fast Fourier transform can be used. In the following, we use the DFT approach.

Approximating the Fourier transform by the weighted sum gives

$$I(\mathbf{k}, \omega) = C_1 \sum_{l=1}^{N} \sum_{j=1}^{N} w_{jl} \Gamma_{jl} V(\mathbf{B}_{jl}, \omega) \exp\{-i\mathbf{B}_{jl} \cdot \mathbf{k}\} \quad (55)$$

where $C_1$ is defined as

$$C_1 = \frac{2\rho c}{T} \left[ \sum_{l=1}^{N} \sum_{j=1}^{N} w_{jl} \Gamma(\mathbf{B}_{jl}) \right]^{-1},$$

and **w** is a weight function. The weighting function is used to offset the nonuniform sampling of the baseline vectors; there are far more small station spacings than large station spacings.

Radio astronomers often use a two-dimensional Gaussian taper function

$$\Gamma_{jl} = \exp\{-|\mathbf{B}_{jl}|^2 / 2\sigma^2\}, \quad (56)$$

where $\sigma$ depends on the maximum station spacing. The two standard weighting functions suggested in radio astronomy are the *natural weight* given by $w_{jl} = 1/N^2$ and the *uniform weight* given by $w_{jl} = 1/KN_g$, where $K$ is the number of visibilities measured within a given grid cell in the baseline vector plane, and $N_g$ is the number of grid cells. The natural weight is better than the uniform weight for improving the signal to noise ratio of weak signals. However, the natural weight gives high weights to the small station spacings and results in a reduced resolution. With uniform weights, the response is less dominated by the array configuration and the resolution is improved.

C. APPLICATIONS TO STRONG MOTION SEISMOLOGY

In the notation of Section II, the intensity is given by

$$I(\mathbf{k}, \omega) = 2C_1 \sum_{l=1}^{N} \sum_{j=1}^{N} w_{jl}\Gamma_{jl}S_{jl}(\omega)\overline{U}_j(\mathbf{k})U_l(\mathbf{k}). \tag{57}$$

The factor of 2 is included because the summation in Eq. (21) considers only positive frequencies. Equation (57) is similar to the form of the power spectrum given in Eq. (18). The main differences are the use of the taper function and that the weights $w_{jl}$ in Eq. (57) are dependent on the differential station location while the weights $W_j$ in Eq. (18) are only dependent on the individual station. We define a modified cross-spectral matrix, $\mathbf{S}^*$, whose elements are

$$S_{jl}^*(\omega) = w_{jl}\Gamma_{jl}S_{jl}(\omega), \tag{58}$$

and in matrix notation, Eq. (57) becomes

$$I(\mathbf{k}, \omega) = 2C_1\mathbf{U}^T(\overline{\mathbf{k}})\mathbf{S}^*(\omega)\mathbf{U}(\mathbf{k}). \tag{59}$$

If the natural weight is used, ($w_{jl} = 1/N^2$), and the taper function is unity ($\sigma^2 = \infty$), then the intensity is simply related to the CV statistic

$$I(\mathbf{k}, \omega) = (2\rho c/T)P^{CV}(\mathbf{k}, \omega). \tag{60}$$

An example of an intensity calculation is given in Section V.

## V. Computational Results

In this section, we present some computational results demonstrating the algorithms outlined in the previous sections and linking array analysis with the synthesis strong ground motions. The SMART 1 array recordings (see Section I) are used in the examples.

A. COHERENCE

Consider the spatial coherence of the S waves recorded at the SMART 1 array (see Fig. 1) during event 5. The mean Fourier amplitude spectrum, normalized at 1 Hz and averaged over all 27 radial components for the

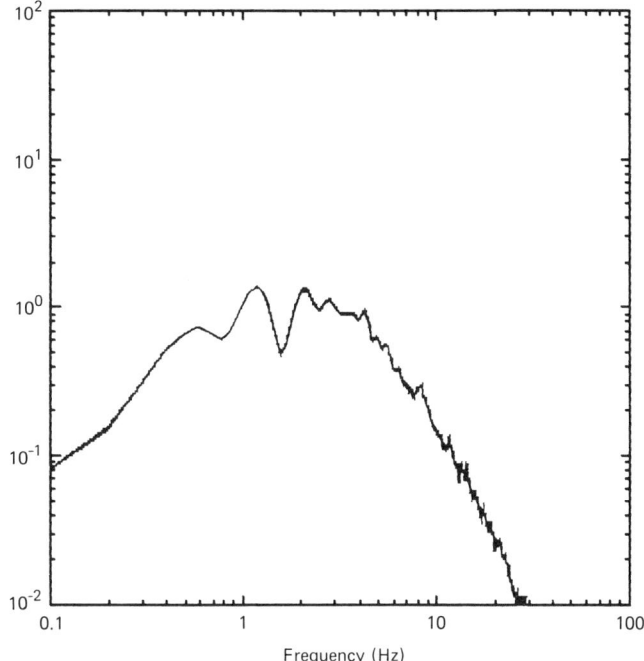

**FIG. 6.** The mean Fourier amplitude spectrum normalized at 1 Hz and averaged over all 27 components for the S wave during SMART 1, event 5.

S-wave window from event 5, is shown in Fig. 6. The signal is well above the digitization noise level over the frequency band 0.5–6 Hz. The spatial coherence of the phasing is examined over this bandwidth using the CV statistic (see Section III).

Figure 7 shows the CV wave number spectra for the radial component at 0.5 to 6 Hz using the middle, inner, and central stations (maximum separation of 2 km). Each wave number plot is for a 1-Hz bandwidth with the center frequency indicated next to each plot. A dominant peak at the azimuth of the source region is evident at frequencies up to 1.5 Hz. At 2 Hz, there is a rapid decay in the spectral peak, and the coherence remains low at the higher frequencies. The coherency (square root of the coherence) is plotted in Fig. 8. The rapid decay in the coherency at 2 Hz is evident. Above 4 Hz, the spectral peaks are not significantly different from noise at the 95% confidence level (shown by the horizontal line).

The measured coherency suggests that in this case, the low frequency part of the seismogram (< 1.5 Hz) could be modeled using deterministic methods as outlined in Chapter 5. As the frequency increases, however, the

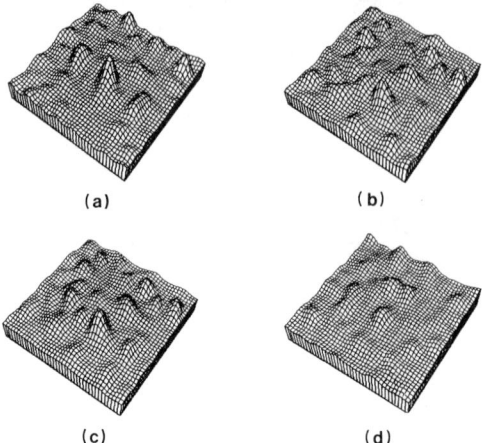

**FIG. 7.** Conventional wave number spectra at center frequencies of (a) 1 Hz, (b) 2 Hz, (c) 4 Hz, and (d) 6 Hz (see text).

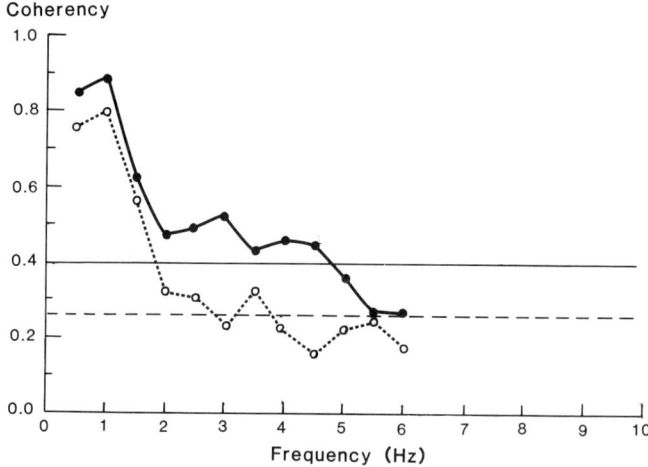

**FIG. 8.** Coherency of the S waves. The solid line is for 17 stations of SMART 1 (middle, inner, and central stations), and the dashed line is for all 27 stations (event 5). The horizontal lines indicate the approximate 95% confidence level of noise for the 17 and 27 station configurations.

percentage of incoherent energy increases. At high frequencies (here above 3 Hz) the recorded motion is dominated by incoherent energy. A practical approach to modeling would be to generate a suite of synthetic ground motions in which the Fourier phase of the incoherent energy is varied in a statistical manner while the Fourier phase of the coherent energy remains unchanged.

Incoherence is important when modeling ground motions for input into a large engineered structure. Dynamic analysis requires ground-motion input at numerous nodes across the base of the structure. The spatial variation of the Fourier phase of the ground motion may not be adequately modeled by simple deterministic wave propagation assumptions.

B. RUPTURE VELOCITY

As mentioned in Section I, rupture velocity is a key source parameter used in generating synthetic, strong ground motions. It is important to have some precise observational estimates for the range of rupture velocities that may be expected.

Estimates have been found, using array recordings for the rupture velocity during SMART 1, event 5, by means of frequency–wave-number analysis. In Section V, A the recorded accelerograms were seen to contain predominately coherent energy propagating from the source region at frequencies up to 1.5 Hz during the S-wave window. These accelerograms can be studied in detail at frequencies below 1.5 Hz for evidence of a moving source.

In this analysis, let the words "rupture" and "rupture front" refer to the position of the source centroid of coherent, high-frequency energy. This meaning is similar to that used by Spudich and Cranswick (1984), who note that, based on the conclusion of Madariaga (1977, 1983) the high-frequency energy radiated by a crack originates at the crack tip.

Temporal changes in the frequency–wave-number spectra for event 5 show a time-dependent azimuth of peak power. If this is interpreted as a measurement of the position of the moving rupture front, the time-dependent rupture speed can be estimated from the time-dependent azimuth $\theta(t)$ found from the wave number spectra. Let time at the source be denoted $\tau$ and time at the receiver denoted $t$. Setting the center of coordinates at the array center, $t$ and $\tau$ are related by

$$t = \tau + t_0 + \frac{|\mathbf{y}(\tau)| - |\mathbf{y}(0)|}{c}, \qquad (61)$$

where $t_0$ is the propagation time from the hypocenter to the array center, $\mathbf{y}(\tau)$ is the position of the rupture front at time $\tau$, and $c$ is the average wave

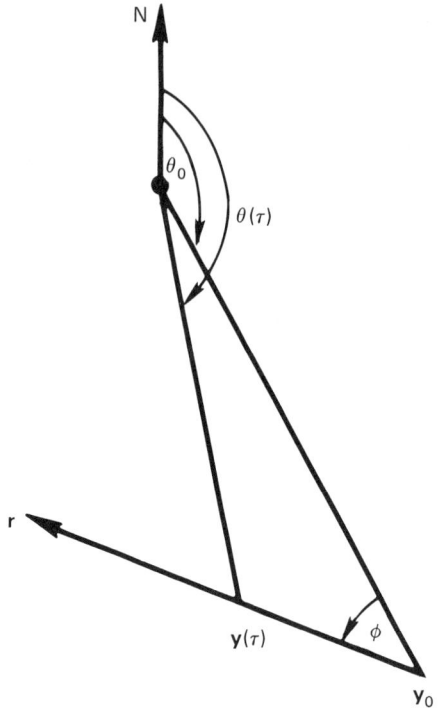

FIG. 9. The fault-array geometry, event 5, SMART 1.

velocity at the source region. For example, a wave leaving the source at time $\tau$ from position $\mathbf{y}(\tau)$ arrives at the center of the array at time $t$. Assuming a unilateral rupture in direction $\hat{\mathbf{r}}$ and a laterally homogeneous velocity structure, by simple geometry (Fig. 9)

$$\mathbf{y}(\tau) = \frac{|\mathbf{y}(0)| \sin[\theta(t) - \theta_0]}{\sin[\theta(t) - \theta_0 + \varphi]} \hat{\mathbf{r}} + \mathbf{y}(0), \qquad (62)$$

where

$$\varphi = \cos^{-1}\{\mathbf{y}(0) \cdot \hat{\mathbf{r}}/|\mathbf{y}(0)|\}. \qquad (63)$$

The rupture direction $\hat{\mathbf{r}}$ can be estimated from the aftershock hypocenter distribution or fault trace if either is known. For the case of a constant rupture direction, the relation between $t$ and $\tau$ is

$$\frac{dt}{d\tau} = 1 + \frac{1}{c} \frac{d|\mathbf{y}(\tau)|}{dL} \frac{dL}{d\tau}, \qquad (64)$$

where $L(t)$ is the rupture length at time $t$ and is given by

$$L(t) = \frac{|\mathbf{y}(0)| \sin[\theta(t) - \theta_0]}{\sin[\theta(t) - \theta_0 + \varphi]}. \tag{65}$$

Frequency–wave-number analysis gives the azimuth $\theta(t)$ as a function of time at the array. This azimuth can be converted to a function of time at the source $\theta(\tau)$ using the approximation

$$\tau_j = t_j - t_0 - 1/c\left\{\left[|\mathbf{y}(0)|^2 + L_j^2 - 2|\mathbf{y}(0)|L_j \cos\varphi\right]^{1/2} - |\mathbf{y}(0)|\right\}, \tag{66}$$

where

$$L_j = \frac{|\mathbf{y}(0)| \sin(\theta_j - \theta_0)}{\sin(\theta_j - \theta_0 + \varphi)}, \tag{67}$$

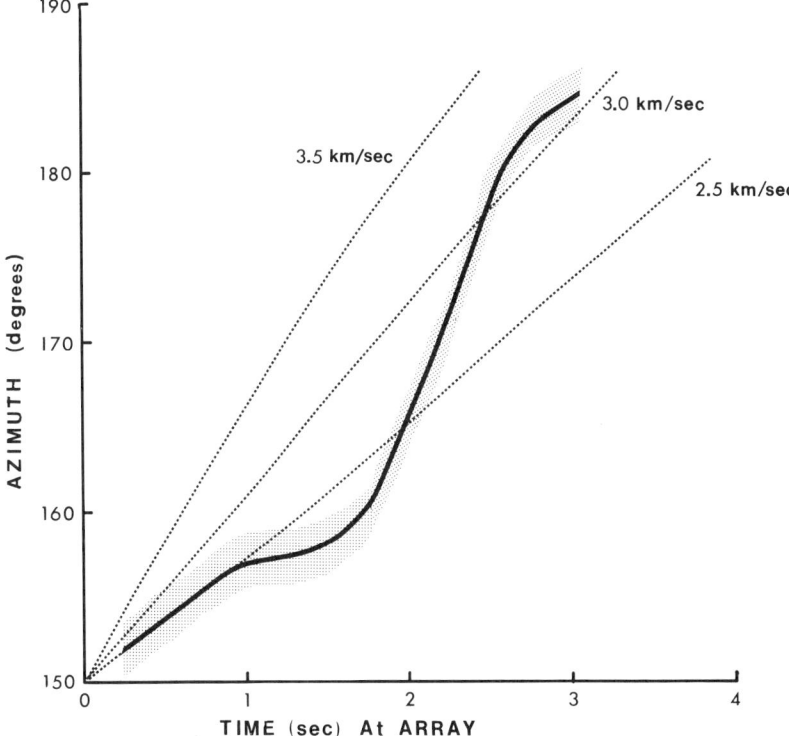

**FIG. 10.** Time-dependent azimuths estimated from the S-wave spectra, event 5. The shaded regions indicates an estimate of one standard error and the dotted lines show the expected azimuths for various constant rupture velocities.

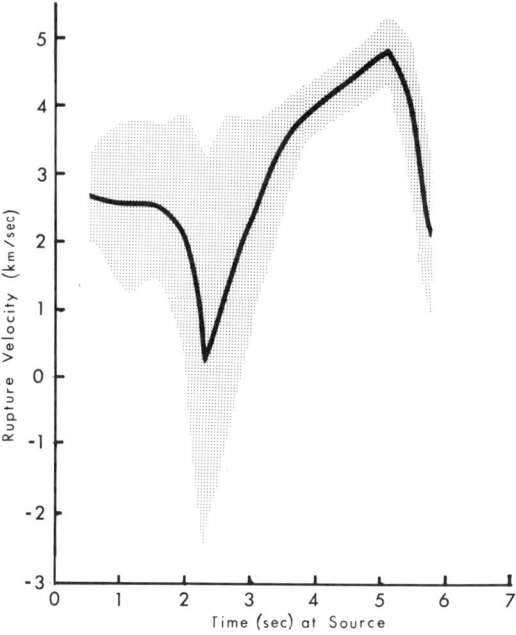

FIG. 11. The time-dependent rupture velocity estimated from the S-wave azimuthal variation, event 5. The shaded region is an estimate of one standard error. The initial acceleration of the rupture front is not observed, but the deceleration after 5.5 sec may correspond to the stopping of the rupture.

and $L_0 = 0$. The estimate of the rupture speed is simply

$$v_r(\tau_j) = (L_j - L_{j-1})/(\tau_j - \tau_{j-1}). \tag{68}$$

The S waves from event 5 were analyzed using Eq. (68) for evidence of a moving source over a 0.8-Hz frequency band centered at 1 Hz (Abrahamson, 1985). This is within the frequency band of coherent energy. The computed values of the (1-Hz) S-wave azimuths smoothed over 1-sec windows are shown in Fig. 10 as a function of time at the array. The time-dependent azimuths expected for various constant rupture velocities are shown for reference. The shaded region indicates one standard error.

There is a clear clockwise (east to west) rotation of the azimuth of the peak power. The time-dependent rupture speed is estimated from these azimuths and is plotted in Fig. 11. The (two-sided) standard error is shown by the shaded region. This estimate of the standard error is a measure of the reading error and does not include systematic uncertainties due to the method itself. For this reason, the standard errors in Figs. 10 and 11 are minimum estimates.

Although the initial acceleration of the rupture front cannot be resolved in this example, the rupture speed near its initiation is approximately 2.5 km/sec. The rupture appears to decelerate sharply after 2 sec, but this deceleration is not significant because of the range of the standard error. After 3.5 sec, the rupture accelerates up to $4.9 \pm 0.5$ km/sec. The mean is thus above the shear wave velocity of 3.5 km/sec. At 5.5 sec, the rupture decelerates down to 1.6 km/sec. This deceleration may indicate the end of the rupture. Beyond 5.5 sec, the S-wave amplitude at 1 Hz decreases rapidly and surface waves dominate the wave number spectra obscuring the end of the rupture. The total inferred rupture length obtained by integrating the rupture velocity is about 19 km. This rupture length is slightly less than the 25-km long rupture indicated by the aftershock distribution.

Super-shear rupture speeds have been proposed by other authors. A short episode of super-shear rupture has been postulated for the 1979 Imperial Valley earthquake (Archuleta, 1982; Spudich and Cranswick,

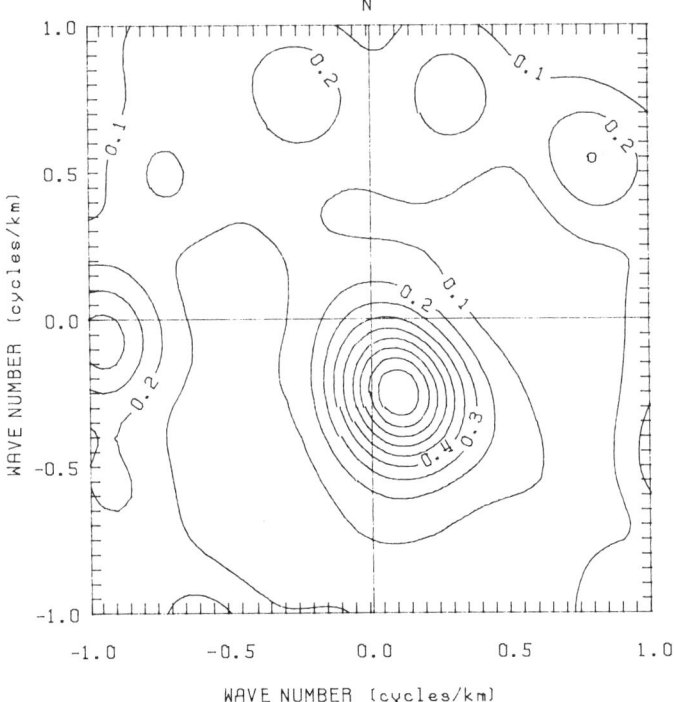

FIG. 12. Seismic intensity of the S waves, event 5, estimated by synthesis mapping, using SMART 1. The spectrum is scaled by $3 \times 10^4$ W/m$^2$.

1984). The inferred super-shear rupture speed shown in Fig. 11, however, may be due to violations of the assumptions of the method. For example, the rupture direction $\hat{r}$ was assumed to be constant. If the fault has curvature, then the inferred rupture velocities may be significantly in error. Similarly, the velocity structure is assumed to be laterally homogeneous, but this is clearly not the case in northeastern Taiwan, where the crust thickens rapidly from the east to the west.

C. INTENSITY

Array estimates of the seismic intensity $I(\mathbf{k}, \omega)$ can be estimated using synthesis mapping. As an example, we have estimated the seismic intensity for the S waves recorded at the SMART 1 array during event 5. As before, a narrow frequency band centered at 1 Hz was selected. The uniform weighting scheme (see Section IV) was used with a grid size of 200 m × 200 m and the taper function (Eq. 56) was adopted with $\sigma^2 = 2000$ m. The intensity, $I(\mathbf{k}, \omega)$, estimated using Eq. (59) is shown in Fig. 12. The contours have been scaled by a factor of $3 \times 10^4$ W/m². As expected, the shape of the intensity wave number spectrum is similar to the shape of the CV spectrum (see Fig. 3a).

Measurements of this kind can be useful both for construction of realistic wave synthetics and for engineering purposes, where estimation of the seismic energy flux across the foundation of a structure is needed.

ACKNOWLEDGMENTS

We are grateful to Dr. D. Brillinger and Mr. R. Darragh for comments on the text. The research incorporated was supported by NSF grant ECE 8417856.

REFERENCES

Abrahamson, N. A. (1985). Estimation of seismic wave coherency and rupture velocity using the SMART 1 strong motion array recordings. Ph.D. dissertation, Univ. of California, Berkeley.
Aki, K., and B. Chouet (1975). Origin of Coda waves: source attenuation and scattering effects. *J. Geophys. Res.* **80**, 3322–3342.
Archuleta, R. (1982). Analysis of the near-source and dynamic measurements from the 1979 Imperial Valley Earthquake. *Bull. Seismol. Soc. Am.* **72**, 1927–1956.
Backus, G. E. (1977). Seismic sources with observable glut moment of spatial degree two. *Geophys. J. R. Astron. Soc.* **51**, 27–45.
Bolt, B. A. (1981). The interpretation of strong motion seismograms. *U.S. Army Eng. Waterways Exp. Stn., Misc. Rep. No.* 17, S-73-1, 1–125.

Bolt, B. A., C. H. Loh, J. Penzien, Y. B. Tsai, and Y. T. Yeh (1982). Preliminary report on the SMART 1 strong motion array in Taiwan. *EERC Rep.* **UCB / EERC-82 / 13**.

Bolt, B. A., N. A. Abrahamson, and Y. T. Yeh (1984). The variation of strong ground motion over short distances. *World Conf. Earth Eng., 8th, San Francisco, Calif.* pp. 183–189.

Boore, D. M., and W. B. Joyner (1978). The influence of rupture incoherence on seismic directivity. *Bull. Seismol. Soc. Am.* **64**, 555–570.

Bracewell, R. B. (1973). "The Fourier Transform and its Applications." McGraw-Hill, New York.

Brillinger, D. R. (1981). "Time Series Data Analysis and Theory," Expanded Ed. Holden-Day, San Francisco, California.

Brillinger, D. R. (1985). A maximum likelihood approach to frequency-wavenumber analysis. *IEEE Trans. Acoust. Speech Signal Process.* **ASSP-33**, 1076–1085.

Bullen, K. E., and B. A. Bolt (1985). "Introduction to the Theory of Seismology," 4th Ed. Cambridge Univ. Press, London and New York.

Capon, J. (1969). High resolution frequency-wavenumber spectrum analysis. *Proc. IEEE* **57**, 1408–1418.

Capon, J. (1970). Applications of detection and estimation theory to large array seismology. *Proc. IEEE* **58**, 760–770.

Capon, J. (1973). Signal processing and frequency wavenumber spectrum analysis for a large aperture seismic array. *Methods Comput. Phys.* **13**, 1–59.

Capon J., and N. R. Goodman (1970). Probability distributions for estimators of the frequency-wavenumber spectrum. *Proc. IEEE* **57**, 1785–1786.

Capon, J., R. J. Greenfield, and R. J. Kolker (1967). Multidimensional maximum-likelihood processing of a large aperture seismic array. *Proc. IEEE* **55**, 192–211.

Chu, S. T. (1984). Statistical estimation of the parameters of a moving source from array data. Ph.D. Thesis, Univ. of California, Berkeley.

Das, S., and K. Aki (1977a). Fault plane with barriers: a versatile earthquake model. *J. Geophys. Res.* **82**, 5658–5670.

Das S., and K. Aki (1977b). A numerical study of two-dimensional spontaneous rupture propagation. *Geophys. J. R. Astron. Soc.* **50**, 643–668.

Douze, E. J., and S. J. Laster (1979). Statistics of semblance. *Geophysics* **44**, 1999–2003.

Enochson, L. D., and N. R. Goodman (1965). Gaussian approximations to the distribution of sample coherence. Tech. Rep. AFFDL-TR-65-57, Wright-Patterson Air Force Base.

Hanks, T. C. (1975). Strong ground motion of the San Fernando California earthquake: ground displacements. *Bull. Seismol. Soc. Am.* **65**, 193–225.

Hartzell, S. H., and T. H. Heaton (1983). Inversion of strong ground motion and teleseismic waveform data for the fault rupture of history of the 1979 Imperial Valley, California earthquake. *Bull. Seismol. Soc. Am.* **73**, 1553–1584.

Hartzell, S. H., and D. V. Helmberger (1982). Strong motion modeling of the Imperial Valley Earthquake of 1979. *Bull. Seismol. Soc. Am.* **72**, 571–599.

Haskell, N. A. (1966). Total energy and energy spectral density of elastic wave radiation from propagating faults. Part II. A statistical source model. *Bull. Seismol. Soc. Am.* **56**, 125–140.

Iwan, W. D., ed. (1978). *Strong-Motion Earthquake Instrum. Arrays*: Proc. Int. Workshop Strong-Motion Earthquake Instrum. Arrays, Honolulu, Hawaii.

Kostrov, B. V. (1966). Unsteady propagation of longitudinal shear cracks. *J. Appl. Math. Mech. (Engl. Transl.)* **30**, 1241–1248.

Liu, H. L., and T. Heaton (1984). Array analysis of the ground velocities and accelerations from the 1971 San Fernando, California, earthquake. *Bull. Seismol. Soc. Am.* **74**, 1951–1968.

McLaughlin, K. L. (1983). Spatial coherency of seismic waveforms. Ph.D. Thesis, Univ. of California, Berkeley.
Madariaga, R. (1977). High frequency radiation from crack (stress drop) models of earthquake faulting. *Geophys. J.* **51**, 625–651.
Madariaga, R. (1983). High frequency radiation from dynamic earthquake fault models. *Ann. Geophys.* **1**, 17–23.
Mikumo. T., and T. Miyatake (1979). Earthquake sequence on a frictional fault model with non-uniform strengths and relaxation times. *Geophys. J. R. Astron. Soc.* **59**, 497–522.
Moffet, A. T. (1968). Minimum redundancy linear arrays. *IEEE Trans. Antennas Propag.* **AP-16**, 172–175.
Neidell, N. S., and M. T. Tanner (1971). Semblance and other coherency measures for multichannel data. *Geophysics* **36**, 482–497.
Niazi, M. (1982). Source dynamics of the 1979 Imperial Valley Earthquake from near-source observations of ground acceleration and velocity. *Bull. Seismol. Soc. Am.* **72**, 1957–1968.
Olson, A. H., and R. Aspel (1982). Finite faults and inverse theory with applications to the 1979 Imperial Valley Earthquake. *Bull. Seismol. Soc. Am.* **72**, 1969–2002.
Penzien, J. (1970). Applications of random vibration theory. *In* "Earthquake Engineering" (R. L. Wiegel, ed.) , Prentice-Hall, New Jersey.
Reid, H. F. (1911). The elastic-rebound theory of earthquakes. *Bull. Dep. Geol. Univ. Calif.* **6**, 412–44.
Spudich, P., and E. Cranswick (1984). Direct observation of rupture propagation during the 1979 Imperial Valley earthquake using a short baseline accelerometer array. *Bull. Seismol. Soc. Am.* **74**, 2083–2114.
Thompson, A. R., B. G. Clark, C. M. Wade, and P. J. Napier (1980). The very large array. *Astrophys. J., Suppl. Ser.* **44**, 151–167.

CHAPTER 3

# Numerical Modeling of Realistic Fault Rupture Processes

T. Mikumo

*Disaster Prevention Research Institute*
*Kyoto University*
*Uji, Kyoto, Japan*

T. Miyatake

*Earthquake Research Institute*
*University of Tokyo*
*Tokyo, Japan*

## I. Introduction

Strong ground motions observed in the near-field during large earthquakes depend primarily on the complex, dynamical rupture processes of the fault and on heterogeneous crustal structures around the source, propagation path, and recording sites. For the purpose of predicting strong motions from a large earthquake, these effects should be fully taken into consideration.

A number of theoretical studies have been developed to predict the waveforms of strong ground motions in the near field. The earliest and most fundamental works (Aki, 1968; Haskell, 1969) are based on simple kinematic dislocation sources in an infinite homogeneous medium (Maruyama, 1963; Burridge and Knopoff, 1964), and later works were based on double-couple point sources in a half-space (see, e.g., Kawasaki *et al*., 1973,

1975) incorporating the free-surface effect. It has been found that these simple models are useful for simulating long-period, direct body waves and also converted-body and Rayleigh waves in the latter case. Observed strong motions, however, usually include high-frequency seismic waves, which may be amplified by reverberations and scattering at many small-scale geological boundaries within the real crustal structure. Theoretical approaches have recently attempted to incorporate the seismic wave response of more realistic earth structures. The response of horizontally layered crustal structures, due to a double-couple point source or Green's function, has been calculated by various analytical techniques, such as generalized ray theory (see, e.g., Helmberger, 1968; Helmberger and Malone, 1975; Heaton and Helmberger, 1977, 1978), the Cagniard–deHoop method for a series of multiply reflected rays (Sato, 1977, 1978), reflectivity methods (Fuchs and Mueller, 1971; Kennett, 1974; Kennett and Kerry, 1979), direct frequency-domain integration (Herrmann and Nuttli, 1975a,b; Herrmann, 1977; Wiggins et al., 1977), the normal mode method (see, e.g., Kawasaki, 1978; Swanger and Boore, 1978), the discrete wave number method (Bouchon, 1979, 1982), the reflectivity method with the Cagniard–deHoop integration (Sato and Hirata, 1980), and by the discrete wave number/finite-element method (Alekseev and Mikhailenko, 1980; Hron and Mikhailenko, 1981; Olson, 1982). Some of these methods generate the complete response of the medium, including all types of body and surface waves.

However, the earth's real structure around fault zones is more complicated than assumed, since it includes lateral heterogeneities, attenuative properties, and local topographical site effects. Hence, complete modeling with the given methods would be extremely difficult. To overcome this difficulty, a semi-empirical approach has been presented by Hartzell (1978) to model strong ground motions from a large earthquake by using the aftershock records as an empirical earth response. This approach is based on the idea that if the source of a foreshock or aftershock can be approximated as a point source on the main shock fault, then the ground motion associated with the small event may be regarded as a Green's function, which involves all the propagating effects from the source to the receiver. This approach has been employed to predict the velocity spectra, peak accelerations, and the maximum duration of ground motions from large earthquakes with complex multiple events (see, e.g., Kanamori, 1979; Hadley and Helmberger, 1980). Hartzell's method has been extended by convolving the phase-delayed records with a specific time function to correct for the difference in the source time function and other fault parameters between the main shock and smaller events distributed over a two-dimensional fault plane (Irikura and Muramatu, 1982; Imagawa and Mikumo, 1982; Irikura, 1983).

In all of these the kinematic dislocation models, however, some simplified assumptions are made rather arbitrarily on the process of rupture propagation over the fault plane and the displacement time history, without physical considerations concerning the stress conditions on and around the fault. For a more comprehensive understanding of complex ground motions in the near field, it is essential to model how the rupture initiates, spreads, and stops on the fault and how dynamic slip motions are developed under shearing stresses.

Another approach along this line has been to deal with the faulting process as a propagating shear crack by taking account of the initial stress field, frictions, and cohesive forces on the fault. The first study of this kind was the pioneering work made by Kostrov (1964, 1966). During the last decade, dynamic crack problems have been solved analytically and numerically for two-dimensional, in-plane and antiplane shear cracks growing at a constant velocity or with self-similar propagation under different fracture criteria and boundary conditions (Burridge, 1969, 1973; Burridge and Halliday, 1971; Hanson *et al.*, 1971, 1974; Ida and Aki, 1972; Ida, 1973; Takeuchi and Kikuchi, 1973; Fossum and Freund, 1975; Husseini *et al.*, 1975; Andrews, 1975, 1976a,b; Yamashita, 1976; Das and Aki, 1977a,b; Stoeckl, 1977). Some of these authors have calculated near- and far-field displacements and the spectra of seismic waves radiated from finite cracks, but the results from the two-dimensional crack studies are incomplete when compared with the observations. Thus, three-dimensional modeling of the spontaneous rupture process is needed, incorporating the distribution of static frictional strengths, sliding frictions, and applied initial shear stress on the fault. Early work in this field includes rupture propagation with constant subsonic speeds (Kostrov, 1964; Richards, 1976; Madariaga, 1976; Archuleta and Frazier, 1978) or in quasi-three-dimensional models (Yamashita, 1976; Miyatake, 1977; Mikumo and Miyatake, 1978). Much progress has recently been made to provide complete solutions for three-dimensional dynamic shear crack problems.

In Section II, we describe an application of kinematic dislocation models to the synthesis of strong ground motion, which includes the representation theorem and current methods of theoretical and semi-empirical synthesis, with some examples of the calculated results from recent earthquakes. A stochastic fault model that involves a random distribution of dislocations, slip velocities, slip angles, and rupture velocities over the fault is also discussed in relation to the problem of radiation of high-frequency seismic waves.

In Section III, we deal with dynamic shear cracks with an emphasis on numerical modeling of complete three-dimensional, spontaneous rupture processes not only in an infinite, homogeneous medium but also in a homogeneous half-space or horizontally layered medium. In this section, we describe the wave equations and boundary conditions, together with their finite-difference expressions, fracture criteria, and the distribution of frictional strength and sliding friction on the fault. Some numerical results for rupture propagation, slip displacements, and ground motions in the near field, which have been calculated for the cases of uniform and strongly heterogeneous fault strengths, are provided to help in understanding the dynamical problems.

## II. Synthesis from Kinematic Dislocation Models

### A. Representation Theorem

In Sections II.A, B, and C, we describe, mainly following Imagawa *et al.* (1984), the methods for theoretical and semi-empirical synthesis of ground motions based on the representation theorem for a kinematic dislocation model. The displacement field due to a slip dislocation over a fault $S$ can be expressed by (see, e.g., Burridge and Knopoff, 1964)

$$u^i(y, t) = \int_0^t d\tau \iint_S s(x, \tau) \cdot g^i(x, t - \tau, y) \, dS, \quad (1)$$

where $u^i(y, t)$ is the $i$th component of displacement at a location $y$ and time $t$, $s(x, t)$ is the slip discontinuity across the fault surface at a position $x$ at $t = \tau$, and $g^i(x, t - \tau, y)$ is the stress tensor at position $x$ and time $t$ due to an impulsive point force applied in the $i$th direction at the location $y$. The component $g^i(x, t - \tau, y)$ may also be taken as the $i$th component of displacement at the receiver position $y$ due to a point dislocation at $x$ on the fault and hence may be regarded as a Green's function or the impulse response of the medium. Equation (1) thus implies that the displacement at any point in the medium is expressed by the space–time convolution of the slip distribution with the Green's function integrated over the fault surface. When the direction of the displacement discontinuity is the same everywhere on the fault surface, the slip can be written as $s(x, t) = m^i \Delta u(x, t)$, where $m^i$ is the second-order moment tensor for a point shear dislocation, which is specified by the direction of the fault plane, and $\Delta u(x, t)$ is the scalar slip or source function. If we further assume that the position $x$ on the fault plane starts to slip at time $t'$ specified by the rupture propagation, $t$ can be replaced by $t - t'$, and Eq. (1) may also be expressed by

$$u^i(y, t) = \int_0^{t-t'} d\tau \iint_S m^i \Delta u(t - t' - \tau) \cdot g^i(x, t - \tau, y) \, dS. \quad (2)$$

## B. Theoretical Synthesis

To perform the integration [Eq. (2)] over the fault, the entire fault surface with dimensions $S = L \times W$ should be divided into $N$ segments, each of which has an elementary size of $\Delta S_j = L_j \times W_j$. The displacement given by Eq. (2) may be rewritten as

$$u^i(t) = \sum_{j=1}^{N} m_j^i \Delta u(t - t_j) * g_j^i(t) \Delta S_j$$

$$= \sum_{j=1}^{N} m_j^i \Delta u(t) * g_j^i(t - t_j) \Delta S_j, \quad (3)$$

where the subscript $j$ refers to the $j$th fault segment, the asterisk ($*$) indicates the convolution operator, and $t_j$ is the phase delay time from the onset of initial rupture. The delay time $t_j$ is expressed by $t_j = r_j/v + R_j/c$ where $r_j$ is the distance between the nucleation point of rupture and the $j$th segment on the fault, and $v$ and $c$ are the rupture velocity over the fault and the average wave velocity in the medium, respectively. If the distance $R_j$ along the ray path between each of the divided segments $\Delta S_j$ and the receiver position $y$ is much greater than the linear dimension of the segments, i.e., $\lambda R_j/2 \gg L_j^2$ [see, e.g., Aki and Richards (1980, Chap. 8), Eq. (14.12)], the far-field terms involved in the Green's function dominate over the other terms, and the displacement $u^i(y, t)$ should have a waveform proportional to $\iint_S \Delta \dot{u}(x, t - t') \, dS$. In this case, $\Delta u(t)$ in Eq. (3) may be replaced by $\Delta \dot{u}(t)$, and $g_j^i(t)$ reduces to the Green's function only for the far field.

If the slip function $\Delta u(t)$ has a ramp functional form, it may be written as

$$\Delta u(t) = (Dt/\tau_R) \cdot H(t), \quad t < \tau_R$$
$$= D, \quad t \geq \tau_R, \quad (4)$$

where $D$ is the final displacement, $\tau_R$ the rise time, and $H(t)$ the Heaviside unit step function. The synthetic seismogram $f^i(t)$ can be obtained by convolving $u_j^i(t)$ with the impulse response $h(t)$ of the recording system used, so that

$$f^i(t) = u^i(t) * h(t). \quad (5)$$

In order to calculate the Green's function $g_j^i(t)$, the discrete wave number/finite-element method (abbreviated as the DWFE method) presented recently by Olson (1982), and which is similar to that given by Alekseev and Mikhailenko (1980), is used. The DWFE method yields the complete elastic response of a vertically heterogeneous medium by combining separable solutions of the elastic equations for the horizontal dependence of the wave motion in terms of a Fourier–Bessel series, with finite-element and finite-difference solutions for the vertical and time dependence (Olson, 1982). The Green's function $g_j^i(t)$ obtained from this method includes all types of body and surface waves, and also near-field terms. The Green's functions are then convolved with the slip function multiplied by the second-order moment tensor and spatially integrated over the fault surface to obtain the theoretical ground displacements, as given by Eq. (3).

## C. Semi-Empirical Synthesis

In the semi-empirical synthesis, the records of small events, such as foreshocks and aftershocks, are used as empirical Green's functions instead of calculating the theoretical functions. If the source of the small event can be approximated as a point source located in the far field and if the source duration is short enough, then $\Delta \dot{u}(t)$ approximates a delta function $\delta(t)$, and the observed seismogram reduces to $f_A(t) \propto m_j^i \cdot g_j^i(t) * h(t)$, which includes only the propagation effects and the instrumental response (Harztell, 1978). If, however, the minor event has a small but finite source area and a finite rise time, then

$$f_{A_j}^i(t) = m_{A_j}^i \Delta \dot{u}_{A_j}(t) * g_j^i(t) \cdot \Delta S_{A_j} * h_A(t), \tag{6}$$

where the subscript $A$ denotes the quantities belonging to the minor event that occurred at the $j$th location on the fault. The fault area of the main shock is then divided into a finite number of segments with a unit dimension equal to the source area of the small events, i.e., $L_j = L_A$ and $W_j = W_A$. The main shock seismogram may be synthesized from Eqs. (3) and (5) as

$$f_M^i(t) = h_M(t) * \sum_{j=1}^{N} m_j^i \Delta \dot{u}_{M_j}(t) * g_j^i(t - t_j) \Delta S_j, \tag{7}$$

where the subscript $M$ indicates the case of the main shock. Taking the Fourier transform of Eqs. (6) and (7), we have

$$F_M^i(\omega) = H_M(\omega) * \sum_{j=1}^{N} m_j^i \Delta \dot{U}_{M_j}(\omega) \cdot G_j^i(\omega) e^{-j\omega t_j} \Delta S_j$$

$$= \sum_{j=1}^{N} Q_j(\omega) \cdot F_{A_j}(\omega) e^{-j\omega t_j} \tag{8}$$

$$Q_j(\omega) \equiv \left[ \Delta \dot{U}_{M_j}(\omega) / \Delta \dot{U}_{A_j}(\omega) \right] \left[ H_M(\omega) / H_A(\omega) \right], \tag{9}$$

since $m_j^i \cdot G_j^i(\omega) \cdot \Delta S_j = m_{A_j}^i \cdot G_j^i(\omega) \cdot \Delta S_{A_j}$. We have assumed that the

## 3. NUMERICAL MODELING

main shock and aftershocks were recorded with two different types of instruments with different amplifications and frequency responses at a single station.

From Eq. (8), the synthetic seismogram for the main shock may be given by

$$f_M^i(t) = \sum_{j=1}^{N} q_j(t) * f_{A_j}(t - t_j), \qquad (10)$$

where

$$q_j(t) = \int_{-\infty}^{\infty} Q_j(\omega) e^{-i\omega t} d\omega \qquad (11)$$

(Mikumo, 1981a; Imagawa and Mikumo, 1982). If the slip displacement and the rise time of the main shock are uniformly distributed over the fault plane and we use aftershocks with identical slips and rise times, then $Q_j(\omega)$, $q_j(t)$, and $f_{A_j}(t)$ turn out to be common over the fault plane and may be replaced by $Q(\omega)$, $q(t)$, and $f_A(t)$, respectively. In this case, Eq. (10) can be rewritten as

$$f_M^i(t) = q(t) * \sum_{j=1}^{N} f_A(t - t_j), \qquad (12)$$

where $q_j(t)$ or $q(t)$ are correction functions for the differences in the source time function between the main shock and the smaller events and between the instrumental responses used for recording the two different types of earthquakes.

The ratio $\Delta \dot{U}_M(\omega)/\Delta \dot{U}_A(\omega)$ can in principle be calculated from Eq. (4), which should include the ratios of the final displacements and the rise times for the main shock and the smaller event used. One possible way of estimating these ratios is to rely on empirical scaling relations in the fault parameters between larger and smaller events. If we use the empirical similarity conditions proposed by Kanamori and Anderson (1975), the scaling relations may be written as (Irikura and Muramatu, 1982; Irikura, 1983)

$$L_M/L_A = W_M/W_A = D_M/D_A = \tau_M/\tau_A = (M_{0_M}/M_{0_A})^{1/3}. \qquad (13)$$

Both of the ratios $D_M/D_A$ and $\tau_M/\tau_A$ will be obtained from the moment ratio $M_{0_M}/M_{0_A}$, which may be estimated from the average spectral amplitudes over low frequencies (Irikura, 1983) or again from an empirical formula (Kanamori and Anderson, 1975)

$$\log(M_{0_M}/M_{0_A}) = 1.5(M_M - M_A).$$

The third and fourth terms in Eq. (13) give $D_M/\tau_M = D_A/\tau_A$, actually assuming the same slip velocity for the larger and smaller events. If this is really the case, we obtain, using Eq. (4),

$$\Delta u_{M_j}(t) \simeq \sum_{k=1}^{n} \Delta u_{A_j}[t - (k-1)\tau_A] \quad \text{for } t \leq \tau_M, \quad (14)$$

and

$$\Delta \dot{U}_{M_j}(\omega) \simeq \sum_{k=1}^{n} \Delta \dot{U}_{A_j}(\omega) e^{-i\omega(k-1)\tau_A}, \quad (15)$$

where $n$ is taken to be the integer closest to $(M_{0_M}/M_{0_A})^{1/3}$. Further, if the same instrument could be used to obtain the records of both the main shock and smaller events, $H_M(\omega) = H_A(\omega)$ and hence $Q_j(\omega) \simeq \sum_{k=1}^{n} e^{-i\omega(k-1)\tau_A}$. In this case, Eq. (10) then reduces to

$$f_{M_j}^i(t) \simeq \sum_{j=1}^{N} \sum_{k=1}^{n} f_{A_j}[t - t_j - (k-1)\tau_A], \quad (16)$$

where $N \simeq n \times n$. This is a special case of the more general expression of Eq. (10) and corresponds exactly to the formula derived independently by Irikura (1983). It has been reported for many earthquakes, however, that the slip velocity depends on the local stress drop on the fault surface, and also that the average stress drop estimated for a number of events scatters in a range between 1 and 100 bars. These observations do not always support the simple assumption of Eq. (16). The use of Eq. (10) would be more appropriate to detailed waveform analysis.

### D. Some Examples and Discussion

In this section, some practical examples of the theoretical and semi-empirical synthesis are described; these were obtained from two recent moderate-size earthquakes in Japan. The first one is the 1984 western Nagano earthquake ($M = 6.8$) that occurred in a central part of the Honshu region of Japan. The focal mechanism solution and spatial distribution of a large number of aftershocks indicate that this earthquake was caused mainly by right-lateral strike-slip motion over a fault plane with dimensions of 12 km × 8 km and a dip of 79° trending in the N 81° E direction (Mikumo *et al.*, 1985a). Strong ground motions from the main shock have been well recorded by low-magnification seismographs ($T_0 = 6.0$ sec, $h = 0.6$, and $V = 1$) at more than 16 stations belonging to the Japan Meteorological Agency (JMA) located in central Japan within epicentral distances of less than 120 km. Figure 1 shows two horizontal component

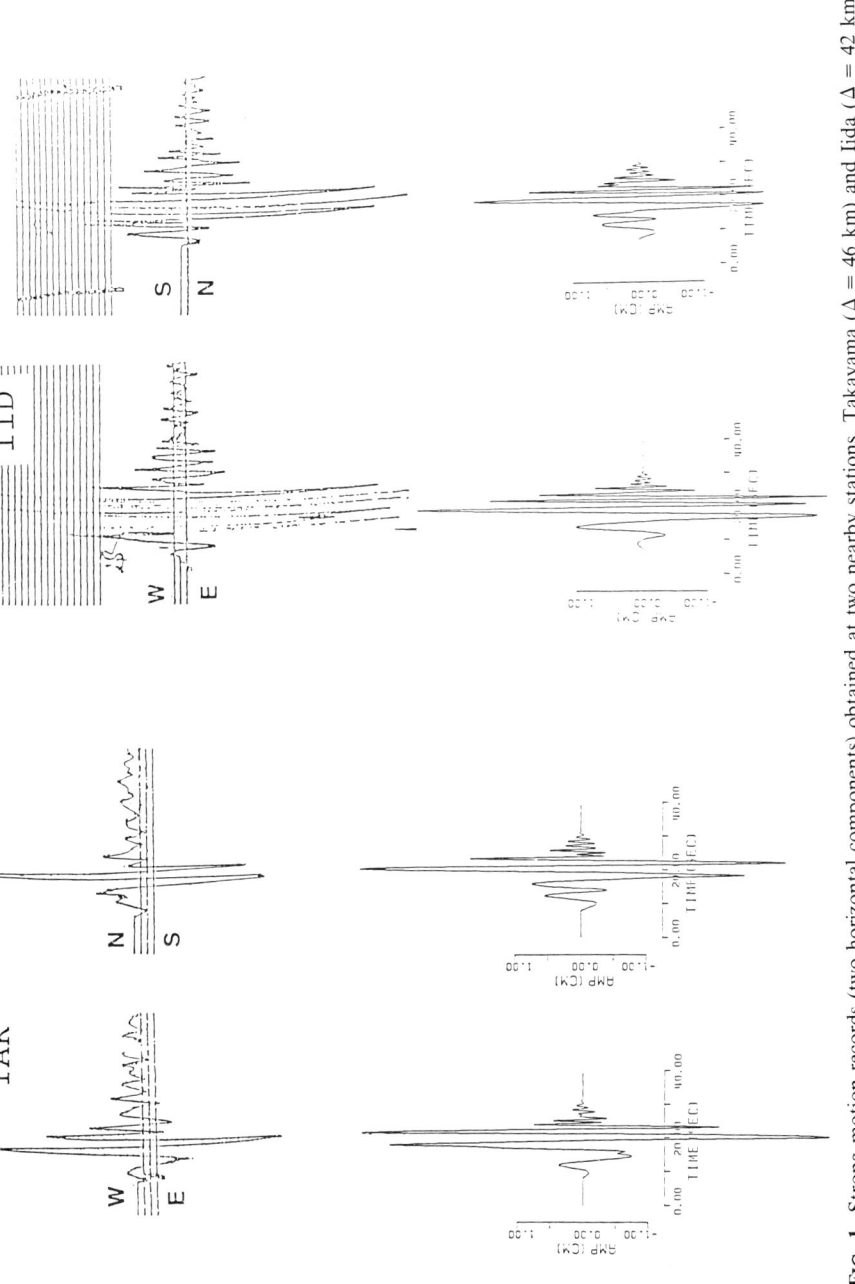

**Fig. 1.** Strong motion records (two horizontal components) obtained at two nearby stations, Takayama ($\Delta = 46$ km) and Iida ($\Delta = 42$ km), during the 1984 Western Nagano earthquake ($M = 6.8$), central Honshu, Japan, and the theoretical, synthetic seismograms. (After Mikumo et al., 1985a.)

records observed at two nearby stations, Takayama ($\Delta = 46$ km) and Iida ($\Delta = 42$ km), together with the corresponding synthetic seismograms. The synthetic seismograms in these cases were obtained from the method of theoretical synthesis described in Section II.B: Theoretical Green's functions have been calculated by the DWFE method for the crustal structure appropriate to this region and for each divided fault segment with dimensions of 2 km × 1 km, integrated over the entire fault plane with the phase delay time due to rupture propagation and then convolved with the slip function. The fault rupture was assumed to initiate at the hypocenter with a depth of 4 km and spread radially at a constant velocity. The fault parameters, including rise time ($\tau$), rupture velocity ($v$), and average final displacement ($\overline{D}$), were successively varied as a forward process to match the synthetics with the records. The estimated parameters are $\tau \simeq 1$ sec, $v \simeq 2.5$ km/sec, $\overline{D} = 1$ m, and the seismic moment $M_0 = 2.9 \times 10^{25}$ dyn cm (Mikumo et al., 1985a). It can be seen that there is satisfactory agreement in the general waveforms of body and surface waves, including their absolute amplitudes, between the observed and theoretical seismograms at the two stations. Theoretical seismograms have also been calculated for many other stations.

Several large aftershocks from this earthquake were recorded at the same stations with velocity-type strong motion seismographs. [The frequency response of the seismographs is flat over the frequency range between 0.02 and 20 Hz of ground velocity and the velocity sensitivity ranges between 1 V/100 kine to 1 V/0.5 kine (Muramatu, 1977).] The synthetic seismograms (Muramatu and Ohnuma, 1985) calculated from the semi-empirical synthesis using the records of a large aftershock ($M = 5.3$), which was located at the eastern side of the aftershock zone, are shown in Fig. 2. Comparison with Fig. 1 indicates that the main shock records at the two stations may also be well simulated by this synthesis.

Some of the best synthetics from the semi-empirical model came from the 1980 Off-Izu Peninsula earthquake ($M = 6.7$), which occurred off the southern coast of central Japan. Its aftershocks, observed within 5 days, were distributed for a length of about 15 km along the N 15° W direction and at depths between 12 and 19 km (Imoto et al., 1981), which suggest the fault dimensions for the main shock. The main shock hypocenter was located near the bottom center of the aftershock zone. The fault-plane solution indicates that this earthquake had a left-lateral strike-slip mechanism. The rise time and average rupture velocity during the main shock faulting were estimated as 1.0 sec and 3.2 km/sec, respectively (Muramatu and Irikura, 1982). The seismic waveforms from several moderate-size aftershocks, as well as that from the main shock, have been well recorded

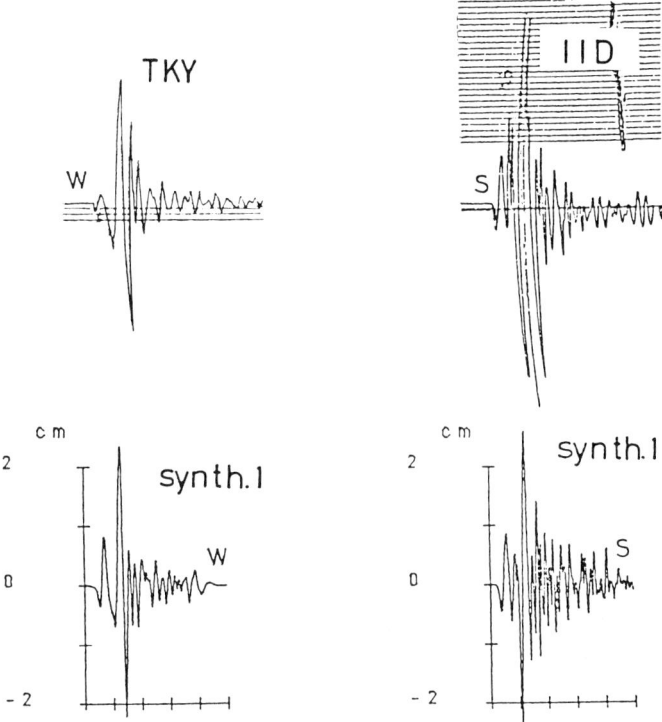

FIG. 2. Comparison between the strong motion records and the semi-empirically synthesized seismograms from aftershock records. (After Muramatu and Ohnuma, 1985.)

by the velocity-type strong motion semismographs (Muramatu, 1977) at three stations in the near field. The observed records and the corresponding synthetic seismograms estimated from the semi-empirical synthesis are shown in Figs. 3 and 4 (Muramatu and Irikura, 1982; Irikura, 1983). The upper two traces in Fig. 3 are the NS-component records of an aftershock ($M = 4.9$) located south of the main shock epicenter and of a foreshock with the same magnitude located north of it, both of which were obtained at a station SMC ($\Delta = 69$ km) located at a medium distance. The third trace indicates the synthetic seismograms obtained using Eq. (16) from the two minor events, and the last one is the main shock record. A comparison between the observed and predicted seismograms shows very good agreement in their waveforms with frequencies up to ~ 1 Hz. Figure 4 shows the records from two aftershocks ($M = 4.6$ and $M = 4.9$), the synthetic seismogram, and the main shock record obtained at a nearby station JIS ($\Delta =$

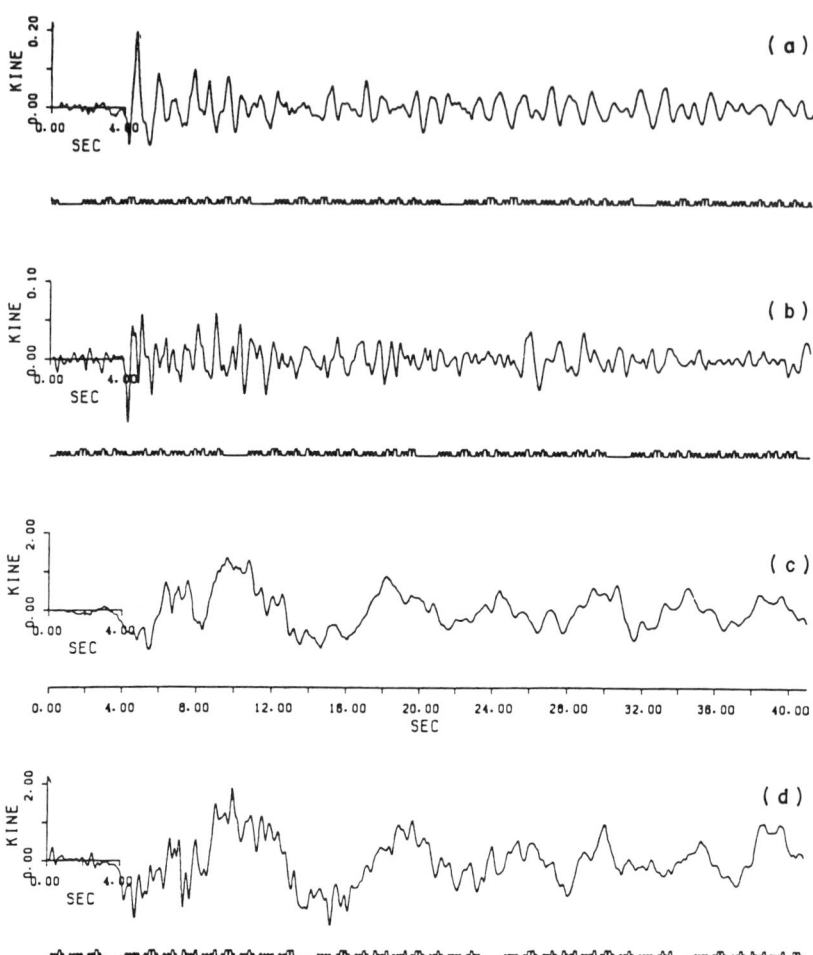

**FIG. 3.** Comparison between the observed and semi-empirically synthesized seismograms from the 1980 Off-Izu Peninsula earthquake ($M = 6.7$). Traces (a) and (b) show the NS-component records from an aftershock and a foreshock obtained at the SMC station ($\Delta = 69$ km), trace (c) indicates the synthesized seismogram from the above two shocks, and trace (d) is the main shock record. (After Irikura, 1983.)

20 km), west of the main shock epicenter (Irikura, 1983). Again, there is a good agreement between the observed and predicted seismograms. The examples shown suggest that the semi-empirical modeling for strong ground motions using smaller events is useful under some favorable conditions.

Both the theoretical and semi-empirical synthesis methods have some problems. In the theoretical synthesis, the crustal structure, particularly of

## 3. NUMERICAL MODELING

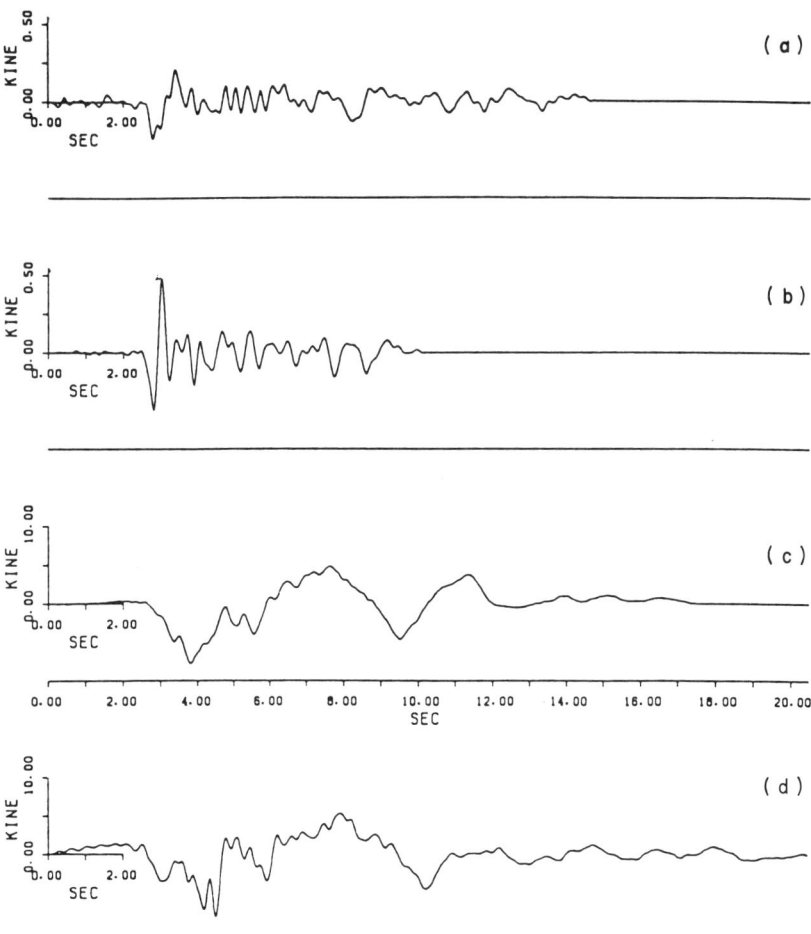

FIG. 4. Comparison between the observed and semi-empirically synthesized seismograms for the JIZ station ($\Delta = 20$ km), where the four traces are arranged in the same way as in Fig. 3. (After Irikura, 1983.)

low-velocity surficial layers, which generate large-amplitude converted body and surface waves, are not always well known. Incorrect estimates for the surficial structure would lead to unrealistic waveforms. If small-scale lateral heterogeneities, including topographical site effects, exist in the area between the source and receiver, there would be some variations of the seismic waveforms. In the calculations shown previously, attenuation properties have not been taken into account. Thus, theoretical calculations may not always generate complete Green's functions for realistic structures.

In the semi-empirical synthesis, the small shocks used should be large enough to excite the earth's response sufficiently above the noise level and also should have the same focal mechanism as the main shock. In some cases, however, the records from the minor event might be affected by its complicated rupture process, if the event had an appreciable source dimension. Also smaller shocks that occur on the main shock fault often have a radiation pattern that is considerably different from that of the main shock. It should also be noted that Green's functions change their waveforms considerably with the source depth, particularly in a complex crustal structure. This implies that an observed record from a minor event, even if it does not involve the radiation effects, should only be regarded as representing a Green's function for its focal depth. It is, therefore, extremely important to use the records of minor events that are as well distributed at different depths and horizontal locations over the fault plane as possible. However, there would be some difficulties in applying the semi-empirical method to the case of large earthquakes with an extensive fault area, since the above restrictions on the selection of suitable minor shocks are quite severe.

### E. Stochastic Fault Model

Strong-motion records often involve large-amplitude, high-frequency seismic waves superimposed on lower frequency waves. The results shown in the previous sections suggest that the high-frequency waves cannot be well simulated either by theoretical synthesis or by semi-empirical synthesis. One possible explanation of the generation of the high-frequency waves may be some amplification effects due to reverberations and scattering of seismic waves within small-scale, inhomogeneous structures, but the source of the waves should be the complex rupture process of heterogeneous faults.

As a first step in explaining how the high-frequency waves are radiated from the fault, we introduce a stochastic fault model with spatially nonuniform displacements, slip angles, slip velocities (or rise times), and rupture velocities (Mikumo, 1976). In this model, these fault parameters are assumed to be functions of position $(\xi, \eta)$ on the fault plane and are distributed independently, in a normal random fashion, with variable standard deviations about their respective mean values. (This randomness should be closely related spatially from the viewpoint of the dynamic rupture process.) The distribution of these fault parameters is given in this

## 3. NUMERICAL MODELING

table:

| Fault parameter | Distribution | Mean | Standard deviation |
|---|---|---|---|
| Final displacement | $D(\xi, \eta)$ | $\overline{D}$ | $\sigma_D$ |
| Slip angle | $\alpha(\xi, \eta)$ | $\overline{\alpha}$ | $\sigma_\alpha$ |
| Rise time | | $\overline{\tau}$ | |
| Slip velocity | $D(\xi, \eta)/\tau$ | $\overline{D}/\overline{\tau}$ | |
| Rupture velocity | $v(\xi, \eta)$ | $\overline{v}$ | $\sigma_v$ |

The mean values will be the controlling factors for long-period seismic waves and static displacements that come from the gross features of the faulting process. For a normal random distribution of these parameters,

$$D(\xi, \eta) = \left(\overline{D}/\sqrt{2\pi\sigma_D}\right) \exp\left[-(D_i - \overline{D})^2/2\sigma_D^2\right], \quad (17)$$

$\alpha(\xi, \eta)$ and $v(\xi, \eta)$ follow the same distribution as given in the preceding table, and all of these parameters are assumed to be two-dimensionally distributed on each segment over the fault plane. The rupture time at any point $(\xi, \eta)$ or $(r, \theta)$ on the fault is

$$T(r, \theta) = \int \frac{dr}{v(r, \theta)}, \quad (18)$$

assuming the rupture starts at $(0, 0)$ and propagates radially over the fault. We further allow some fluctuations in the source time function, following Haskell (1964):

$$\Delta u(t) = \begin{cases} 0, & t < T \\ (\overline{D}/\tau) \cdot [t - \sin(2m\pi t/\tau)/2m\pi t], & T \leq t < T + \tau \\ D, & t \geq T + \tau. \end{cases} \quad (19)$$

Strong motions in the near field from this stochastic model have been calculated for a number of different cases based on the fault parameters for the 1969 Central Gifu earthquake in Japan (Mikumo, 1973). These cases, shown in Figs. 5(a)–(f), include the following:

1. Only the final displacements are randomly distributed with $\sigma_D/\overline{D}$ ranging between 0 and 0.75 [case (b)]; the other parameters are fixed.

2. Only the slip angles are randomly distributed with $\sigma_\theta$ ranging between 0 and 30° [case (c)]; the other parameters are fixed.

3. Only the rupture velocities are randomly varied with $\sigma_v/\overline{v}$ ranging between 0 and 0.60 [cases (d), (e), and (f)]; the other parameters are fixed.

4. Two or three of the parameters are randomly distributed within the indicated ranges.

5. Only the source time function has some fluctuations with $m = 1$ to 3.

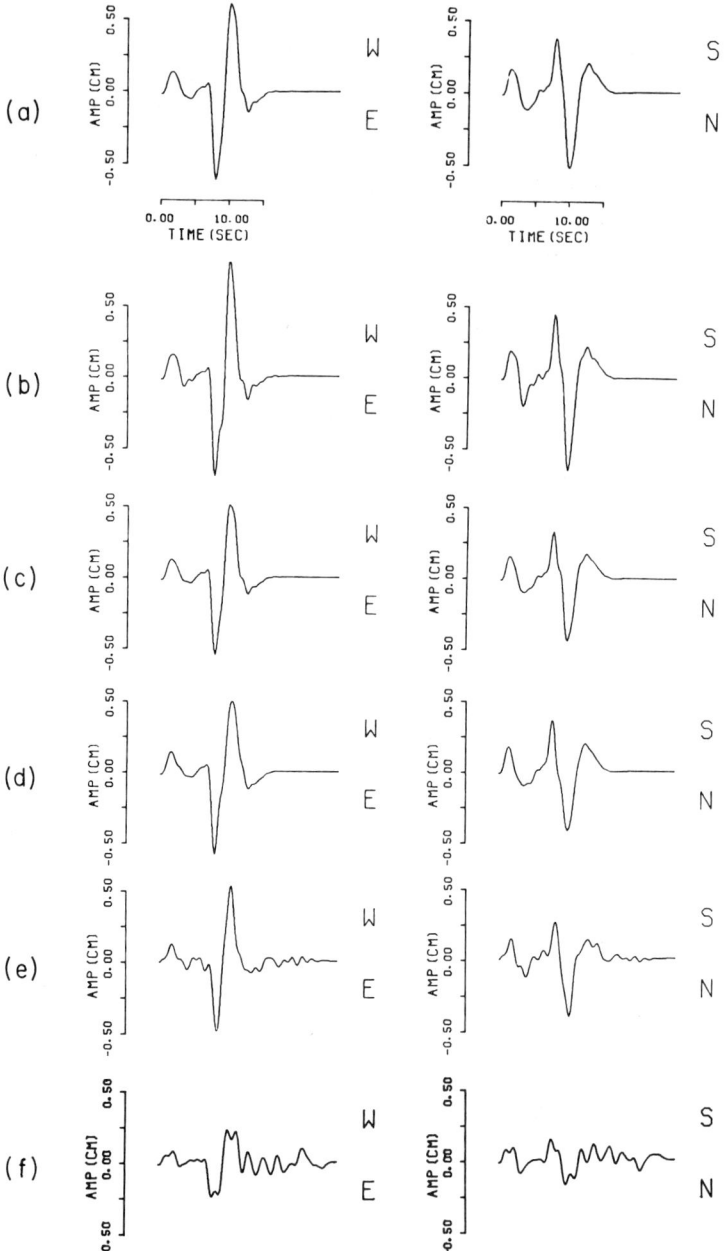

FIG. 5. Synthetic seismograms calculated for several different assumed cases with randomly distributed fault parameters in a stochastic dislocation model. For details, see text. (After Imagawa and Mikumo, 1982.)

Figure 5 shows the synthetic seismograms calculated for each one of these cases with its largest variance (Imagawa and Mikumo, 1982). In these cases, the unit dimension of the divided fault segments has been taken as 2 km × 1 km or 1 km × 1 km within the entire fault area of 20 km × 10 km. Comparisons of these traces clearly indicate that large variations in the fault displacements [case (b)] and slip angles [case (c)] on these fault segments do not produce major effects on the waveforms. High-frequency components tend to be gradually increased as $\sigma_v/\bar{v}$ increases from 0.2 to 0.6 [cases (d) to (f)], suggesting that rapid variations in the rupture velocity can radiate high-frequency waves. The fluctuation in the source time function

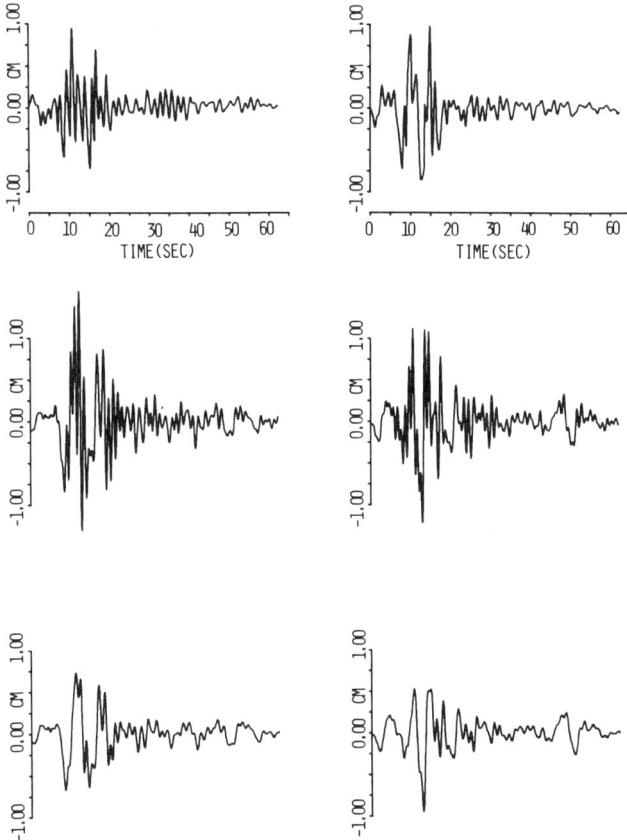

FIG. 6. Strong motion records obtained at the GIF station from the 1969 Central Gifu earthquake ($M = 6.9$) and the semi-empirically synthesized seismograms from a stochastic fault model, using four aftershock records. (After Imagawa and Mikumo, 1982.)

from a smooth ramp function also yields high-frequency waves (not shown here), as expected. This last case should not be regarded as being caused by an independent factor, because the source of the fluctuation should in principle be variations in the rupture velocity on the fault. Cases (g) to (j) (not shown in figure) also show the synthetic seismograms calculated for variable rupture velocities on fan-shaped fault segments $\Delta r \times \Delta \phi$, where $\Delta r = 2$ to 3 km, and $\Delta \phi = 30$ to $45°$. Case (j) produces high-frequency waves and preserves low-frequency components, while case (f) loses low-frequency waves to a considerable extent.

The above calculations do not involve the effects of earth structures. Figure 6 shows an example [case (k)] of the synthetic seismograms, which incorporate the empirical Green's functions derived from four aftershock records together with an assumed incoherent rupture propagation (Imagawa and Mikumo, 1982).

These results demonstrate that the most effective factors in the generation of high-frequency waves are accelerations or decelerations of the incoherent rupture propagation on the fault. Similar conclusions have also been suggested from other studies (Madariaga, 1977; Boore and Joyner, 1978; Campillo, 1983).

## III. Three-Dimensional Modeling of Spontaneous Fault Rupture Processes

### A. Dynamic Shear Crack Problems

As mentioned in the introduction, theoretical modeling of the faulting process has been practiced since the middle 1960s to deal with dynamic shear crack problems. This section briefly reviews some of these studies, particularly two-dimensional spontaneous shear cracks and current three-dimensional crack models.

#### 1. Two-Dimensional Models

Two- and three-dimensional problems of propagating shear cracks were studied by Kostrov (1964). He obtained self-similar solutions of a circular crack propagating with a constant velocity in an infinite, homogeneous medium and also gave a simple but very useful expression for the displacement time history and the stress field.

In 1966, Kostrov also solved the problem of spontaneous propagation of antiplane shear cracks with semi-infinite length by using the dynamic Griffith's fracture criterion for the energy balance at the advancing crack tip. He found that the rupture propagates at variable velocities under uniform applied stress and concentrated load. This was the first study that dealt with spontaneous rupture propagation. Kikuchi and Takeuchi (1970, 1972) and Takeuchi and Kikuchi (1973) extended Kostrov's method to a finite length antiplane shear crack and investigated the propagation and stopping mechanisms of the crack by introducing a zero-stress drop condition into the problem. The stopping mechanism of a semi-infinite crack was also investigated by Husseini *et al.* (1975) based on Kostrov's method, who calculated two different cases for a limited stress and surface energy arrest.

Ida and Aki (1972), Ida (1973), and Andrews (19766a,b) introduced a slip-weakening model in which the frictional stress near the crack tip decreases as a linear function of fault slip. The slip-weakening fracture criterion, along with a finite-difference method applied to in-plane shear cracks (Andrews, 1976b) subjected to a uniform stress, indicates that the maximum rupture speed could reach *P*-wave velocity in the medium. If a transition zone for slip weakening is small, however, the criterion reduces to Griffith's criterion. A similar conclusion for the rupture velocity has been obtained for self-similar, in-plane shear cracks lacking cohesion (Burridge, 1973).

Das (1976) and Das and Aki (1977a,b) demonstrated that Irwin's criterion, based on the stress intensity factor at the crack tip, can be approximately translated to a finite-stress criterion and calculated antiplane and in-plane shear crack propagation on the faults with uniform and nonuniform strengths, by using a numerical boundary integral technique. They proposed (Das and Aki, 1977b) the "barrier model" in which the critical stress level varies over the fault, showing that the presence of unbroken barriers can effectively excite high-frequency seismic waves and can slow down the rupture propagation. This model has played an important role, along with the "asperity model" (Kanamori, 1981), in seismic source theory. "Barriers" in this case are defined as high-strength portions that do not break or that strongly resist slip during the main rupture but sometimes break at a later time as multiple shocks and aftershocks. The term "asperity" implies large, strongly coupled portions of the fault surface that yield large stress drops during the main rupture.

2. Three-Dimensional Models

Although theoretical studies on three-dimensional, dynamic shear cracks can also be traced back to Kostrov (1964), no significant progress was made until the late 1970s. Richards (1976) obtained an analytical solution for three-dimensional elliptical shear cracks growing steadily with a subsonic speed in an infinite prestressed medium. Madariaga (1976) solved, by a finite-difference method, the problem of a circular shear crack that propagates with a fixed rupture velocity and stops suddenly. By using a finite-element method, Archuleta and Frazier (1978) calculated shallow strike-slip fault motions expected from a semi-circular shear crack that initiates at the ground surface, propagates with a constant speed, and abruptly stops. Besides these, several other studies on three-dimensional cracks propagating with constant rupture velocity have been made by using different techniques (Das, 1980; Burridge and Moon, 1981; Day, 1982a). Some simple formulas have been presented (Day, 1982a) for the relationship between the fault parameters, such as the rise time, final displacement, and fault dimensions.

For spontaneous, three-dimensional shear cracks, Yamashita (1976) and Mikumo and Miyatake (1978) have made numerical simulations of rupture propagation under the finite-stress criterion on quasi-three-dimensional models, assuming a gradually varying initial stress field and various types of nonuniform distributions of static frictional strength, respectively. It has been shown (Mikumo and Miyatake, 1978) that strongly heterogeneous faults could yield incoherent rupture propagation with decelerated velocities, generate high-frequency seismic waves, and leave unbroken patches on the fault. Some of these results will be described later. In 1980, Miyatake (1980) extended the previous model by solving a complete three-dimensional crack propagation under Irwin's criterion for a dynamic rupture processes on homogeneous and heterogeneous faults and for seismic waves radiated from the fault. Das (1981) also solved the problem of spontaneously propagating shear cracks in an infinite medium by applying the boundary integral equation technique under a critical stress level fracture criterion and obtained results for the rupture velocity similar to Mikumo and Miyatake (1978). She also obtained a relation between the average fault slip and average stress drop for a rectangular fault. Numerical simulations for the same problem were also performed by Day (1982b) and Virieux and Madariaga (1982) under conditions of both uniform and nonuniform prestress, with the slip-weakening and Irwin's fracture criterion, respectively. These provided some results for spatial variations of rupture velocity and slip velocity function. Many of the important results will be presented in Sections III. E, G, H, and I, together with those from recent studies (Mikumo *et al.*, 1985b).

B. Some Considerations on Realistic Rupture Processes

We assume that the actual rupture process on the fault may be described as follows: the rupture nucleates in a very small region where a number of small cracks are developed within part of a preexisting fault zone or newly created weak zone under increasing tectonic stress. The rupture initiates suddenly when the tectonic shear stress becomes greater than the weakest shear strength of the fault zone materials. Immediately, stress is concentrated in some part of the adjacent region causing successive fault slips if the concentrated stress plus the applied tectonic stress exceeds the shear strength. The rupture spreads rapidly over the fault and finally stops at the fault edges or in regions where the shear strength is higher than the stress. In this case, the velocity of the spontaneous rupture propagation will be specified by the elastic constants of fault zone materials and by the difference between the shear stress and the fracture strength. If either the fault has heterogeneous elastic properties and strength or the shear stress is not homogeneously distributed over the fault, the rupture will not propagate uniformly in space and time. As discussed in Section II.B, this type of incoherent rupture would cause episodic fault motion and would radiate high-frequency seismic waves.

In the present section, we model the above fault rupture process as a three-dimensional, dynamic shear crack propagating in an infinite, homogeneous medium; homogeneous half-space; in a horizontally layered half-space; or in more complex heterogeneous structures. The fracture criteria for rupture propagation, the form of the distribution of the strength, shear stress, and some related problems are also discussed.

### C. Fault Rupture Process in an Infinite, Homogeneous Medium

We first describe the fault rupture process in an infinite, homogeneous medium following Miyatake (1980). The fault plane is taken here as the $xy$ plane ($z = 0$) of a Cartesian coordinate system as shown in Fig. 7, assuming that the uniform initial shear stress $\sigma_0$ ($T_0$ in Fig. 7) is applied parallel to the fault plane. All three displacement components ($u$, $v$, and $w$) and six stress components ($\sigma_{xx}$, $\sigma_{xy}$, $\sigma_{xz}$, $\sigma_{yy}$, $\sigma_{yz}$, and $\sigma_{zz}$), which are caused by dynamic faulting at any point in the medium, should satisfy the equations of motion, and hence we have the wave equations

$$(\lambda + \mu)\frac{\partial^2 u}{\partial x^2} + \mu \nabla^2 u + (\lambda + \mu)\frac{\partial^2 v}{\partial x \partial y} + (\lambda + \mu)\frac{\partial^2 w}{\partial x \partial z} = \rho \frac{\partial^2 u}{\partial t^2},$$

$$(\lambda + \mu)\frac{\partial^2 u}{\partial x \partial y} + (\lambda + \mu)\frac{\partial^2 v}{\partial y^2} + \mu \nabla^2 v + (\lambda + \mu)\frac{\partial^2 w}{\partial y \partial z} = \rho \frac{\partial^2 v}{\partial t^2}, \quad (20)$$

$$(\lambda + \mu)\frac{\partial^2 u}{\partial x \partial z} + (\lambda + \mu)\frac{\partial^2 v}{\partial z \partial y} + (\lambda + \mu)\frac{\partial^2 w}{\partial z^2} + \mu \nabla^2 w = \rho \frac{\partial^2 w}{\partial t^2},$$

where $\lambda$ and $\mu$ are Lamé's elastic constants and $\rho$ is the density of the medium. When we consider two regions occupying $z > 0$ and $z < 0$, $u$ and

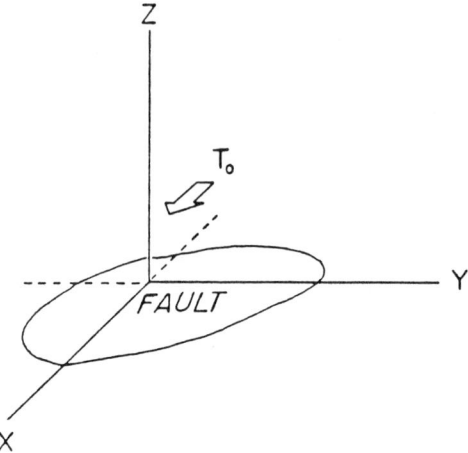

FIG. 7. Fault geometry and the applied shear stress.

$v$ are discontinuous and $\sigma_{zz}$ is continuous across the plane $z = 0$. The displacements $u$ and $v$ and the stress $\sigma_{zz}$ are odd functions of $z$. This implies that $u$ and $v$ tend to the one-half of the displacement discontinuities and $\sigma_{zz} = 0$ as $z \to 0$. It is therefore sufficient to treat the present problem in the half-space $z > 0$ because of symmetry. The fault is initially subject to the shear stress $\sigma_0$ working in the $x$ direction, i.e., $\sigma_{xz} = \sigma_0$. In dynamic problems, once fault slip initiates at any point on the fault surface, the slip motion is resisted by the sliding frictional stress $\sigma_d$ (dynamic friction), and the driving stress $\sigma_{xz}$ reduces to $\sigma_0 - \sigma_d$.

The boundary conditions in the present problem are described in Eq. (21). On the fault plane ($z = 0$), we have

$$\sigma_{xz} = \mu\left(\frac{\partial u}{\partial z} + \frac{\partial w}{\partial x}\right) = \sigma_0 - \sigma_d,$$

$$\sigma_{yz} = \mu\left(\frac{\partial v}{\partial z} + \frac{\partial w}{\partial y}\right) = 0, \qquad (21)$$

$$\sigma_{zz} = \lambda\left(\frac{\partial u}{\partial x} + \frac{\partial v}{\partial y} + \frac{\partial w}{\partial z}\right) + 2\mu\frac{\partial w}{\partial z} = 0.$$

At $z = 0$ outside the fault surface, $\sigma_{zz} = 0$, and $u$ and $v$ are also taken to be zero since no fault slip occurs there. The $w$ component perpendicular to the fault plane takes nonzero values over any point at $z = 0$.

Now, we solve Eq. (1) numerically by the finite-difference scheme under the boundary conditions of Eq. (21). The finite-difference grids are taken in

3. NUMERICAL MODELING        113

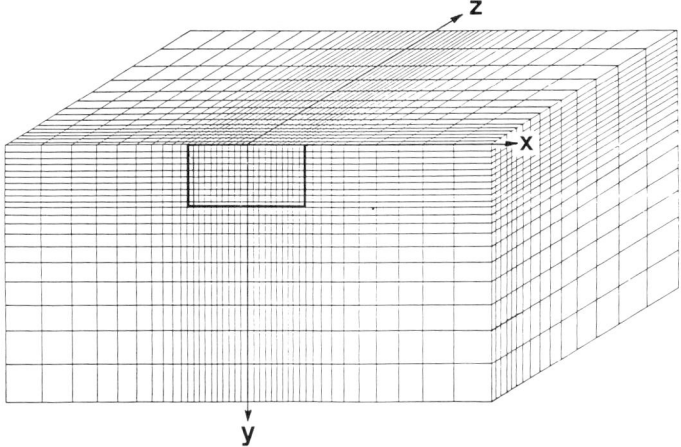

FIG. 8. Schematic view of the finite-difference grids used in the present study.

a three-dimensional space in such a way as given by Eq. (22) (Miyatake, 1980), with constant finer intervals within and just above the fault plane and with gradually increasing intervals outside and far above the plane, as illustrated in Fig. 8. Thus,

$$\Delta x_i = \Delta x_0 \text{ for } i < N_x, \quad \Delta x_i = (1 + \alpha_x) \Delta x_{i-1} \text{ for } i \geq N_x,$$

$$\Delta y_i = \Delta y_0 \text{ for } i < N_y, \quad \Delta y_i = (1 + \alpha_y) \Delta y_{i-1} \text{ for } i \geq N_y, \quad (22)$$

$$\Delta z_i = \Delta z_0 \text{ for } i < N_z, \quad \Delta z_i = (1 + \alpha_z) \Delta z_{i-1} \text{ for } i \geq N_z,$$

where $\alpha_x$, $\alpha_y$, and $\alpha_z$ are numerical constants, and $\Delta x_i$, $\Delta y_i$, and $\Delta z_i$ are the $i$th grid intervals in the $x$, $y$, and $z$ directions, respectively. The outer boundaries of the model space are far enough away so that compressional waves reflected back from the boundaries do not arrive onto the fault until after the time that the rupture propagation reaches the fault edges. An alternative way to eliminate the effects of reflected waves from these artificial boundaries would be to introduce viscous damping of normal and shear stress components along the boundaries (see, e.g., Lysmer and Kuhlemeyer, 1969) or absorbing boundary conditions (see, e.g., Smith, 1974; Clayton and Engquist, 1977; Suzuki and Hakuno, 1984; Berg, 1984), but we do not use these conditions here. The wave equations [Eq. (20)] can be rewritten for the finite-difference grids given in Eq. (22) as (Miyatake,

1980, with a few corrections due to misprints)

$$(\lambda + 2\mu)\left(\frac{u^t(i+1,j,k)}{\Delta x_{i+1}\cdot(\Delta x_{i+1}+\Delta x_i)} - \frac{u^t(i,j,k)}{\Delta x_{i+1}\cdot\Delta x_i} + \frac{u^t(i-1,j,k)}{\Delta x_i\cdot(\Delta x_{i+1}+\Delta x_i)}\right)$$

$$+\mu\left(\frac{u^t(i,j+1,k)}{\Delta y_{j+1}\cdot(\Delta y_{j+1}+\Delta y_j)} - \frac{u^t(i,j,k)}{\Delta y_{j+1}\cdot\Delta y_j} + \frac{u^t(i,j-1,k)}{\Delta y_j\cdot(\Delta y_{j+1}+\Delta y_j)}\right)$$

$$+\mu\left(\frac{u^t(i,j,k+1)}{\Delta z_{k+1}\cdot(\Delta z_{k+1}+\Delta z_k)} - \frac{u^t(i,j,k)}{\Delta z_{k+1}\cdot\Delta z_k} + \frac{u^t(i,j,k-1)}{\Delta z_k\cdot(\Delta z_{k+1}+\Delta z_k)}\right)$$

$$+\frac{(\lambda+\mu)}{2}\frac{[v^t(i+1,j+1,k) - v^t(i+1,j-1,k) - v^t(i-1,j+1,k) + v^t(i-1,j-1,k)]}{(\Delta x_{i+1}+\Delta x_i)(\Delta y_{j+1}+\Delta y_j)}$$

$$+\frac{(\lambda+\mu)}{2}\frac{[w^t(i+1,j,k+1) - w^t(i+1,j,k-1) - w^t(i-1,j,k+1) + w^t(i-1,j,k-1)]}{(\Delta x_{i+1}+\Delta x_i)(\Delta z_{k+1}+\Delta z_k)}$$

$$= (\rho/2\,\Delta t^2)(u^{t+\Delta t}(i,j,k) - 2u^t(i,j,k) + u^{t-\Delta t}(i,j,k)),$$

$$(\lambda + 2\mu)\left(\frac{v^t(i,j+1,k)}{\Delta y_{j+1}\cdot(\Delta y_{j+1}+\Delta y_j)} - \frac{v^t(i,j,k)}{\Delta y_{j+1}\cdot\Delta y_j} + \frac{v^t(i,j-1,k)}{\Delta y_j\cdot(\Delta y_{j+1}+\Delta y_j)}\right)$$

$$+\mu\left(\frac{v^t(i+1,j,k)}{\Delta x_{i+1}\cdot(\Delta x_{i+1}+\Delta x_i)} - \frac{v^t(i,j,k)}{\Delta x_{i+1}\cdot\Delta x_i} + \frac{v^t(i-1,j,k)}{\Delta x_i\cdot(\Delta x_{i+1}+\Delta x_i)}\right)$$

$$+\mu\left(\frac{v^t(i,j,k+1)}{\Delta z_{k+1}\cdot(\Delta z_{k+1}+\Delta z_k)} - \frac{v^t(i,j,k)}{\Delta z_{k+1}\cdot\Delta z_k} + \frac{v^t(i,j,k-1)}{\Delta z_k\cdot(\Delta z_{k+1}+\Delta z_k)}\right)$$

$$+\frac{(\lambda+\mu)}{2}\frac{[u^t(i+1,j+1,k) - u^t(i+1,j-1,k) - u^t(i-1,j+1,k) + u^t(i-1,j-1,k)]}{(\Delta x_{i+1}+\Delta x_i)(\Delta y_{j+1}+\Delta y_j)}$$

$$+\frac{(\lambda+\mu)}{2}\frac{[w^t(i,j+1,k+1) - w^t(i,j-1,k+1) - w^t(i,j+1,k-1) + w^t(i,j-1,k-1)]}{(\Delta y_{j+1}+\Delta y_j)(\Delta z_{k+1}+\Delta z_k)}$$

$$= (\rho/2\,\Delta t^2)(v^{t+\Delta t}(i,j,k) - 2v^t(i,j,k) + v^{t-\Delta t}(i,j,k)),$$

$$(\lambda + 2\mu)\left(\frac{w^t(i,j,k+1)}{\Delta z_{k+1}\cdot(\Delta z_{k+1}+\Delta z_k)} - \frac{w^t(i,j,k)}{\Delta z_{k+1}\cdot\Delta z_k} + \frac{w^t(i,j,k-1)}{\Delta z_k\cdot(\Delta z_{k+1}+\Delta z_k)}\right)$$

## 3. NUMERICAL MODELING

$$+ \mu \left( \frac{w^t(i+1, j, k)}{\Delta x_{i+1} \cdot (\Delta x_{i+1} + \Delta x_i)} - \frac{w^t(i, j, k)}{\Delta x_{i+1} \cdot \Delta x_i} + \frac{w^t(i-1, j, k)}{\Delta x_i \cdot (\Delta x_{i+1} + \Delta x_i)} \right)$$

$$+ \mu \left( \frac{w^t(i, j+1, k)}{\Delta y_{j+1} \cdot (\Delta y_{j+1} + \Delta y_j)} - \frac{w^t(i, j, k)}{\Delta y_{j+1} \cdot \Delta y_j} + \frac{w^t(i, j-1, k)}{\Delta y_j \cdot (\Delta y_{j+1} + \Delta y_j)} \right)$$

$$+ \frac{(\lambda + \mu)}{2} \frac{[u^t(i+1, j, k+1) - u^t(i+1, j, k-1) - u^t(i-1, j, k+1) + u^t(i-1, j, k-1)]}{(\Delta z_{k+1} + \Delta z_k)(\Delta x_{i+1} + \Delta x_i)}$$

$$+ \frac{(\lambda + \mu)}{2} \frac{[v^t(i, j+1, k+1) - v^t(i, j-1, k+1) - v^t(i, j+1, k-1) + v^t(i, j-1, k-1)]}{(\Delta y_{j+1} + \Delta y_j)(\Delta z_{k+1} + \Delta z_k)}$$

$$= (\rho/2 \Delta t^2)(w^{t+\Delta t}(i, j, k) - 2w^t(i, j, k) + w^{t-\Delta t}(i, j, k)). \quad (23)$$

The corresponding boundary conditions [Eq. (21)] are given by

$$\mu \left( \frac{u^t(i, j, 2) - u^t(i, j, 1)}{\Delta z_0} + \frac{w^t(i+1, j, 2) - w^t(i-1, j, 2)}{2 \cdot \Delta x_0} \right) = \sigma_0 - \sigma_d,$$

$$\mu \left( \frac{v^t(i, j, 2) - v^t(i, j, 1)}{\Delta z_0} + \frac{w^t(i, j+1, 2) - w^t(i, j-1, 2)}{2 \cdot \Delta y_0} \right) = 0,$$

$$\lambda \left( \frac{u^t(i+1, j, 2) - u^t(i-1, j, 2)}{2 \cdot \Delta x_0} + \frac{v^t(i, j+1, 2) - v^t(i, j-1, 2)}{2 \cdot \Delta y_0} \right)$$

$$+ (\lambda + 2\mu) \left( \frac{w^t(i, j, 2) - w^t(i, j, 1)}{\Delta z_0} \right) = 0.$$

(24)

From the finite-difference equations of Eq. (23), $u^{t+\Delta t}(i, j, k)$, $v^{t+\Delta T}(i, j, k)$, and $w^{t+\Delta t}(i, j, k)$ (for $k \neq 1$) can be solved at each time step that is successively advanced by a finite time interval $\Delta t$. The three displacement components on the fault surface (for $k = 1$) are obtained through the boundary conditions [Eq. (24)], which are again put into Eq. (23). Through this procedure, we observe that the rupture propagating spontaneously on the fault surface depends on the fracture criteria described in Section III.D, and that seismic waves are radiated from the fault.

The accuracy of the numerical solutions has been checked in the case of the rupture propagating with a prescribed constant velocity, and the solutions compared with a self-similar solution by Kostrov (1964) for circular crack expansion. It has been confirmed (Miyatake, 1980) that the obtained spatial distribution of fault displacements and the form of the source time function show reasonable agreement with those from Kostrov's analytical solution, if we choose the parameters $\Delta x_0 = \Delta y_0 = \Delta z_0 \equiv \Delta h = 1$ km, $\Delta t = 0.05$ sec, $V_P = 5.5$ km/sec, $V_S = 3.2$ km/sec, and $\rho = 2.75$ g cm$^{-3}$ and a rupture velocity of 3.0 km/sec, where $V_P$ and $V_S$ are P- and S-wave velocities, respectively. The choice of these parameters satisfies the stability conditions for two-dimensional, in-plane wave propagation and also for the three-dimensional problem, i.e., $\Delta t / \Delta h \leq (v_P^2 + v_S^2)^{-1/2}$ (Aki and Richards, 1980, Chap. 13), or $v_S \cdot \Delta t / \Delta h \leq \frac{1}{2}$ for a Poisson's ratio of $\frac{1}{4}$.

## D. Fracture Criteria

Since the fracture of some materials occurs when the applied stress exceeds its material strength, it appears reasonable to introduce a fracture criterion based on this simple idea into the present model. Actually, such a finite-stress criterion has been employed in two-dimensional crack problems (see, e.g., Burridge and Halliday, 1971; Burridge, 1973). In numerical calculations of these crack problems, however, the calculated stress at the crack tip would depend on the grid interval used in the finite-difference or finite-element models, since the crack tip is a singular point. For this reason, it may not be adequate in the numerical modeling of the rupture process to rely *a priori* on the simple concept of comparing a critical stress level directly with fault strength; more physical considerations are needed.

In fracture mechanics, there are two different types of fracture criteria: (1) Griffith's criterion and (2) Irwin's criterion, both of which may be used in theoretical seismology (Aki and Richards, 1980). Griffith's fracture criterion implies that the fracture occurs when the supply of mechanical energy from the surrounding material exceeds the consumed surface energy at the crack tip. This criterion has been used in several cases of two-dimensional cracks (see, e.g., Kostrov, 1966; Fossum and Freund, 1975; Husseini *et al.*, 1975). Under this criterion, there still remains a stress singularity at the advancing crack tip. In real materials, however, there should be an upper bound or shear stress that the medium can support, and it would be possible to remove the apparent mathematical singularity by introducing either cohesion or inelastic deformation into the rupture front (Andrews, 1976b). Irwin's criterion is that the crack will extend when the stress intensity factor at the crack tip exceeds the cohesive force appropriate to the material. These two criteria are the same for a static case or at the initiation of a crack but yield slightly different results for dynamic cracks. It was found that the rupture velocity calculated from Griffith's criterion is somewhat faster than that from Irwin's criterion (Das and Aki, 1977a). A major difference between the two criteria is the material constant adopted,

## 3. NUMERICAL MODELING

which is the surface energy in the former and the specific stress intensity factor in the latter.

For numerical solutions, particularly of a complex rupture process, it would be advantageous to incorporate the finite-stress criterion into calculations. Das and Aki (1977a) showed that this criterion may be approximately translated into Irwin's criterion based on the critical intensity factor. In this case the crack tip is supposed to lie between two neighboring grid points, and the stress difference between these points is compared with a critical stress value $S_c$ (Aki and Richards, 1980). The stress concentration at the crack tip may be represented by

$$\sigma = K/\sqrt{2\pi x}, \tag{25}$$

where $x$ is the distance from the crack tip, and $K$ is the stress intensity factor for two-dimensional cracks. The average stress over the grid outside the crack tip is

$$\bar{\sigma} = \frac{1}{\Delta x} \int_0^{\Delta x} \frac{K}{\sqrt{2\pi x}} \, dx = \frac{2K}{\sqrt{2\pi \Delta x}}. \tag{26}$$

From this criterion, the critical average stress $S_c$ may also be related to Irwin's critical intensity factor $K_c$,

$$S_c = \sigma_s - \sigma_d = K_c/\sqrt{2\pi \Delta x}, \tag{27}$$

where $\sigma_s$ and $\sigma_d$ are the static strength and the sliding frictional stress, respectively. It has thus been demonstrated that the finite-stress criterion corresponds approximately to Irwin's criterion and hence may be used for numerical solutions. In Sections III. E, G, H, and I of this chapter, the criterion will be incorporated into a model of a complex rupture process. Since $K_c$ is related to the critical crack half-length $L_c$ by $K_c = (\sigma_0 - \sigma_d)\sqrt{\pi L_c}$ in two-dimensional shear cracks, $S_c/(\sigma_0 - \sigma_d) = \sqrt{2L_c/\Delta x}$ (Aki and Richards, 1980). If a parameters $S$ is introduced in numerical solutions by $1 + S = S_c/(\sigma_0 - \sigma_d)$, then

$$S = (\sigma_s - \sigma_d)/(\sigma_0 - \sigma_d) - 1 = (\sigma_s - \sigma_0)/(\sigma_0 - \sigma_d). \tag{28}$$

The parameter $S$, which is sometimes called "the stress jump factor," has often been used as a criterion for the initiation of rupture (Andrews, 1976b; Das and Aki, 1977a).

For three-dimensional shear cracks, however, it seems rather difficult to describe a complete fracture criterion on the basis of the stress-intensity factor. Irwin's criterion for this case may be expressed as (Cherepanov, 1979)

$$f(K_{\text{II}}, K_{\text{III}}) = 0,$$

where $K_{II}$ and $K_{III}$ are the stress-intensity factors for two-dimensional, in-plane and antiplane shear cracks, respectively. The function $f(K_{II}, K_{III})$ may be defined from experimental data or some physical considerations. Das (1981), Miyatake (1980), and Virieux and Madariaga (1982) assumed the form

$$f(K_{II}, K_{III}) = K_c^2 - K_{II}^2 - K_{III}^2 = 0. \tag{29a}$$

This form has an advantage for numerical simulations, because $K_{II}^2 + K_{III}^2$ can be easily calculated from the $xz$ component of the applied shear stress $\tau_{xz}$. Another possible form would be

$$f(K_{II}, K_{III}) = K_c^2 - (1 - \nu) \cdot K_{II}^2 - K_{III}^2 = 0, \tag{29b}$$

which is derived from the energy balance between static energy release due to a crack extension and the surface energy (i.e., Griffith's criterion in a static case). To derive Eq. (29b), $(1 - \nu)K_{II}^2 + K_{III}^2$ has to be calculated by $(1 - \nu) \cdot \tau_{xz}^2 \cdot \cos^2 \phi + \tau_{xz}^2 \cdot \sin^2 \phi$, where $\phi$ is the angle of rupture propagation with respect to the reference axis. The second formula is not easy to apply to our three-dimensional model because of the difficulty of calculating $\phi$.

### E. Some Results for Homogeneous Faults

In this section, we discuss the fault rupture process on homogeneous faults, by taking account of the fracture criteria described in Section III.D and presenting some results (Miyatake, 1980). The main purpose is to investigate a basic relationship between the velocity of rupture propagation and the difference between the frictional strength of the fault and the initial shear stress. In this section, all parameters specifying the medium, grid spacings, extent of the model space, and time increment are taken to be the same as in Section III.C. Two examples of spontaneous rupture propagation on homogeneous faults with different values of constant strength (i.e., 210 and 250 bars) are shown in Fig. 9. In these cases, the initial shear stress is 200 bars, and the sliding frictional stress is 100 bars, which is assumed to be independent of slip displacement and slip velocity. The latter value is somewhat lower than results from laboratory friction experiments on rock materials. The stress jump factor $S$ in these cases is 0.1 and 0.5, respectively; and the ratios of the critical half-length of the initial crack to the grid interval $L_c/\Delta h$ are taken as 1.2 and 2.1, respectively. Figure 9(a) shows the rupture times traveled along the $x$ and $y$ axes taken parallel and perpendicular to the applied shear stress. The travel-time diagrams show that the rupture moves slowly at an initial stage, then gradually increases its propagation speed and finally gains its terminal velocities in the two

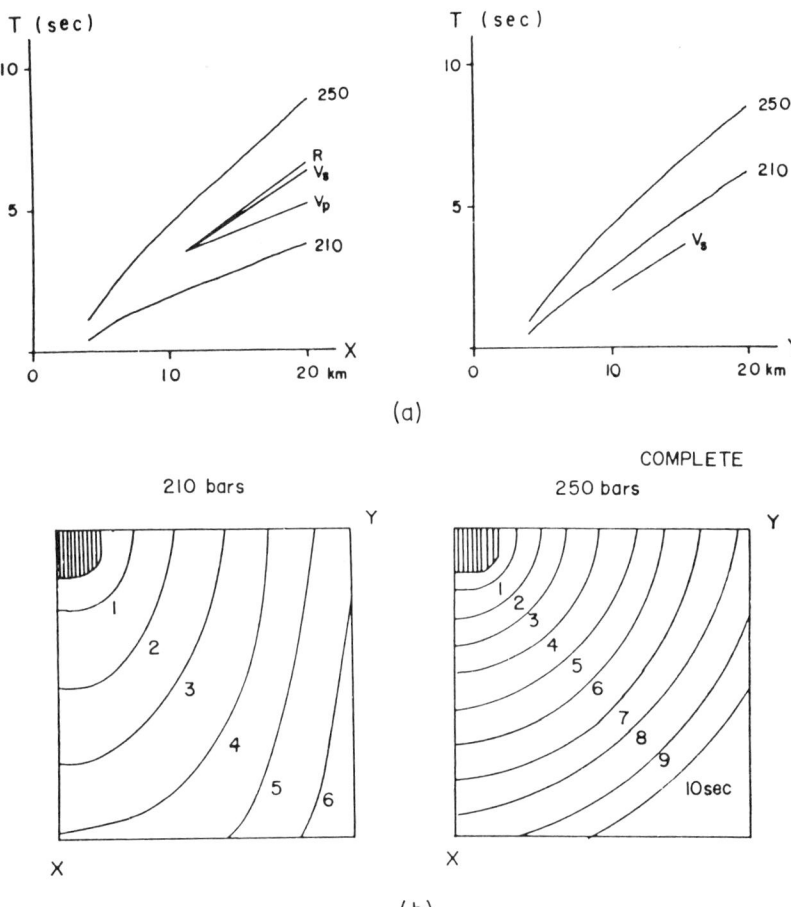

FIG. 9. Mode of rupture propagation on a homogeneous fault with a uniform strength in an infinite medium. The x and y axes are taken parallel and perpendicular to the direction of the applied shear stress. (a) Rupture time of the crack tip in the two directions; (b) rupture fronts on the fault plane. (After Miyatake, 1980.)

directions. This situation, characterized by the slow initial growth of the rupture, would be more obvious if the sliding frictional stress had some velocity or displacement dependence due to cohesive properties of the fault surface, as has been shown in two-dimensional, in-plane shear cracks (Andrews, 1976b) and also in a quasi-three-dimensional problem (Mikumo, 1981b).

Another important feature [Fig. 9(b)] is that for a lower strength (210 bars) the rupture takes more time in the $y$ direction than in the $x$ direction.

This implies that the rupture in this case propagates with nearly elliptically shaped fronts traveling with a velocity close to that of P waves in the direction parallel to the shear stress and with a velocity close to that of S waves in the direction perpendicular to it. This has also been shown in a quasi-three-dimensional model (Mikumo and Miyatake, 1978) and later confirmed by other workers (Das, 1981; Virieux and Madariaga, 1982). For a higher strength (250 bars), however, the rupture velocities along the two directions reduce to close to the velocity of Rayleigh waves, indicating nearly circular rupture fronts. It is also found from additional numerical calculations that a sharp decrease of the rupture velocities in the $x$ direction occurs for $S > 0.3$ ($\sigma_s > 230$ bars) in the case of $L_c/\Delta h = 1.6$. For two-dimensional, in-plane shear cracks, it has been suggested analytically, using a cohensionless fracture criterion, that the admissible crack speed could be P-wave velocity for a weak static-friction level (Burridge, 1973). It has also been shown numerically that the rupture velocity changes from sub-Rayleigh to super-shear-velocities and approaches P-wave velocity as the crack length increases for some range of $L_c/L$ in the case of $S < 1.63$ (Andrews, 1976b). Our results for the spontaneous rupture velocity along the $x$ direction were derived from a complete three-dimensional model and appear to agree essentially with those of Andrews. A major and important difference from the two-dimensional model and also from earlier three-dimensional models with a fixed rupture velocity (see, e.g., Richards, 1976; Archuleta and Frazier, 1978) is that the rupture spreads spontaneously at different velocities in different directions with respect to the direction of the applied shear stress in the case of a relatively low fault strength.

This mode of rupture propagation may be explained by the increase of stress associated with the arrival of different types of waves, with respect to the level of fault strength. On each point on the fault located in the direction of the initial shear stress, the stress at the crack tip should increase as soon as P waves arrive there from the initiation point of rupture. If the fault strength is relatively low, the rupture immediately occurs there due to the stress increase. However, if the strength is high enough, it would not occur at that time but would later break when large-amplitude Rayleigh waves arrive. In the direction perpendicular to the initial stress, on the other hand, a large stress increase is caused by the arrival of S waves and the rupture occurs at this time.

### F. Spatial Variations of Fracture Strength and Sliding Frictional Stress on the Fault

As discussed in Sections III.D and III.E, the fault rupture process depends entirely on the static and dynamic (sliding) frictional stresses on the fault surface, $\sigma_s$ and $\sigma_d$, and the initial tectonic shear stress $\sigma_0$. In other words, the process is specified by the critical stress difference or the effective fracture strength defined as $\sigma_s - \sigma_0$ and the dynamic stress drop defined as $\sigma_0 - \sigma_d$. It is natural to suppose that these parameters are not homogeneous but rather nonuniformly distributed over the fault surface. In this section, we shall discuss the properties of fault materials and physical quantities that could affect the above parameters and also look for possible sources of their spatial variations.

The fracture strength depends strongly on material properties, which include the cohesive force, surface energy, etc. The cohesive properties of the fault surface associated closely with the sliding frictional stress have been explained, on the basis of laboratory experiments, by the role of indentation and ploughing of asperities* on the surface (Scholz and Engelder, 1976), and hence should depend on the critical dimension of the contact of the asperities (Dieterich, 1978). The creep of indentation asperities, resulting from stress corrosion cracking or hydrolytic weakening, would lead to stick-slip instability in frictional sliding. The static frictional strength and the sliding frictional stress also depend on some physical quantities, such as the normal stress, pore pressures, and ambient temperature around the fault zone. Laboratory friction experiments on intact rocks (Byerlee, 1970, 1978) indicate a well-known empirical relationship between the shear strength $\tau_s$ and the normal stress $\sigma_n$, which can be written as

$$\tau_s = \tau_{s0} + \mu_s \cdot \sigma_n, \qquad (30)$$

where $\tau_{s0}$ is found to be nearly zero and $\mu_s = 0.85$ for $0.02$ kb $< \sigma_n < 2$ kb, and $\tau_{s0} = 0.5$ kb and $\mu_s = 0.60$ for $2$ kb $< \sigma_n < 17$ kb. It has been shown that this relation [Eq. (30)] may not be strongly affected by rock type, temperatures (up to 500°C), and strain rates. A similar relation has also been suggested for the sliding friction of granite $\tau_k$ with $\tau_{k0} = 0.25$ kb and $\mu_k = 0.47$ for $2$ kb $< \sigma_n < 12$ kb (Byerlee, 1978). In real fault zones, however, the shear strength will reduce to a considerable extent due to pore fluid pressures $p$ in the upper crust, and hence $\sigma_n$ should be replaced by the effective normal stress $\sigma_n - p$ or $(1 - \lambda)\sigma_n$ where $\lambda = p/\sigma_n$. The behavior of $\sigma_s$ and $\sigma_d$ on the fault surface would correspond as a first approximation to $\tau_s$ and $\tau_k$ estimated from the laboratory experiments that are strongly dependent on the normal stress and pore pressures. If this is the case, the ratio of the sliding frictional stress to the static frictional strength takes an almost constant value of 0.75 (Byerlee, 1978) irrespective of the effective normal stress, but the value would vary with increasing temperatures (Sibson, 1977).

*This means the "asperity" widely used in rock mechanics and does not mean the "asperity" model of seismology.

Next, we consider several possible sources that would yield spatial variations in the fault strength and sliding frictions. These may be attributed mainly to three different sources: inhomogeneous environments, a heterogeneous medium, and nonuniform fault geometry.

The first one includes possible lateral and depth variations of the applied stress, pore pressures, and temperatures over the fault provided that the fault zone is long and extends down to some depth. Although their lateral variations are not well known from observations, the variations with depth have been investigated in some detail. In a shallower section of the crust, the normal stress $\sigma_n$ and, hence, the static frictional strength $\sigma_s$ of rock materials should increase down to a certain depth with increasing hydrostatic pressures. The rate of increase of the shear strength with depth depends on the coefficient of static friction $\mu_s$ and the ratio of pore pressures to hydrostatic pressures as indicated by Eq. (30). Below a certain depth in the middle crust, on the other hand, the strength should decrease with depth, because rock materials there become ductile due to the increase of temperature. For this ductile regime, the shear strength may be roughly estimated from experimental relations between the strength, strain rate, and temperature (e.g., Sibson, 1982, 1984; Meissner and Strehlau, 1983). The second source may be the heterogeneous structure or inhomogeneous elastic properties of the medium, such as low-velocity fault zones or gouge layers intervened between hard rocks and more complex three-dimensional structures around fault zones. These will have some effect on the strength of the fault zones. These effects on the faulting process will be discussed in Section III.I.

The third possible source that yields lateral variations of fault strength and frictions is nonuniform fault geometry, including local bending and branching of the fault plane. Some simple calculations for a corrugated fault surface with a small bend show that variations of dynamic stress drop can be expressed as $\Delta(\Delta\sigma) = 4\pi\mu_d\sigma_0 h/\lambda$, where $\mu_d$ is a coefficient of dynamic friction, $\sigma_0$ is the applied shear stress, and $h$ and $\lambda$ are the amplitude and wavelength of fault bend, respectively (Miyatake, 1985). For $\sigma_0 = 300$ bars, $\lambda/h = 0.1$, and $\mu_d = 0.5$, $\Delta(\Delta\sigma)$ is expected to be 180 bars, which could yield large variations of stress drops along the fault. For larger fault bends, it would be more difficult to estimate the effects because of large-amplitude and short-wavelength indentation or ploughing of asperities. Local branching of a fault needs much more surface energy than the extension of a simple fault. The stress intensity factor at the crack tip tends to decrease, since it must create more surface. This could act as a stopping mechanism of rupture propagation (Mansinha, 1964). The importance of fault geometry on the fault rupture process has also been emphasized recently by King and Nabelek (1985).

3. NUMERICAL MODELING 123

These sources of spatial variations, their physical parameters, and their detailed distributions around the fault zone are not well known. More laboratory experiments and *in situ* measurements of these quantities are needed to estimate the fault strength.

G. SOME RESULTS FOR HETEROGENEOUS FAULTS

Although it is extremely difficult to obtain definite information from observations on spatial variations of the frictional strength on a real fault, there is some evidence for the form of the frequency distribution of the strength from laboratory experiments on rock materials. It has been shown that the tensile strength of intact rock samples of granite indicates a Weibull distribution (Yamaguchi and Nishimatsu, 1967), which is similar to a normal random distribution with a certain variance.

It has also been suggested on the basis of laboratory experiments that the fracture strength of rock materials could become bimodal under increasing stress; this may be due to the formation of two different sizes of asperities with different strengths along two symmetry axes of crystals in response to the growth of microcracks (Spetzler *et al.*, 1982).

From this evidence, it may be reasonable to suppose that fault materials should have a bimodal or more complex distribution of strengths rather than a Weibull distribution. A second reason is that the fault surface should have different degrees of contact possibly due to the nonuniform fault geometry and normal stress mentioned in Section III.F, that is, tightly locked portions will have high degrees of contact due to indentation and ploughing asperities (Scholz and Engelder, 1976), while the existence of fluid pore pressures and crushed gouge materials will make a loose contact. Thus, there may be strong and weak portions on the fault surface with different frictional strengths with a strength distribution that could be multimodal. The high strength portions may be called "strong patches." We assume several different types of heterogeneous strength distributions over the fault and make numerical calculations for the rupture process to take all of the above situations into consideration. In this section, some examples of the results obtained from our earlier quasi-three-dimensional model (Mikumo and Miyatake, 1978) are presented. This model is not completely three-dimensional since the medium is bounded in the $z$ direction by a fixed boundary not very far above the fault plane, and hence the calculated fault displacements are somewhat smaller than those in a complete three-dimensional model. The fault dimensions are 40 km × 40 km. The applied shear stress is 198 bars and the sliding frictional stress is assumed to be 100 bars. The only difference between the two cases shown in Fig. 10 is that the static frictional strength in case (a) ranges between 200 and 326 bars indicating a

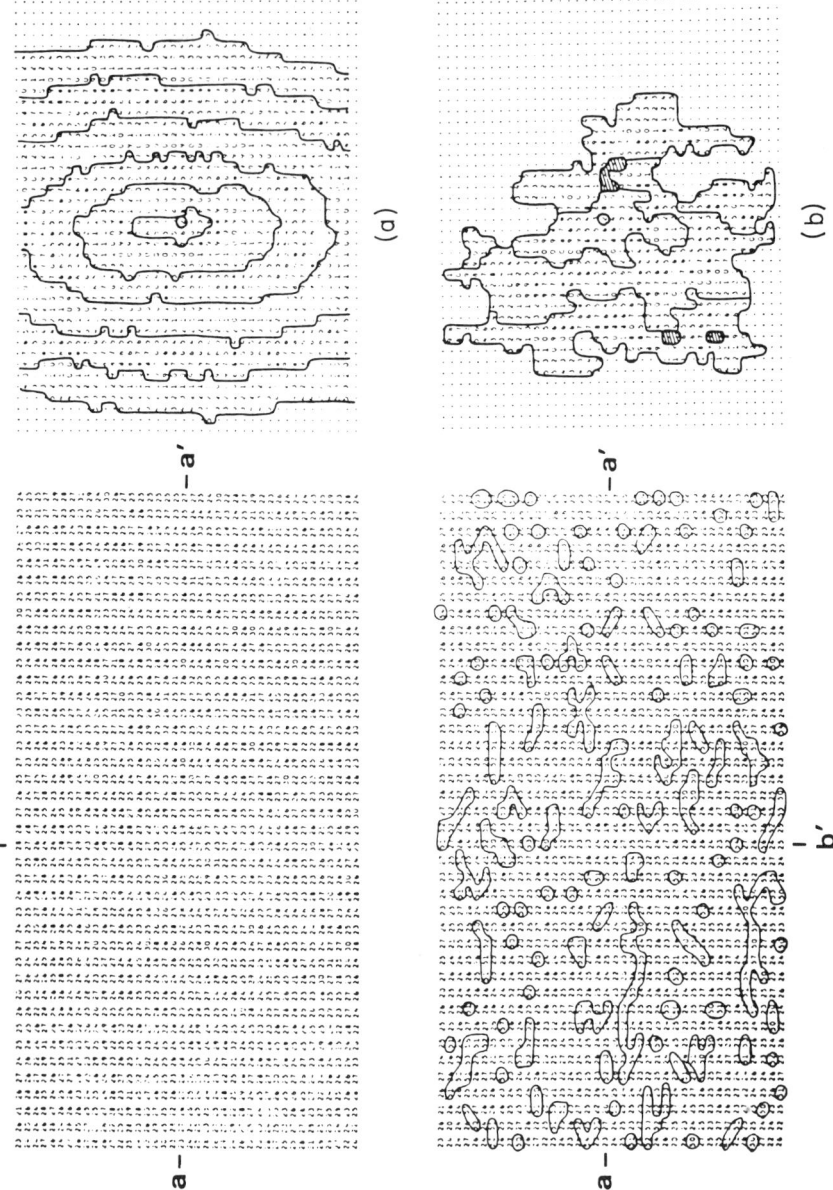

Fig. 10. Assumed distributions of frictional strength on the fault and the mode of rupture propagation. (a) The case for a weakly heterogeneous fault; (b) the case for a strongly heterogeneous fault. (After Mikumo and Miyatake, 1978.)

weakly heterogeneous distribution represented by a single Weibull distribution, while case (b) has strongly nonuniform strengths ranging from 200 to 590 bars represented by a bimodal Weibull distribution. The left side of Fig. 10 shows the assumed spatial distribution of strength. Case (b) includes a number of strong patches with higher strengths within the encircled portions. The right side of Fig. 10 gives the rupture fronts shown at every second in case (a) and at every 2 sec in case (b). Case (a) shows that the rupture extends faster in the direction parallel to the applied shear stress (pointing downward), which is almost the same as in Fig. 9(b) for a homogeneous fault but with rupture fronts slightly deviated from elliptic shapes. Case (b) indicates a very unusual pattern of rupture propagation: the rupture velocity is decelerated to an appreciable extent by high-strength inclusions distributed over the fault surface, and the shape of the rupture fronts is no longer elliptic. Several portions with extremely high strengths, which are shown by small shaded areas, remain unruptured, resisting high-stress concentrations. The rupture propagation finally stops at the stage shown, leaving large surrounding areas and several bay-shaped areas unbroken.

The final displacements and remnant stresses after the completion of the rupture are shown in Fig. 11, which gives their distributions along the a–a' and b–b' axes on the fault. Case (b) shows large variations of displacements including the unruptured portions that did not slip, while the final displacements in case (a) indicate a smooth distribution as expected. In case (a), the initial stress drops to the level of sliding frictional stress, and the stress drop is almost constant inside the faulted area. The remnant stresses in case (b) have large variations, and there are still high-stress concentrations above the initial stress level in and around the unbroken regions. These unbroken regions could be the source of aftershocks and eventual large earthquakes at a later stage.

A possible mechanism of aftershock occurrence would be time-delayed stress corrosion or weakening of the fracture strength due to the concentrated stress for a long time range. Another conceivable mechanism may be time-dependent recovery of the shear stress and strength on the fault. Mikumo and Miyatake (1979) introduced a visco-elastic stress recovery process into the three-dimensional fault in a standard three-parameter solid, taking into account an experimental formula (Dieterich, 1978) for the strength recovery. The results indicate that aftershocks with high stress drops occur in and around the regions left unruptured during the main faulting, while those with low stress drops take place successively in spaced regions so as to fill the gaps that have not yet been ruptured since the main shock, due to minor stress readjustments.

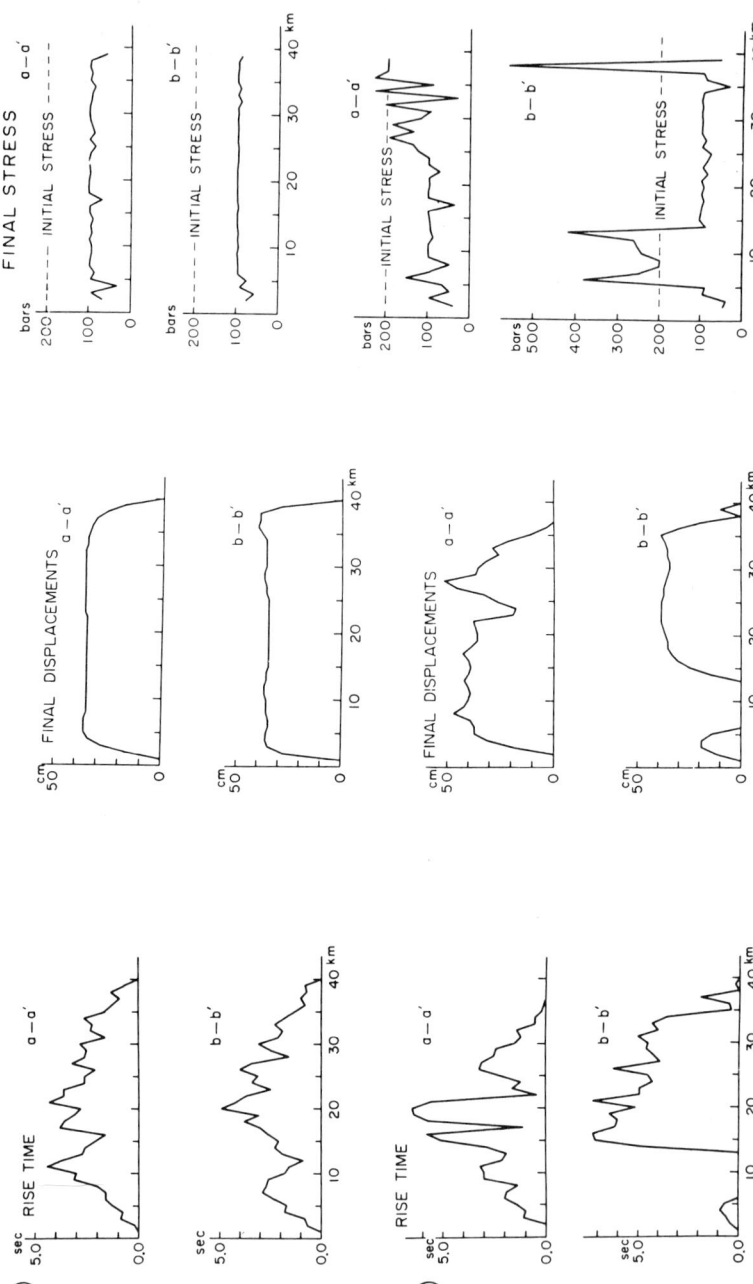

Fig. 11. Distribution of rise time, final displacements, and remnant stresses along the a–a' and b–b' axes shown in Fig. 10. (After Mikumo and Miyatake, 1978.)

## H. THE FAULT RUPTURE PROCESS IN A HOMOGENEOUS HALF-SPACE

In this section, the rupture process on a vertical strike-slip fault in a homogeneous half-space is discussed (Miyatake, 1980), and the effects of the free surface and the fault dimensions under realistic conditions are considered. The results may be compared with observations made in relatively uniform structures.

The geometry of the model space is the same as shown in Fig. 8, where the $xy$ plane ($z = 0$) is the fault plane, and the $xz$ plane ($y = 0$) is taken here as the free surface. The boundary conditions to be added to this case are those for the free surface at $y = 0$:

$$\sigma_{yy} = \lambda \cdot \left( \frac{\partial u}{\partial x} + \frac{\partial v}{\partial y} + \frac{\partial w}{\partial z} \right) + 2\mu \frac{\partial v}{\partial y} = 0,$$

$$\sigma_{yx} = \mu \left( \frac{\partial v}{\partial x} + \frac{\partial u}{\partial y} \right) = 0, \qquad (31)$$

$$\sigma_{yz} = \mu \left( \frac{\partial w}{\partial y} + \frac{\partial v}{\partial z} \right) = 0.$$

The corresponding finite-difference equations can be written as

$$(\lambda + 2\mu) \cdot \left( \frac{v(i,2,k) - v(i,1,k)}{\Delta y_1} \right)$$
$$+ \lambda \cdot \left( \frac{u(i+1,1,k) - u(i-1,1,k)}{\Delta x_{i-1} + \Delta x_i} \right.$$
$$\left. + \frac{w(i,1,k+1) - w(i,1,k-1)}{\Delta z_{k-1} + \Delta z_k} \right) = 0,$$

$$\mu \cdot \left( \frac{u(i,2,k) - u(i,1,k)}{\Delta y_1} + \frac{v(i+1,1,k) - v(i-1,1,k)}{\Delta x_{i-1} + \Delta x_i} \right) = 0, \qquad (32)$$

$$\mu \cdot \left( \frac{w(i,2,k) - w(i,1,k)}{\Delta y_1} + \frac{v(i,1,k+1) - v(i,1,k-1)}{\Delta z_{k-1} + \Delta z_k} \right) = 0.$$

The displacements at the free surface, $u(i, 1, k)$, $v(i, 1, k)$ and $w(i, 1, k)$ at time $t$ can be calculated from the boundary conditions of Eq. (31) and then put into the wave equation [Eq. (23)] for $t + \Delta t$. An alternative way is to estimate fictitious displacements at hypothetical grid points just above the free surface, which will be described in Section III.I.

Several examples of the rupture processes shown here (Miyatake, 1980) are those calculated for the faults located at depths between 2 and 12 km ($2 \leq y \leq 12$ km) and extended for 20 km ($-10 \leq x \leq 10$ km). In this case, the fault does not break the ground surface, and therefore $u$ and $v$ on and just above the fault surface (for $z = 0$ and $y < 2$ km) are taken to be zero. Four different strength distributions are assumed on the fault surface: the uniform strength, the existence of a square-shaped strong patch within a weak-strength region, a random distribution of strength as in cases (a) and (b) in Section III.G, and the case with a beltlike strong patch. The second type of strength distribution corresponds to a blocklike geological structure on the fault surface.

The first case is a homogeneous fault with a uniform strength of 240 bars ($S = 0.4$). The fault rupture is assumed to initiate at the middepth on the left edge of the fault. Figure 12 shows the displacement distribution on the fault at every second, indicating that the spontaneous rupture spreads radially at an initial stage up to $\sim 5$ sec when its front reaches the upper and lower side edges and extends almost unilaterally after this time toward the longer side. The rupture velocity in this case is found to be somewhat smaller than that for the strength of 250 bars ($S = 0.5$) in the infinite medium described in Section III.E. The lower velocity may be attributed to the smaller dimension of the fault in this case because the stress intensity factor reduces to some extent. The fault displacements are also smaller than expected from two-dimensional crack theory, since the rupture extension is restricted by the fault width rather than by the fault length. This conclusion has been confirmed by Das (1981). The rupture process is almost completed at about 12 sec and does not appear to develop thereafter.

The second example is for an inhomogeneous fault with a square-shaped strong patch in its central part. The strength of the patch and the surrounding region are taken as 400 bars ($S = 2$) and 240 bars ($S = 0.4$), respectively. It can be seen from Fig. 13 that the rupture propagates in a way similar to the case of the homogeneous fault up to $\sim 6$ sec and then it goes around the central strong patch, leaving the patch unbroken. The patch works as a barrier for some time but starts to slip after the rupture reaches the right fault edge and finally completely breaks due to stress concentration. This rupture would correspond to the generation of multiple shocks on the fault surface of large earthquakes. However, if the strength of the

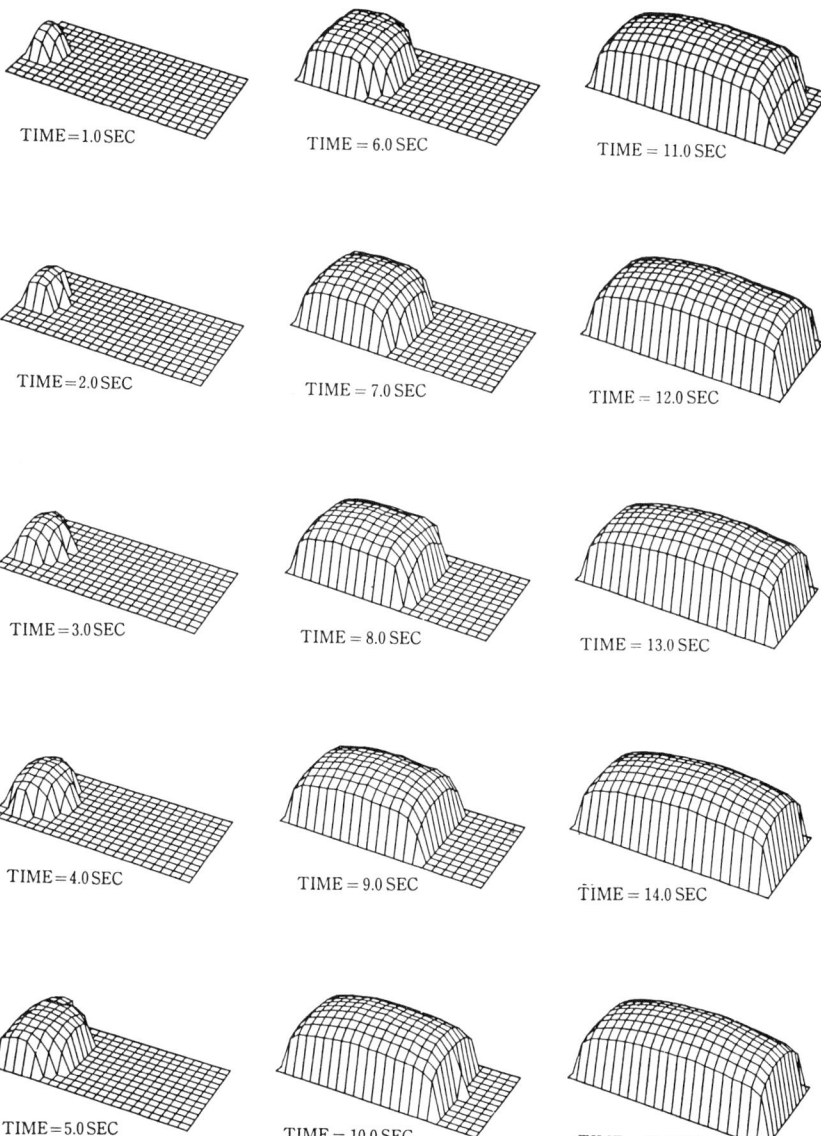

**FIG. 12.** Rupture propagation and the displacement distribution on the fault with a uniform strength in a homogeneous half-space. (After Miyatake, 1980.)

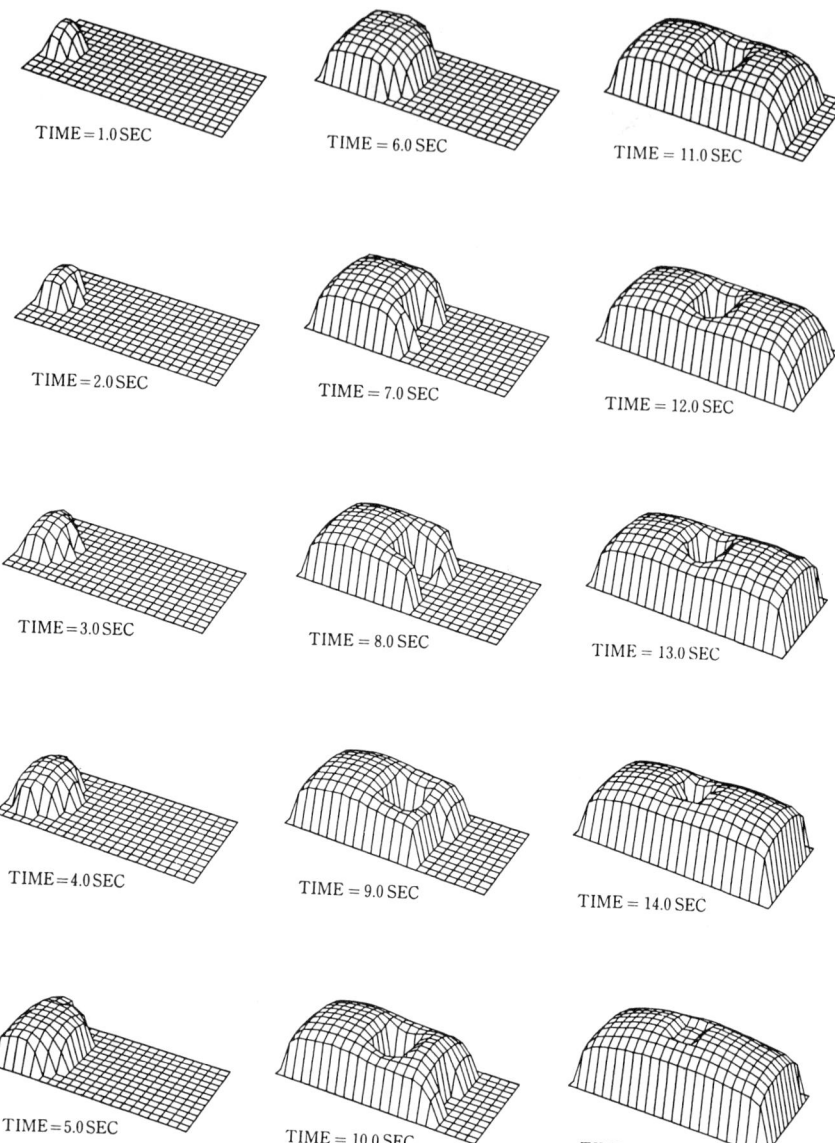

**FIG. 13.** Rupture propagation and the displacement distribution on a heterogeneous fault with a square-shaped strong patch in the central part. (After Miyatake, 1980.)

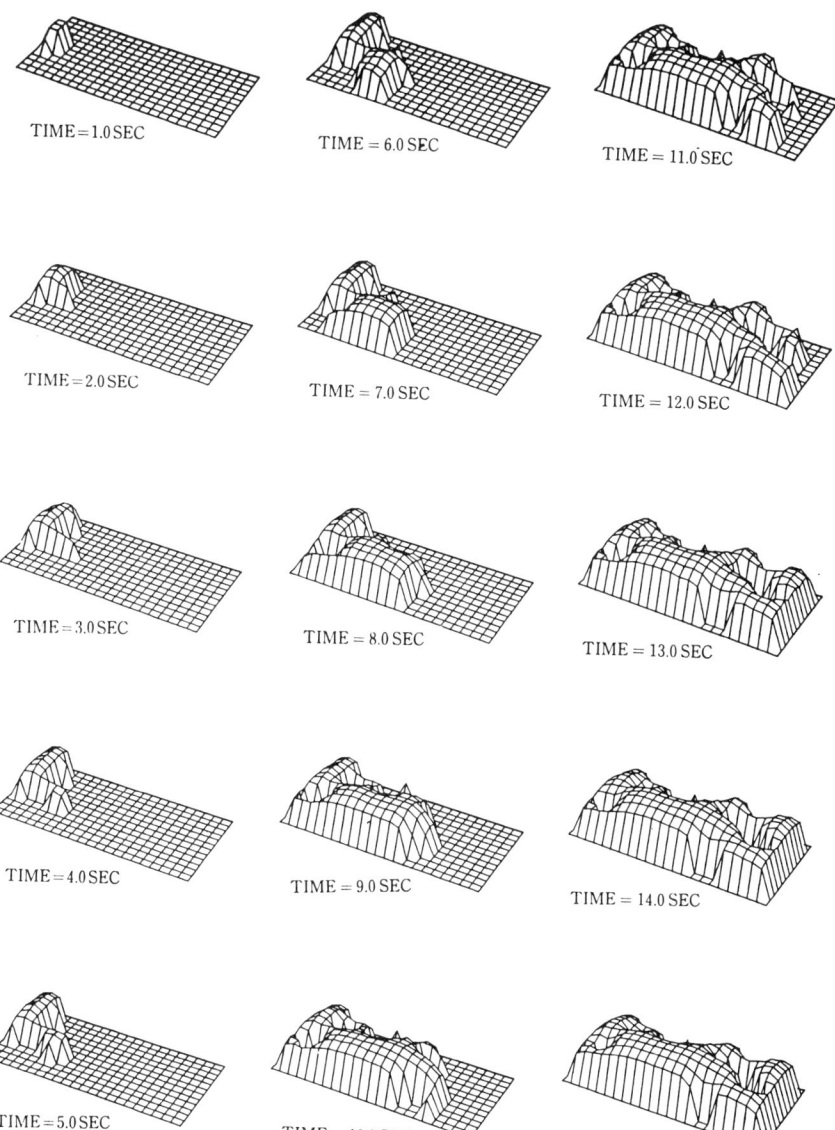

**FIG. 14.** Rupture propagation and the displacement distribution on a strongly heterogeneous fault with multimodal strength distribution. (After Miyatake, 1980.)

central patch is considerably larger than assumed here, the patch would not be broken up to the final stage.

The third case includes heterogeneous strengths that are represented by simple Gaussian, bimodal and multimodal distributions. For a weakly heterogeneous fault with a Gaussian strength distribution, the rupture propagation and the fault displacements are almost the same as in the first case for a homogeneous fault. Figure 14 shows the rupture process for a strongly heterogeneous fault with a multimodal strength distribution, including 2 large and 4 small strong patches with the highest strength of 450 bars ($S = 2.5$) and the lowest one of 200 bars in the rest of the region. The rupture extends downward up to 3 sec and is blocked by a large strong patch located at the left bottom. Then it moves toward the right side up to 9 sec, leaving unbroken this patch and another strong patch at the upper right portion, and reaches the right fault edge at ~ 13 sec. After 30 sec, there remain 5 unbroken strong patches as barriers. These unbroken barriers will generate aftershocks or eventual large shocks some time later, as briefly mentioned in Section III.G.

The fourth example (Miyatake, 1984) is shown in Fig. 15, which gives the patterns of rupture propagation on a fault located at $2 \leq y \leq 8$ km. In case (a), the strength is uniform (235 bars, $S = 0.35$) over the fault, and in case (b) the fault is assumed to include a beltlike, high-strength (300 bars, $S = 1.0$) zone with a 3-km-width in the central part. The rupture fronts in

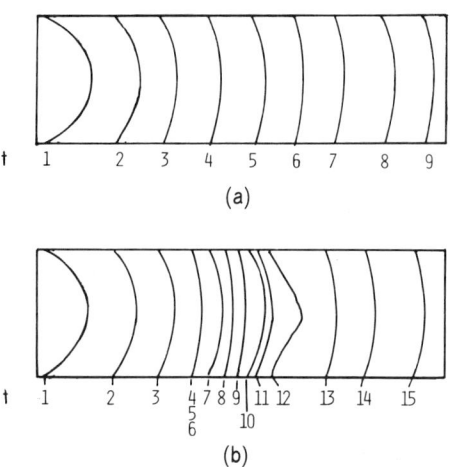

FIG. 15. Patterns of rupture propagation on two different types of faults in a homogeneous half-space; (a) the case with uniform strength, (b) the case with a beltlike strong patch in the central part of the fault. (After Miyatake, 1984.)

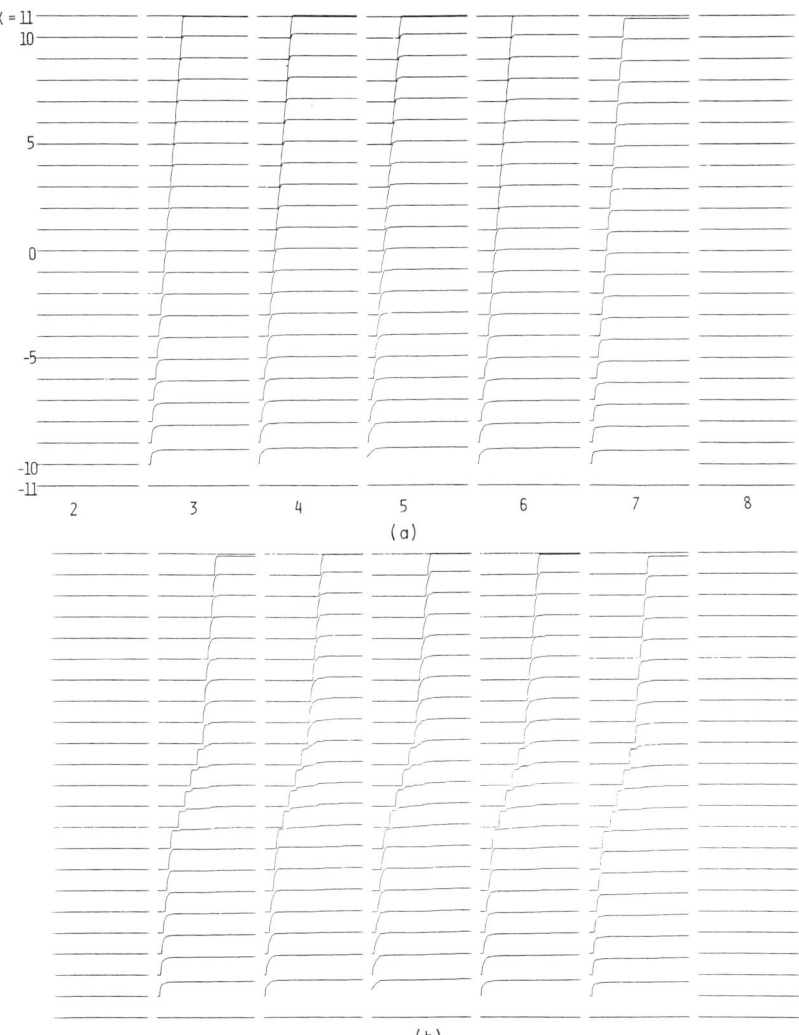

**FIG. 16.** Displacement time functions (the $x$ component) at different locations on the fault. Cases (a) and (b) correspond to the two different types of faults shown in Fig. 15. (After Miyatake, 1984.)

case (b) indicate remarkable deceleration effects around the central part due to the existence of the high-strength zone. Figures 16(a) and (b) show the displacement time functions at different locations on the fault plane in the above two cases. These are the $u$ components parallel to the fault, which are arranged in the $x$ direction as ordinate and in the $y$ direction as abscissa, respectively. A comparison between the two cases clearly indicates the decelerated rupture propagation and episodic displacements in case (b). The distributions of the final displacements and rise times in the homogeneous fault in case (a) depends on the fault width rather than the fault length.

More results for the case when the fault rupture breaks the ground surface are presented in Section III.I, which presents the rupture process in a horizontally layered structure.

## I. Fault-Rupture Process in a Horizontally Layered Half-Space and in a More Heterogeneous Structure

In this section, we deal with the rupture process on a vertical strike-slip fault embedded in a horizontally layered and more heterogeneous structure (Mikumo et al., 1985b). Calculations are made for several cases when the rupture initiates at the base or a middepth of the fault and when it breaks out on the ground surface or does not reach the surface. These cases in heterogeneous crustal structures may be more realistic for comparison with the results of observations inferred from strong ground motions in the near field and teleseismic body and surface waves.

The fault geometry considered here is the same as in Section III.H. The medium is horizontally layered in the $y$ direction, with the free surface at $y = 0$. The boundary conditions in this case are the continuity of normal and tangential stresses and the three displacement components at each layer interface,

$$
\begin{aligned}
(\sigma_{yy})_m &= (\sigma_{yy})_{m-1}, & (u)_m &= (u)_{m-1} \\
(\sigma_{yx})_m &= (\sigma_{yx})_{m-1}, & (v)_m &= (v)_{m-1} \\
(\sigma_{yz})_m &= (\sigma_{yz})_{m-1}, & (w)_m &= (w)_{m-1}.
\end{aligned}
\quad (33)
$$

A possible way to solve this problem numerically is to introduce the notion of fictitious displacements at the layer interface (Alterman and Karal, 1968; Kelley et al., 1976). The displacements correspond to those that would be obtained by supposing that an upper layer with a lower velocity extended down by one grid interval $\Delta y_i$ into the lower medium with a higher velocity. These fictitious displacements $\tilde{u}_{m-1}$, $\tilde{v}_{m-1}$, and $\tilde{w}_{m-1}$ can be

## 3. NUMERICAL MODELING

obtained from Eq. (33) as

$$\tilde{u}_{m-1}(i, j+1, k)$$
$$= u_{m-1}(i, j, k) + (\mu_m/\mu_{m-1})[u_m(i, j+1, k) - u_{m-1}(i, j, k)]$$
$$+ (\Delta y_j/\Delta x_{i-1}) \cdot (\mu_m/\mu_{m-1})[v_{m-1}(i, j, k) - v_{m-1}(i-1, j, k)],$$
$$\tilde{w}_{m-1}(i, j+1, k)$$
$$= w_{m-1}(i, j, k) + (\mu_m/\mu_{m-1})[w_m(i, j+1, k) - w_{m-1}(i, j, k)]$$
$$+ (\Delta y_j/\Delta z_{k-1}) \cdot (\mu_m/\mu_{m-1})[v_{m-1}(i, j, k) - v_{m-1}(i-1, j, k)],$$
$$\tilde{v}_{m-1}(i, j+1, k)$$
$$= v_{m-1}(i, j, k) + (\mu_m/\mu_{m-1})[v_m(i, j+1, k) - v_{m-1}(i, j, k)]$$
$$+ \frac{\Delta y_j}{\Delta x_{i-1}} \cdot \frac{\lambda_m + 2\mu_m}{\lambda_{m-1} + 2\mu_{m-1}} \cdot [u_{m-1}(i, j, k) - u_{m-1}(i-1, j, k)]$$
$$+ \frac{\Delta y_j}{\Delta z_{k-1}} \cdot \frac{\lambda_m + 2\mu_m}{\lambda_{m-1} + 2\mu_{m-1}} \cdot [w_{m-1}(i, j, k) - w_{m-1}(i-1, j, k)]. \tag{34}$$

These terms are put into the wave equation [Eq. (23)], where $u_m(i, j+1, k)$, $v_m(i, j+1, k)$, and $w_m(i, j+1, k)$ at the $m$th interface $y = h_m$ are replaced by $\tilde{u}_{m-1}$, $\tilde{v}_{m-1}$, and $\tilde{w}_{m-1}$, respectively. Likewise, one can introduce fictitious displacements at the free surface by extending artificial grid points to a distance $\Delta y_1$ above the surface.

From Eq. (34), we have

$$\tilde{u}_1(i, 0, k) = u_1(i, 2, k) + \frac{2 \Delta y_1}{\Delta x_{i-1} + \Delta x_i}$$
$$\cdot [v_1(i+1, 1, k) - v_1(i-1, 1, k)],$$
$$\tilde{w}_1(i, 0, k) = w_1(i, 2, k) + \frac{2 \Delta y_1}{\Delta z_{k-1} + \Delta z_k}$$
$$\cdot [v_1(i, 1, k+1) - v_1(i, 1, k-1)],$$
$$\tilde{v}_1(i, 0, k) = v_1(i, 2, k) + \frac{2 \Delta y_1}{\Delta x_{i-1} + \Delta x_i} \cdot \frac{\lambda_1}{\lambda_1 + 2\mu_1}$$
$$\cdot [u_1(i+1, 1, k) - u_1(i-1, 1, k)]$$
$$+ \frac{2 \Delta y_1}{\Delta z_{k-1} + \Delta z_k} \cdot \frac{\lambda_1}{\lambda_1 + 2\mu_1}$$
$$\cdot [w_1(i, 1, k+1) - w_1(i, 1, k-1)]. \tag{35}$$

Here, $u_1(i, j-1, k)$, $v_1(i, j-1, k)$, and $w_1(i, j-1, k)$ for $j = 1$ at the

TABLE I. HORIZONTALLY LAYERED MEDIUM

| $V_P$ (km/sec) | $V_S$ (km/sec) | $\rho$ (g/cm$^3$) | $H$ (km) | No. |
|---|---|---|---|---|
| 3.20 | 2.00 | 2.30 | 2.0 | I |
| 5.50 | 3.25 | 2.65 | 3.0 | II |
| 6.10 | 3.50 | 2.72 | 20.0 | III |
| 6.60 | 3.75 | 2.85 | 8.0 | IV |
| 7.80 | 4.45 | 3.20 | 10.0 | V |
| 8.00 | 4.60 | 3.30 |  | VI |

ground surface included in Eq. (23) are replaced by $\tilde{u}_1$, $\tilde{v}_1$, and $\tilde{w}_1$, respectively. The expressions obtained in Eq. (35) are the extensions of the two-dimensional case (Kelly *et al.*, 1976) to the three-dimensional case.

The horizontally layered structure assumed here (see Table I) is that appropriate to a part of the central Honshu region of Japan, which is almost the same structure as adopted in Section II.D. The fault dimension is taken as 20 km × 12 km, which is assumed to be located at depths $0 \leq y \leq 12$ km in case (a) and $2 \leq y \leq 14$ km in case (b). The first case corresponds to the case when the fault reaches the ground surface. The static frictional strength and sliding frictional stress in this case are assumed to be 205–210 bars and 150 bars over the fault, respectively, and the initial shear stress is 200 bars. The expected stress drop is 50 bars, and the stress jump factor $S$ ranges between 0.05 and 0.10. We calculate here the rupture process for the following three cases for a homogeneous fault embedded in the vertically heterogeneous medium.

Figure 17 shows two different modes of rupture propagation that initiate at depths of $y = 5$ km and (at the bottom) $y = 12$ km of the vertical fault, where the rupture fronts are drawn at an interval of 0.5 sec. The rupture in both cases propagates with nearly elliptic shapes indicating a faster velocity in the direction parallel to the applied shear stress, as has been shown for the cases with relatively low $S$ values in an infinite medium and in a homogeneous half-space. A closer examination indicates that the rupture propagation slows down to an appreciable extent within the uppermost low-velocity layer, as might be expected. These situations are more obvious in case (a) than in case (b). The upward rupture velocity within the layer along the direction perpendicular to the shear stress is found to be that of S waves in the layer. This velocity is slower than the corresponding rupture propagation in a homogeneous half-space. This implies that the rupture velocity depends on elastic properties of the medium involving the fault.

A remarkable difference between the horizontally layered structure and the homogeneous half-space and between the two cases (a) and (b) in the

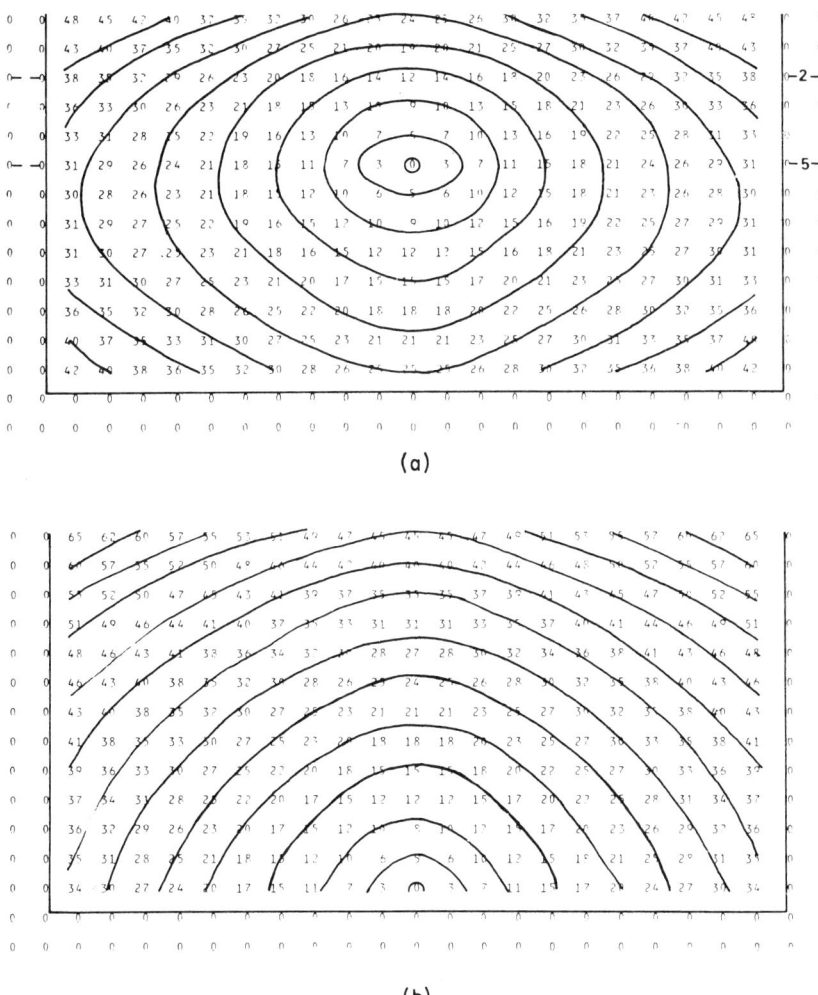

FIG. 17. Mode of rupture propagation on a vertical fault with a uniform strength in a horizontally layered structure; (a) the rupture initiates from a middepth of the fault, and (b) the rupture initiates from the base of the fault. (After Mikumo et al., 1985b.)

layered medium lies in the distribution of the final fault displacements, particularly near the ground surface. The upper graph of Fig. 18 shows the slip distribution in case (a) along the fault length (in the $x$ direction), and the lower graph indicates its distribution with depth (in the $y$ direction). These diagrams clearly show that the slip displacements gradually increase toward the ground surface when the fault breaks through the surface. The

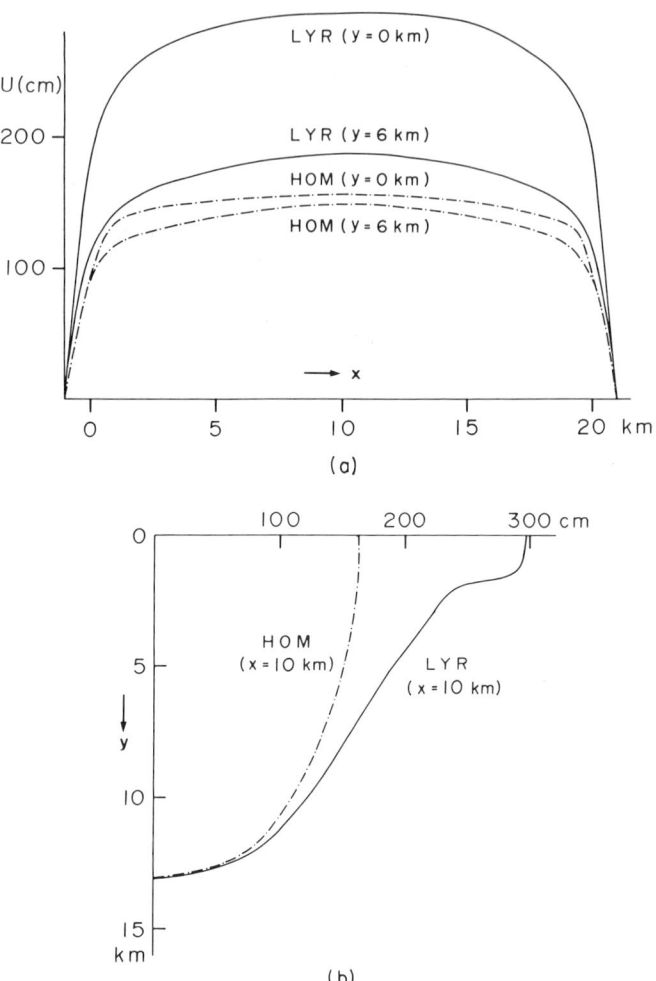

FIG. 18. Distribution of final displacements over the fault when it breaks out the ground surface as shown in Fig. 17. Case (a) indicates the slip displacements along the direction of fault length, and (b) gives their distribution with depth. The terms HOM and LYR imply the cases for the homogeneous half-space and the horizontally layered structure, respectively. (After Mikumo et al., 1985b.)

form of the displacement distribution with depth in the homogeneous half-space is exactly the same as expected from the corresponding analytical solution for two-dimensional antiplane shear cracks (see, e.g., Burridge and Halliday, 1971). The horizontally layered structure with a low-velocity surficial layer accommodates much larger fault slip in an upper crustal section, particularly in the low-velocity surficial layer. Also, the presence of

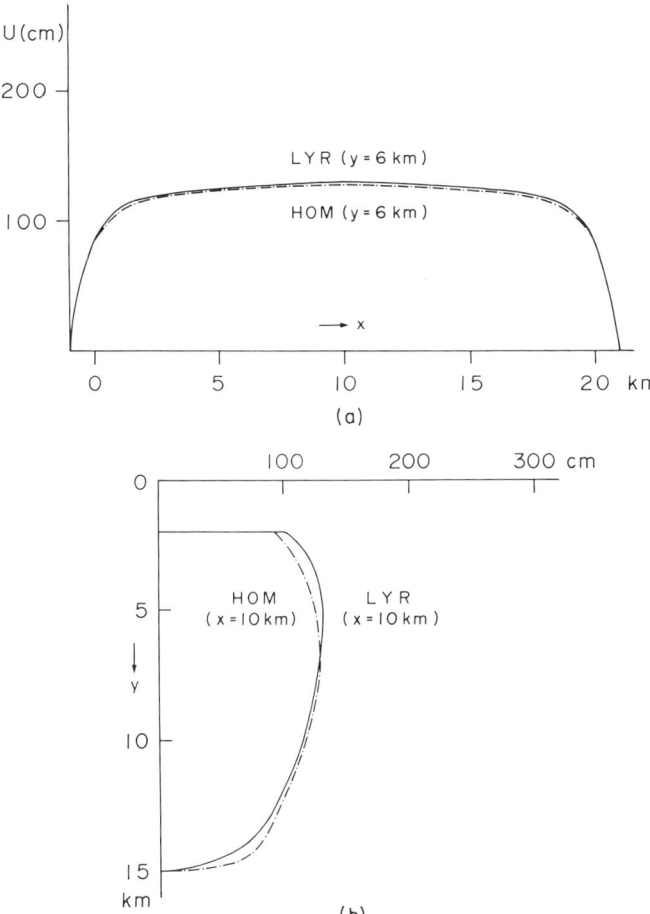

**FIG. 19.** Distribution of final slip displacements over the fault when it does not reach the ground surface. Cases (a) and (b) correspond to those in Fig. 18. (After Mikumo et al., 1985b.)

the low-velocity surficial layer gives larger displacements even in a section below 5 km, which has the same velocity structure as in the homogeneous case. This can be attributed to the low impedance of materials in the uppermost layer.

Figure 19 provides a similar distribution of fault displacements for case (b) when the fault does not reach the ground surface. It can be seen that the maximum slip in this case does not occur at the shallowest portion of the medium, but the distribution is still affected by the existence of the low-velocity layer and the free surface. The difference in the displacements from the homogeneous case is not very large.

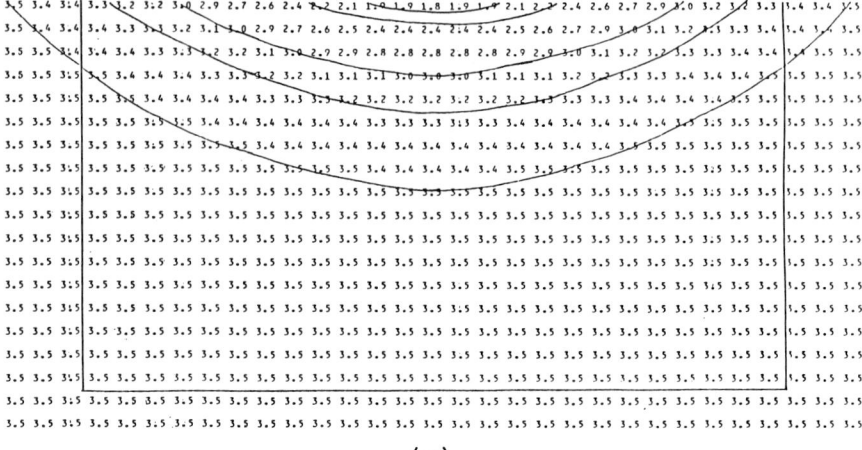

(a)

## SLV

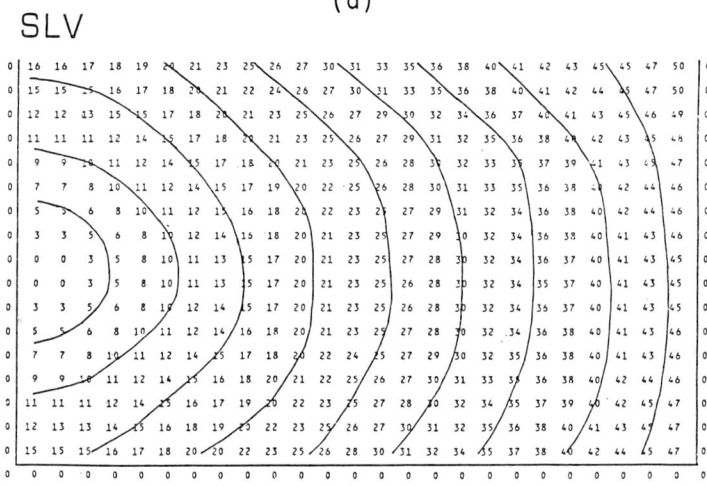

(b)

## SLV

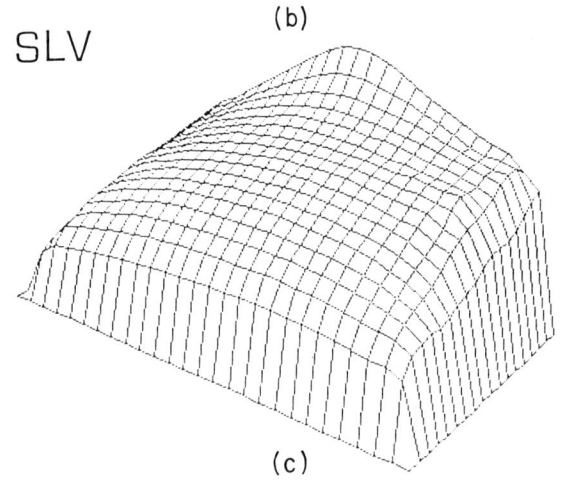

(c)

## 3. NUMERICAL MODELING

For more heterogeneous media, such as three-dimensionally varying structures including vertically and laterally heterogeneous cases, elastic constants $\lambda$ and $\mu$ and density $\rho$ are no longer constant even at the same depth but change in the horizontal directions. The wave equations for a weakly heterogeneous medium with a continuously varying structure have been constructed so as to include the space derivatives of these parameters into the wave equations (Hirahara et al., 1985; Mikumo et al., 1985b). An example of the recent results (Hirahara et al., 1985) for the rupture process in these cases is also presented here. Figure 20(a) shows the velocity structure assumed, which is not horizontally layered but has a downwarping low-velocity zone intervening into the upper crustal section with a normal velocity. The lowest P- and S-wave velocities and density in the uppermost layer are the same as those in Table I, but are assumed to vary gradually with depth. The mode of rupture propagation given in Fig. 20(b) shows that the rupture initiating from a middepth at the left edge of the fault propagates faster downward and appears considerably decelerated in the low-velocity zone. The distribution of final displacements in Fig. 20(c) again indicates appreciably larger slips in the upper section of the fault than in the layered half-space shown in Fig. 18. Thus, the existence of a low-velocity zone yields large fault slips, which provides effects similar to that of a zone with high stress drops. These effects would be greatly enhanced if these two different zones existed together. More results are shown in Hirahara et al. (1985).

The results obtained in Sections III.H and I are for the fault rupture processes that occur under a constant stress drop with depth. However, this assumption might not always be true in the real earth. The fault rupture extending down to middle to lower crustal sections will depend on the properties of crustal materials, which changes from a brittle regime to a ductile one. Mikumo et al. (1985b) took into consideration the depth dependence of the shear stress and the frictional strength of the fault embedded in a horizontally layered structure. One of the results shows that the rupture initiating from a nucleus at the base of the brittle zone can propagate not only upwards but also downwards even into a ductile zone with small stress drops. When the nucleus is evolved in the shallow section where the stress drop is quite small, it is not easy for the rupture to extend into a large scale, as have been suggested by Das and Scholz (1983). If, however, the stress jump factor $S$ is smaller than a certain limit (small

---

FIG. 20. Rupture propagation in a three-dimensional, heterogeneous crustal structure in which the fault does not reach the ground surface; (a) assumed velocity profile around the fault, (b) rupture fronts in the structure shown in (a), and (c) distribution of final slip displacements. (After Mikumo et al., 1985b.)

$\sigma_s - \sigma_0$ and/or large $\sigma_0 - \sigma_d$) in the shallow sections, the rupture could initiate at a very shallow location and extend downwards into the middle crust. Thus, the initiation and propagation of the rupture over a vertical fault extending from shallow to middle sections of the crust depends primarily on the $S$ value and its rate of variations with depth.

The fault zone materials involved in the ductile zone within the middle to lower crustal sections may have rheological or cohesive properties. In this case the sliding frictional stress should be dependent on slip velocity and/or slip displacement, as have been demonstrated by several experimental friction laws by Scholz and Engelder (1976) and Dieterich (1978, 1979) and also by a more refined form including state variable friction laws by Ruina (1983). An attempt was made (Mikumo, 1981b) to incorporate a simplified velocity- and displacement-dependent friction relation into a three-dimensional model,

$$\frac{d\sigma_d}{dt} = B\left(\frac{kv}{D} - \frac{1}{v}\frac{dv}{dt}\right), \quad \text{for } \sigma_d \geq \sigma_{d\,\text{min}} \qquad (36)$$

where $B = \beta \cdot \sigma_n$, $k$, $\beta$, and $\sigma_n$ being numerical constants and the normal stress, and $D$ and $v$ are the slip displacement and velocity, respectively. Equation (36) implies that the sliding frictional stress to resist fault slippage and hence the applied shear stress drops slowly with time for small $B$ values until it reaches the lowest value $\sigma_{d\,\text{min}}$. Numerical calculations show that the rupture velocity remarkably decreases with decreasing $B$ values for a fixed value of $\sigma_s$. The deceleration effects are greatly enhanced as spatial heterogeneities of $\sigma_s$ increase and as the initial shear stress $\sigma_0$ has lower levels with respect to the average strength $\bar{\sigma}_s$. For these cases, the growth of rupture is extremely slow in a region with the dimension as large as ten times the nucleus length. If the ductile zone has such strong cohesive properties due to the increase in temperatures, slow fault movements could take place there and trigger a normal earthquake in the brittle zone. This type of fault rupture would radiate low-frequency waves at an initial stage before higher frequency waves are transmitted to a recording site.

### J. Ground Motions in the Near Field

It now seems possible, in principle, to predict strong ground motions that would be observed in the near field during large earthquakes, incorporating rupture initiation, propagation, and stopping on a fault embedded in a realistic structure. In this section, we deal with seismic waves radiated from three-dimensional, propagating shear cracks in a homogeneous half-

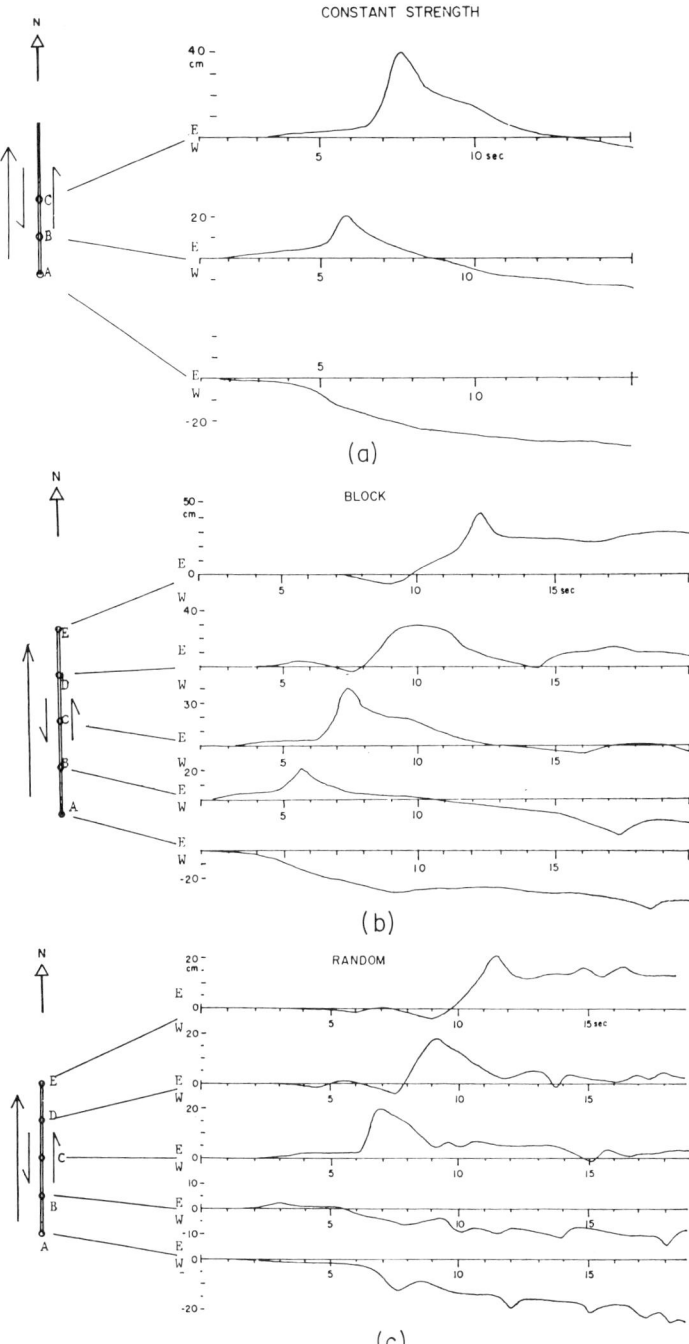

**FIG. 21.** Ground motions just above the fault plane embedded in a homogeneous half-space. These are the $z$ components perpendicular to the fault that does not break out the ground surface. Cases (a), (b), and (c) correspond to the three different cases given in Figs. 12, 13, and 14. (After Miyatake, 1980.)

space and in a horizontally layered structure, following the results in Miyatake (1980, 1984) and Mikumo et al., (1985b).

The ground displacements in these cases are calculated as the displacement components $u(i, 1, k)$, $v(i, 1, k)$, and $w(i, 1, k)$ through the wave equation [Eq. (23)] under the boundary conditions Eq. (32) or Eq. (35) in Sections III.H and III.I. An alternative method to calculate ground displacements at locations far from the fault is to use the distribution of slip functions $s(x, \tau)$ or $m_j^i \cdot \Delta u(t)$ at each fault segment, which have been calculated for the present shear crack model, together with the Green's functions $g_j^i(t)$ and the rupture time $t_j$, as expressed in Eqs. (1) and (3).

Several examples of the ground displacements calculated by the first method are presented here. The first example (Miyatake, 1980), shown in Fig. 21 gives ground displacements just above the fault plane (at $y = 0$ and $z = 0$). Since the strike-slip fault assumed in this case does not reach the ground surface, the displacements have only the $z$ component perpendicular to the fault plane. A high-cut filter with a cut-off frequency of 2 Hz has been applied to remove higher frequency numerical noise. Three cases (a), (b), and (c) in Fig. 21 correspond to three different types of faults with a uniform strength, a square-shaped strong patch, and multimodal heterogeneous strengths, which have been shown in Figs. 12, 13, and 14, respectively. A large-amplitude phase arrives in all cases at 2 to 9 sec after the initiation of rupture. The large phase is due to the arrival of S waves radiated from the passage of the rupture, and is not caused by the fracture of the fault segment nearest to the observation points. The onsets of reverse motions between 14 and 18 sec at points D, C, B, and A in case (b) correspond to the arrival time of S waves reflected from the right edge of the fault after the rupture encircled the central strong patch and stopped at the edge. The strongly heterogeneous fault in case (c) generates high-frequency components superimposed on lower frequency waves. A noticeable difference in the waveform at point B between cases (a) and (c) is probably due to insufficient radiation of S-wave energy from the upper left portion of the fault that remains unruptured in the early stage and finally breaks in the latter case. Thus, the ground motions right above the fault directly reflect the passage of rupture and include large-amplitude S waves. The waveforms become complicated and tend to involve high-frequency waves when the fault has heterogeneous strength distributions.

The second example (Miyatake, 1984) provides the distribution of the displacement waveforms at different locations on the ground surface near the fault. These correspond to the cases given in Figs. 15 and 16. Figure 22 shows the $z$ displacements perpendicular to the fault plane, where the ordinate and abscissa give the locations in the $z$ and $x$ directions. The first large-amplitude pulses observed at $x \geq 5$ km and $0 \leq z \leq 17$ km in cases

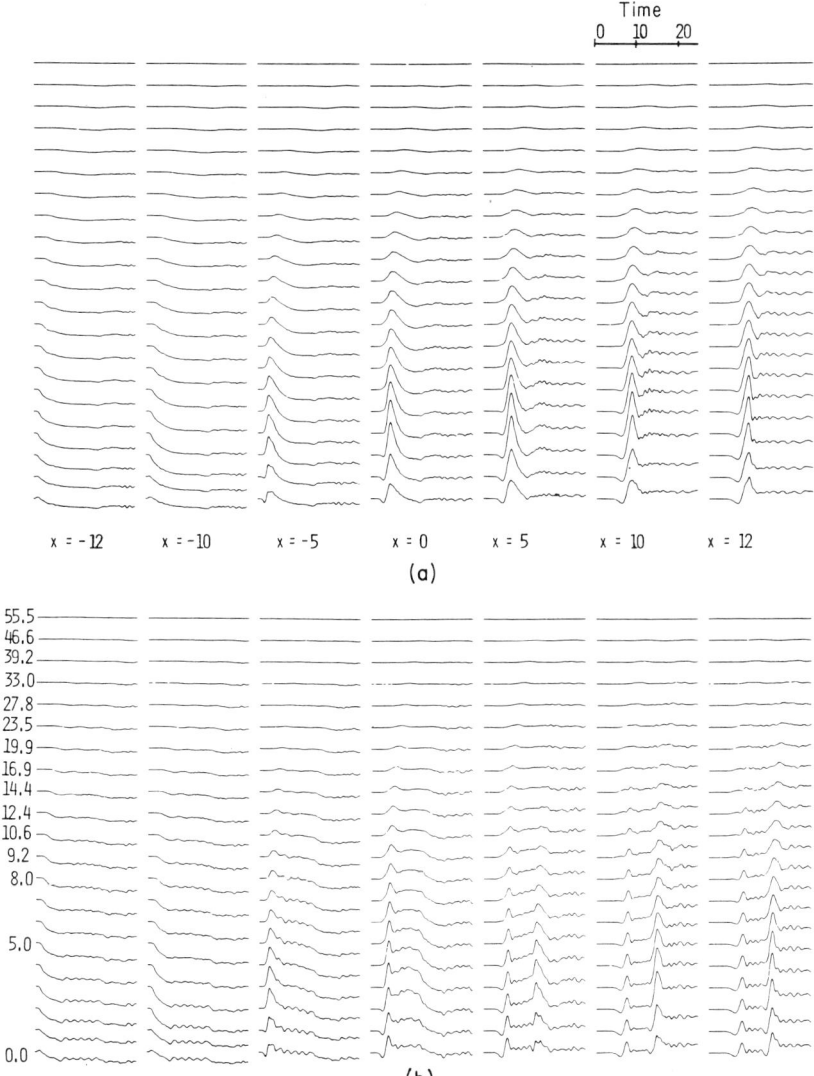

**FIG. 22.** Distribution of the displacement waveforms at different locations on the ground surface. Cases (a) and (b) correspond to the two cases given in Figs. 15 and 16. (After Miyatake, 1984.)

(a) and (b) are due to the arrival of S waves radiated from the propagating rupture. Peculiar slow movements after the first pulse observed at $x = 0$ in case (b) may be due to the healing of the rupture as it gradually decelerated before the high-strength zone. The second pulses observed at $x \geq 5$ km and $0 \leq z \leq 11$ km are also S waves generated when the rupture accelerated beyond the high-strength zone. These examples clearly indicate that the ground motions observed in the near field are strongly affected by rapid variations of rupture velocity, as has been suggested in a stochastic model presented in Section II.E.

The effects of low-velocity surficial layers on the ground motions and the effects of the breakout of fault rupture onto the ground surface are also discussed (Mikumo et al., 1985b). As might be expected from a comparison between the two cases in Fig. 18, the ground motions at locations just above and near to the fault (at $y = 0, z = 0$ to 5 km), particularly in the $x$ component (parallel to the fault), show larger displacements (up to two times) for the case with the low-velocity layers than for the homogeneous half-space. The differences in the amplitudes of the $y$ (vertical) and $z$ components (perpendicular to the fault) computed for the two structures are not as large as the $x$ component.

The break-out effects of the fault rupture on ground motions in the near field can be estimated from a comparison between the ground displacements near the fault in the two cases shown in Figs. 18 and 19. The $x$ component (at $x = 0$, $y = 0$, and $z = 4$ km) from the case when the rupture breaks out is more than twice the corresponding displacement for the case when the upper edge of the fault remains at 2 km below the ground surface. At another location ($x = 10$ km, $y = 0$, and $z = 4$ km) close to the fault edge, the $x$ and $z$ components in the former case are about 50% larger, but the $y$ component (vertical component) is found to be smaller than the corresponding displacements in the latter.

## IV. Summary

We have reviewed current methods and some results for strong motion synthesis from kinematic dislocation sources and with emphasis for numerical modeling of fault rupture processes by three-dimensional dynamic shear cracks.

For the kinematic problem, we described explicitly the methods of theoretical and semi-empirical synthesis based on the representation theorem and presented results obtained by applying the methods to recent moderate-size earthquakes. It has been shown that strong motion records, particularly the low-frequency components, can be satisfactorily modeled

## 3. NUMERICAL MODELING

by the theoretical models with a realistic crustal structure and also by the semi-empirical analysis using aftershock or foreshock records as empirical Green's functions. High-frequency waves, however, cannot be well simulated by either of the two methods unless incoherent rupture propagation is also incorporated. A stochastic fault model with various fault parameters randomly distributed over the fault has been tentatively introduced, and it has been confirmed that the most effective factor for generating high-frequency waves is the acceleration or deceleration of the rupture front.

For the dynamic problem, we emphasized the basic properties and possible complexities of the fault rupture process under realistic conditions rather than constructed synthetic seismograms from the model to match the observed records. The three-dimensional models examined included spontaneous dynamic shear cracks propagating on the fault in an infinite elastic medium, a homogeneous half-space, and a horizontally layered medium and more heterogeneous structures. The fault was assumed to have inherent static frictional strength and sliding frictional stress, which have uniform or heterogeneous properties, being subjected to the initial shear stress. The method used to solve the wave equations for the three-dimensional space, under appropriate boundary conditions and a fracture criterion, was a finite-difference method. For numerical solutions for spontaneous rupture propagation, we used the finite-stress fracture criterion, which is equivalent to Irwin's criterion based on the critical stress-intensity factor.

One of the basic and important features of the faulting process is the mode of rupture propagation. The results clearly indicate that the rupture velocity depends on the difference between the fault strength and the initial stress level, elastic properties of fault zone materials, cohesive properties of the fault surface, and also on heterogeneous strength distribution. The initiation of rupture is specified by the stress jump factor $S = (\sigma_s - \sigma_0)/(\sigma_0 - \sigma_d)$ and the dimension of the nucleus. An important result is that when the difference between the applied stress and the strength is relatively small, the rupture propagates elliptically with a velocity close to that of the P wave along the direction parallel to the shear stress and with an almost S-wave velocity in the direction perpendicular to it. The velocities along the two directions reduce to the Rayleigh velocity when the fault strength is relatively high with respect to the initial stress. If the fault is embedded in an inhomogeneous elastic medium such as in a horizontally layered or three-dimensionally varying structure, the rupture velocity is also controlled by the elastic properties. The presence of low-velocity surficial layers decelerates the rupture propagation and yields large fault displacements in the layers. If the fault surface has strong cohesive properties due to indentation and ploughing of asperities, the sliding frictional stress depends on the slip velocity of fault motion and will yield a slow rupture. This could be the case for a ductile zone in the middle to lower crust due to the increase in temperature. It was also found that rupture propagation is strongly affected by spatial variations and the form of frequency distribution of frictional strength and sliding frictional stress on the fault. If the fault surface has a few large, high-strength zones or if there are many randomly distributed small, high-strength inclusions, the propagation of the rupture is remarkably decelerated or arrested in some extreme cases.

The existence of high-strength zones provides strong deceleration effects and incoherent modes of rupture propagation and hence could radiate high-frequency seismic waves. In some cases, these "strong patches" remain unbroken during the main rupture, and high stresses are concentrated around the zones. The unruptured zones and the large variations of remnant stresses could be the source of aftershocks, which would occur probably because of time-delayed stress corrosion or viscoelastic stress recovery and relaxation combined with creep recovery of the strength. These heterogeneities of fault strengths on the fault could also yield spatial and temporal variations of seimicity when the tectonic shear stress slowly increases over a long time (Mikumo and Miyatake, 1983).

The present model provides reasonable explanations to various aspects of seismic faulting: (1) the initiation, spreading, and stopping mechanisms of rupture; (2) the stick-slip instability of faulting; (3) the radiation of high-frequency seismic waves; (4) the nonuniform distribution of fault displacements; (5) the occurrence of earthquake sequences foreshocks, the main shock sometimes followed by multiple shocks and aftershocks; and (6) the aseismic fault slip at depth.

The models described are a forward process based on kinematic and dynamic faulting models, involving a number of assumptions. Actual rupture processes on the fault in the earth should be closely investigated from observations of seismic waves in the near and far fields. Application of recent inversion techniques from teleseismic body and surface waves (see, e.g., Kikuchi and Fukao, 1985) or from strong ground motions in the near field (see, e.g., Olson and Aspel, 1982; Hartzell and Heaton, 1983; Spudich and Frazier, 1984; Takeo and Mikami, 1985; Fukuyama and Irikura, 1985) are extremely important to clarify the two-dimensional distribution of the final displacements or stress drop and the arrival times of the rupture front on each of the fault segments.

REFERENCES

Aki, K. (1968). *J. Geophys. Res.* **73**, 5359–5376.
Aki, K., and P. G. Richards (1980). "Quantative Seismology: Theory and Methods," Vol. 2. Freeman, San Francisco, California.

Alekseev, A. S., and B. G. Mikhailenko (1980). *J. Geophys.* **48**, 161–172.
Alterman, Z., and F. C. Karal (1968). *Bull. Seismol. Soc. Am.* **58**, 367–398.
Andrews, D. J. (1975). *Bull. Seismol. Soc. Am.* **65**, 163–182.
Andrews, D. J. (1976a). *J. Geophys. Res.* **81**, 3375–3582.
Andrews, D. J. (1976b). *J. Geophys. Res.* **81**, 5679–5687.
Archuleta, R. J., and G. A. Frazier (1978). *Bull. Seismol. Soc. Am.* **68**, 541–572.
Berg, A. P. (1984). *Geophys. J. R. Astron. Soc.* **79**, 3–10.
Boore, D. M., and W. B. Joyner (1978). *Bull. Seismol. Soc. Am.* **68**, 283–300.
Bouchon, M. (1979). *J. Geophys. Res.* **84**, 6149–6156.
Bouchon, M. (1982). *Bull. Seismol. Soc. Am.* **72**, 745–757.
Burridge, R. (1969). *Philos. Trans. R. Soc. London, Ser. A* **265**, 353–381.
Burridge, R. (1973). *Geophys. J. R. Astron. Soc.* **35**, 439–455.
Burridge, R., and G. S. Halliday (1971). *Geophys. J. R. Astron. Soc.* **25**, 261–283.
Burridge, R., and L. Knopoff (1964). *Bull. Seismol. Soc. Am.* **54**, 1874–1888.
Burridge, R., and R. Moon (1981). *Geophys. J. R. Astron. Soc.* **67**, 325–342.
Byerlee, J. D. (1970). *Tectonophysics* **9**, 475–486.
Byerlee, J. D. (1978). *Pure Appl. Geophys.* **116**, 615–626.
Campillo, M. (1983). *Bull. Seismol. Soc. Am.* **73**, 723–734.
Cherepanov, G. P. (1979). "Mechanics of Brittle Fracture," McGraw-Hill, New York.
Clayton, R., and B. Engquist (1977). *Bull. Seismol. Soc. Am.* **67**, 1529–1540.
Das, S. (1976). ScD Thesis, MIT, Cambridge, Massachusetts.
Das, S. (1980). *Geophys. J. R. Astron. Soc.* **62**, 592–604.
Das, S. (1981). *Geophys. J. R. Astron. Soc.* **67**, 325–393.
Das, S., and K. Aki (1977a). *Geophys. J. R. Astron. Soc.* **50**, 643–668.
Das, S., and K. Aki (1977b). *J. Geophys. Res.* **82**, 5658–5670.
Das, S., and C. H. Scholz (1983). *Nature (London)* **305**, 621–623.
Day, S. M. (1982a). *Bull. Seismol. Soc. Am.* **72**, 705–727.
Day, S. M. (1982b). *Bull. Seismol. Soc. Am.* **72**, 1881–1902.
Dieterich, J. H. (1978). *Pure Appl. Geophys.* **116**, 790–806.
Dieterich, J. H. (1979). *J. Geophys. Res.* **84**, 2161–2175.
Fossum, A. F., and L. B. Freund (1975). *J. Geophys. Res.* **80**, 3343–3347.
Fuchs, K., and G. Mueller (1971). *Geophys. J. R. Astron. Soc.* **23**, 417–433.
Fukuyama, N., and K. Irikura (1985). *Tectonophysics.* Submitted.
Hadley, D. M., and D. V. Helmberger (1980). *Bull. Seismol. Soc. Am.* **70**, 617–630.
Hanson, M. E., A. R. Sanford, and R. J. Shaffer (1971). *J. Geophys. Res.* **76**, 3375–3383.
Hanson, M. E., A. R. Sanford, and R. J. Shaffer (1974). *Geophys. J. R. Astron. Soc.* **38**, 365–376.
Hartzell, S. H. (1978). *Geophys. Res. Lett.* **5**, 1–4.
Hartzell, S. H., and T. Heaton (1983). *Bull. Seismol. Soc. Am.* **73**, 1553–1583.
Haskell, N. (1964). *Bull. Seismol. Soc. Am.* **54**, 1811–1841.
Haskell, N. (1969). *Bull. Seismol. Soc. Am.* **59**, 865–908.
Heaton, T. H., and D. V. Helmberger (1977). *Bull. Seismol. Soc. Am.* **67**, 315–330.
Heaton, T. H., and D. V. Helmberger (1978). *Bull. Seismol. Soc. Am.* **68**, 31–48.
Helmberger, D. V. (1968). *Bull. Seismol. Soc. Am.* **58**, 179–214.
Helmberger, D. V., and S. D. Malone (1975). *J. Geophys. Res.* **80**, 4881–4888.
Herrmann, R. B. (1977). Final Rep., Contact DACW39-76-C-0058. Waterways Exp. Stn., Vicksburg, Mississippi.
Herrmann, R. B., and O. W. Nuttli (1975a). *Earthquake Eng. Str. Dyn.* **4**, 49–58.
Herrmann, R. B., and O. W. Nuttli (1975b). *Earthquake Eng. Str. Dyn.* **4**, 59–72.
Hirahara, K., T. Mikumo, and T. Miyatake (1985). *Abstr. Fall Meet. Seismol. Soc. Jpn.* B84.

Hron, F., and B. G. Mikhailenko (1981). *Bull. Seismol. Soc. Am.* **71**, 1011–1029.
Husseini, M. I., D. B. Jovanovich, M. J. Randall, and L. B. Freund (1975). *Geophys. J. R. Astron. Soc.* **43**, 367–385.
Ida, Y. (1973). *J. Geophys. Res.* **78**, 3418–3429.
Ida, Y., and K. Aki (1972). *J. Geophys. Res.* **77**, 3796–3805.
Imagawa, K., and T. Mikumo (1982). *Jishin* **35**, 575–590.
Imagawa, K., N. Mikami, and T. Mikumo (1984). *J. Phys. Earth* **32**, 317–338.
Imoto, M., I. Karakama, and R. Matsu'ura (1981). *Jishin* **34**, 481–493.
Irikura, K. (1983). *Bull. Disaster Prev. Res. Inst., Kyoto Univ.* **33**, 63–104.
Irikura, K., and I. Muramatu (1982). *Proc. Int. Earthquake Microzonation Conf.*, 3rd, Seattle, Wash. **1**, 447–458.
Kanamori, H. (1979). *Bull. Seismol. Soc. Am.* **69**, 1645–1670.
Kanamori, H. (1981). In "Earthquake Prediction" (D. W. Simpson and P. G. Richards, eds.), pp. 1–19. Am. Geophys. Union, Washington, D.C.
Kanamori, H., and D. L. Anderson (1975). *Bull. Seismol. Soc. Am.* **65**, 1073–1095.
Kawasaki, I. (1978). *J. Phys. Earth* **26**, 211–237.
Kawasaki, I., Y. Suzuki, and R. Sato (1973). *J. Phys. Earth* **21**, 251–284.
Kawasaki, I., Y. Suzuki, and R. Sato (1975). *J. Phys. Earth* **23**, 43–61.
Kennett, B. L. N. (1974). *Bull. Seismol. Soc. Am.* **64**, 1685–1696.
Kennett, B. L. N., and J. Kerry (1979). *Geophys. J. R. Astron. Soc.* **57**, 557–583.
Kelly, K. R., R. W. Ward, S. Treitel, and R. M. Alford (1976). *Geophysics* **41**, 2–27.
Kikuchi, M., and Y. Fukao (1985). *Phys. Earth Planet. Inter.* **37**, 235–248.
Kikuchi, M., and H. Takeuchi (1970). *Jishin* **23**, 304–312.
Kikuchi, M., and H. Takeuchi (1972). *Jishin* **24**, 298–303.
King, G., and J. Nabelek (1985). *Science* **228**, 984–987.
Kostrov, B. V. (1964). *J. Appl. Math. Mech. (Engl. Transl.)* **28**, 1077–1087.
Kostrov, B. V. (1966). *J. Appl. Math. Mech. (Engl. Transl.)* **30**, 1241–1248.
Lysmer, R., and R. L. Kuhlemeyer (1969). *J. Eng. Mech. Div., Am. Soc. Civ. Eng.* **95**, 859–588.
Madariaga, R. (1976). *Bull. Seismol. Soc. Am.* **66**, 639–666.
Madariaga, R. (1977). *Geophys. J. R. Astron. Soc.* **51**, 625–651.
Manshina, L. (1964). *Bull. Seismol. Soc. Am.* **54**, 369–376.
Maruyama, T. (1963). *Bull. Earthquake Res. Inst.* **41**, 467–486.
Meissner, R., and J. Strehlau (1982). *Tectonics* **1**, 73–79.
Mikumo, T. (1973). *J. Phys. Earth* **21**, 191–212.
Mikumo, T. (1976). Proj. Rep. No. 21, pp. 191–212. Build. Res. Inst.,
Mikumo, T. (1981a). Rep. Nat. Disaster No. A-56-3, pp. 25–30. Sci. Res. Fund, Minist. Educ.,
Mikumo, T. (1981b). *Geophys. J. R. Astron. Soc.* **65**, 129–153.
Mikumo, T., and T. Miyatake (1978). *Geophys. J. R. Astron. Soc.* **54**, 417–438.
Mikumo, T., and T. Miyatake (1979). *Geophys. J. R. Astron. Soc.* **59**, 497–522.
Mikumo, T., and T. Miyatake (1983). *Geophys. J. R. Astron. Soc.* **74**, 559–583.
Mikumo, T., K. Imagawa, and M. Kato (1985a). *Abstr. IASPEI Symp.* No. 8.
Mikumo, T., K. Hirahara, and T. Miyatake (1985b). *Tectonophysics*. Submitted.
Miyatake, T. (1977). *Jishin* **30**, 449–461.
Miyatake, T. (1980). *J. Phys. Earth* **28**, 565–616.
Miyatake, T. (1984). *Jishin* **37**, 257–267.
Miyatake, T. (1985). *Abstr. IASPEI Symp.* No. 7.
Muramatu, I. (1977). *Jishin* **30**, 191–212.
Muramatu, I., and K. Irikura (1982). *J. Jpn. Soc. Nat. Disaster Sci.* **1**, 29–43. In Jpn.
Muramatu, I., and H. Ohnuma (1985). *Abstr. IASPEI Symp.* No. 8.

Olson, A. H. (1982). Ph.D. Thesis, Univ. of California, San Diego.
Olson, A. H., and R. Aspel (1982). *Bull. Seismol. Soc. Am.* **72**, 1969-2002.
Richards, P. G. (1976). *Bull. Seismol. Soc. Am.* **66**, 1-32.
Ruina, A. (1983). *J. Geophys. Res.* **88**, 10359-10370.
Sato, R. (1977). *J. Phys. Earth* **25**, 43-68.
Sato, R. (1978). *J. Phys. Earth* **26**, 17-37.
Sato, R., and N. Hirata (1980). *J. Phys. Earth* **28**, 145-168.
Scholz, C. H., and J. T. Engelder (1976). *Int. J. Rock Mech. Min. Sci. Geomech. Abstr.* **13**, 149-154.
Sibson, R. H. (1977). *J. Geol. Soc. London* **133**, 191-213.
Sibson, R. H. (1982). *Bull. Seismol. Soc. Am.* **72**, 151-163.
Sibson, R. H. (1984). *J. Geophys. Res.* **89**, 5791-5799.
Smith, W. D. (1974). *J. Comput. Phys.*, **15**, 492-503.
Spetzler, H., H. Mizutani, and F. Rummel (1982). High Pressure Geosci. Res., Ger. Sci. Found.
Spudich, P., and L. N. Frazier (1984). *Bull. Seismol. Soc. Am.* **74**, 2061-2082.
Stoeckl, H. (1977). *J. Geophys.* **43**, 311-327.
Suzuki, T., and M. Hakuno (1984). *Bull. Earthquake Res. Inst.* **59**, 327-359. In Jpn.
Swanger, H. J., and D. M. Boore (1978). *Bull. Seismol. Soc. Am.* **68**, 907-922.
Takeo, M., and N. Mikami (1986). *Tectonophysics*. Submitted.
Takeuchi, H., and M. Kikuchi (1973). *J. Phys. Earth* **21**, 27-37.
Virieux, J., and R. Madariaga (1982). *Bull. Seismol. Soc. Am.* **72**, 345-369.
Wiggins, R. A., J. Sweet, and G. A. Frazier (1977). *Trans. Am. Geophys. Union* **58**, 1193.
Yamaguchi, U., and Y. Nishimatsu (1967). "Introduction to Rock Mechanics." Tokyo Univ. Press, Tokyo. In Jpn.
Yamashita, T. (1976). *J. Phys. Earth* **24**, 417-444.

# CHAPTER 4

# Complete Strong Motion Synthetics

G. F. Panza

*Istituto di Geodesia e Geofisica*
*Università di Trieste*
*Trieste, Italy*
   *and*
*International School for Advanced Studies*
*Trieste, Italy*

P. Suhadolc

*Istituto di Geodesia e Geofisica*
*Università di Trieste*
*Trieste, Italy*

## I. Introduction

The broad field of effort involved in understanding earthquake forces, in analyzing their effects on structures, in developing effective building codes and methods of design, and in assessing earthquake risk to existing construction is termed earthquake engineering. A portion of this effort is research devoted to understanding the nature of strong ground motion.

Usually only strong ground motion from earthquakes of magnitude ~ 5 and greater is of much engineering interest. Furthermore, with only a few exceptions, potentially damaging effects are confined to within several tens of kilometers of the causative fault, even in great earthquakes. Outside of this range, although motion is still perceptible, it is typically associated only with nonstructural damage or damage to particularly weak, poorly engineered structures.

Thus, strong motion seismology is concerned with high-frequency seismic waves from large earthquakes; this type of motion is the least understood area in earthquake seismology. High-frequency waves from small earthquakes are also complicated, but at least they are easy to record and study. The strong motion in the vicinity of large earthquakes, however, is not recorded very often. Nevertheless, significant progress has been made in the past decade in understanding how high-frequency waves are generated during the rupture of an earthquake fault.

Strong motion seismology was initiated by earthquake engineers who developed and operated seismographs to record strong ground motion. Early instruments recorded in analog form; thus, their numerical processing required the tedious process of digitization. Quite recently digital accelerographs have been developed, which will greatly facilitate the analysis of strong motion data.

Digital accelerographs have the advantage that the required digitization of an analog signal is part of the recording process, and subsequent data processing is, therefore, much more convenient. In addition, a memory buffer allows the automatic retention of the previous few seconds of motion, so that the entire record starting before the P-wave arrival can be recorded. Another potential advantage of the digital accelerograph for some applications is the possibility of automatic gain-ranging, so that both the frequent, smaller accelerations and the larger, potentially damaging motions can be recorded accurately.

The availability of digital accelerograms, combined with the increasing demand from structural engineers for more quantitative methods of analysis and interpretation, makes the development of efficient theoretical tools for the construction of synthetic records very important. This demand is complemented by the ongoing international effort to establish a worldwide broadband digital network as proposed in *Project Geoscope* (Romanowicz et al., 1984) and in *Science Plan for a New Global Seismographic Network* (Incorporated Research Institutions for Seismology, 1984).

Recently an increasing number of seismologists have been involved in both collecting and interpreting strong motion data. Various deterministic and stochastic earthquake source models have been applied in their interpretation. Theoretical results on fracture mechanics as well as laboratory results on rock failure have led to a better understanding of the physics of fault rupture. Various attempts have also been made to relate the earthquake rupture process to the observable behavior of earthquake faults. Such a relation is vitally important in evaluating the ground motion expected for earthquake faults mapped by geologists.

This chapter attempts to describe a tool capable of synthesizing strong motion records from a basic understanding of fault mechanics and seismic-wave propagation in the earth.

The mathematical foundation of the approach we follow was firmly established in the early 1960s by Maruyama (1963) and Burridge and Knopoff (1964), who showed that the point-force equivalent of fault slip is a double couple and that the seismogram can be computed by a space-time convolution of the slip function and Green's function. The slip function describes the fault displacement during an earthquake as a function of time and position on the fault plane. Green's function is the response of the earth when an impulsive double couple is applied at a point on the fault plane. The slip function and Green's function express quantitatively the source and the propagation effect, respectively, on seismic motion. In order to find the slip function from an observed seismogram, it is necessary to know the Green's function.

The study of the slip function began from the low-frequency end of the seismic spectrum—the period range longer than ~ 20 sec. Through this low-frequency window it has been possible to determine the "seismic moment," i.e., the moment of the equivalent point double couple. As such, it has become a key parameter in a comparative study of far-field seismic observations with near-field observations on earthquake faults by geologic and geodetic measurements. It has also been used for estimating present-day plate motion and for long-term earthquake prediction. It is a direct measure of the extent of faulting and, therefore, of the damage caused by static deformation, including tsunami damage in the case of a submarine earthquake.

The seismic moment is not directly useful for earthquake engineering, because the strong motion depends crucially on the details of the rupture process, which cannot be seen clearly through the long-period window. It is thus necessary to extend the window to higher frequencies. This is a difficult task because of the complex propagation effects on high-frequency seismic waves involving strong scattering and attenuation.

Usually, the frequency content of the strong ground motion is broadband. That is, the acceleration amplitude spectra are roughly constant over a range of frequencies from 5 to 0.5 Hz or less. At lower frequencies the amplitude decreases with decreasing frequencies. The point where this occurs, the corner frequency $f_0$, depends primarily on the duration of the strong shaking or equivalently on the size of the source, with the longer motions from larger earthquakes delivering more energy at lower frequencies.

At higher frequencies, experience has shown that there is a tendency for the motion to be strongly attenuated for frequencies above 5 to 10 Hz. In this case, the frequency at which attenuation begins, defined as $f_{\max}$ by Hanks (1982), appears to depend primarily on the distance from the source of the motion and on the nature of the local geology. Papageorgiou and Aki (1983), on the other hand, propose that $f_{max}$ originates from the nonelasticity of the fault, and in particular that it varies inversely with the size of the cohesive zone behind the propagating crack tip. As might be expected, records obtained close to the source on competent rock show the most high-frequency motion. Even in this case, the motion attenuates significantly with increasing frequency above $\sim$ 15 to 20 Hz. There is some evidence that measurable motion may be present at frequencies of 30 Hz or higher, but this motion is of interest only in special cases.

An exception to this general description occurs when the surficial deposits at the recording site are very soft. Under these circumstances, the soil deposits can "amplify" the motions at some frequencies and reduce it at others, for example, the accelerograms obtained in Mexico City during the May 11, 1962, and the September 19, 1985, earthquakes. Path effects along the source-to-site distance, of the order of a few hundred kilometers, have helped in reducing the high-frequency motion and increasing the duration of shaking. However, the amplification of motion at periods near 2.5 sec is believed to be caused primarily by the response from the soft soils of the old lake bed on which Mexico City is located (see, e.g., Jennings, 1983).

The construction of synthetic signals of long duration having a broadband frequency content can hardly be done using the ray approach (see, e.g., Chapman, 1985; Červený, 1985). At present, a very suitable tool for this purpose seems to be modal summation (Panza, 1985; Suhadolc and Panza, 1985), which has already been applied at lower frequencies (see, e.g., Liao et al., 1978). Probably the lack of an explicit statement of the details of high-frequency eigenvalue and eigenfunction evaluation has been the main factor delaying large-scale application of multimode, synthetic seismograms to the interpretation of strong motion records. There are essentially two types of computational problems that have recently been overcome: a) removing the loss of precision contained in the original Thomson (1950) and Haskell (1953) technique for the computation of Rayleigh-wave dispersion and (b) achieving the necessary accuracy and efficiency in modal computation at high frequencies, where many modes get very close to each other. To deal with the loss of precision problem, two methods are available —Knopoff's (1964a) method and the method of delta matrices (Pestel and Leckie, 1963; Thrower, 1965; Dunkin, 1965; Watson, 1970). Very recently, as a result of an intensive international cooperation, Schwab et al. (1984) have shown, both for eigenvalue and eigenfunction determinations, that there are no loss of precision problems when existing improvements of the original formulation are used, even for frequencies as high as 10,000 Hz. Examples of highly accurate and efficient computations for quite realistic structural models at high frequencies have recently been given by Panza (1985) and Suhadolc and Panza (1985).

Once the problem of an accurate computation of eigenvalues and eigenfunctions is solved in an efficient way, the inclusion of the effect of point sources can easily be made using the formalism given by Ben-Menahem and Harkrider (1964). These results can be easily generalized to sources with finite dimensions and durations (see, e.g., Ben-Menahem, 1961; Hartzell, 1978; Kanamori, 1979), even in two and three dimensions.

In this paper we will concentrate on Rayleigh modes (P and SV waves), since their treatment presents a rather high degree of complexity and, in our opinion, the effect of such components of motion has not yet received in engineering seismology the necessary consideration. However, all of the results presented here have a general validity and with the proper modifications can be easily extended to the treatment of Love modes (SH waves).

## II. Automatic Computation of Eigenvalues and Eigenfunctions

### A. Perfectly Elastic Media

1. General Background

Knopoff (1964a) has given the solution to problems of elastic wave propagation in multilayered media as the quotient of products of matrices. In the case of SH waves, the matrices are of order two; in the case of P–SV waves, the matrices are of order four. The individual matrix elements are determinants of order two or four in the two cases.

Concerning the determination of the Rayleigh-wave phase velocity using Knopoff's method, it was reported (Schwab, 1970) that with 16 decimal digits carried during computation and 15.4 significant figures required in the computed phase velocities, the number of wavelengths of a layered structure above the homogeneous half-space can be increased to 196 without any loss of precision. To control overflow when the thickness ($H$) of the structure corresponds to a large number of wavelengths ($\lambda$), a simple normalization is required (Schwab et al., 1984). Once the normalization is included, so that large values of $H/\lambda$ can be treated, only these overflow/underflow situations must be avoided: the matrix elements for the layers with $c < \beta_m < \alpha_m$, where $c$ is the phase velocity, $\beta_m$ the S-wave velocity, and $\alpha_m$ the P-wave velocity of the $m$th layer, contain factors of the form (Schwab, 1970)

$$\left\{\begin{array}{c}\sinh\\ \cosh\end{array}\right\} P_m^* \left\{\begin{array}{c}\sinh\\ \cosh\end{array}\right\} Q_m^*, \qquad (1)$$

where

$$P_m^* = -(\omega d_m/c)\sqrt{1 - (c^2/\alpha_m^2)} = +(\omega d_m/c)r_{\alpha_m}^* \quad \text{(real)},$$

and (2)

$$Q_m^* = -(\omega d_m/c)\sqrt{1 - (c^2/\beta_m^2)} = +(\omega d_m/c)r_{\beta_m}^* \quad \text{(real)},$$

where $d_m$ is the thickness of the $m$th layer and $\omega$ is the angular frequency. In the notation used in this paper, the asterisk denotes the imaginary part of a complex quantity. For large values of the arguments, the magnitudes of these factors are approximated by the quantity

$$\tfrac{1}{4}\exp\left[(\omega d_m/c)(r_{\alpha_m}^* + r_{\beta_m}^*)\right], \quad (3)$$

which is always positive and which can be factored out in Eq. (1).

Since our interest in Eq. (1) is limited to changes in sign of the dispersion function, this factor can be deleted when treating layer $m$; consequently, there is no more need to deal with exponentials having arguments above a certain level.

The case $\beta_m < c < \alpha_m$ and large $d_m/\lambda$ can be treated by analogy, and it is possible to delete terms like

$$\tfrac{1}{2}\exp\left[(\omega d_m/c)r_{\alpha_m}^*\right]. \quad (4)$$

The power of this approach has been extensively tested by Schwab et al. (1984).

Once the phase velocity $c$ is obtained for a given angular frequency $\omega$, the group velocity $u$ is obtained from

$$u = \frac{c}{1 - (dc/d\omega)(\omega/c)}, \quad (5)$$

where standard implicit function theory is applied to the dispersion function $F$ to obtain

$$\frac{dc}{d\omega} = -\left(\frac{\partial F}{\partial \omega}\right)_c \bigg/ \left(\frac{\partial F}{\partial c}\right)_\omega. \quad (6)$$

For details see Schwab and Knopoff (1972).

The algorithmic details of the eigenfunction evaluation with Knopoff's method are rather involved, although in principle only a straightforward application of Cramer's rule is required, whereas the details for the original formulation (Haskell, 1953) are quite simple. Full details concerning Knopoff's method are given by Schwab et al. (1984).

Here, we simply remember that $u_m^*$, $w_m$, $\sigma_m^*$, and $\tau_m$ are, respectively, the horizontal and vertical components of displacement and the normal and tangential components of stress at the $m$th interface, the free surface corresponding to $m = 0$.

Since all of the problems connected with the loss of precision at high frequencies have been solved, we propose to use the summation of higher modes of surface waves for the generation of "complete" strong motion synthetics at high frequencies. We will describe in detail the algorithms we have developed for Rayleigh waves, i.e., for synthetizing the radial and vertical component of motion; however, our results can readily be extended to Love waves.

The key point in the use of multimode summation, both for Love and Rayleigh modes, is an efficient computation of the phase velocity for the different modes at sufficiently small frequency intervals $\Delta f$ with sufficient precision. To be efficient it is not advisable to determine at each frequency and for each mode the zeros of the dispersion function using the standard root-bracketing and root-refining procedure (see, e.g., Schwab and Knopoff, 1972). This must be used only when strictly necessary, as for instance at the beginning of each mode. For all other points $i$ of each mode, the phase velocity can be estimated by cubic extrapolation, using the values of the phase slowness $s = 1/c$ and $df/ds$ already determined at frequencies $f_{i-2}$ and $f_{i-1}$. However, the precision that can be reached in this way is not satisfactory, thus the phase velocity value must be refined. This can be done by an iterative cubic fit in the $F$–$c$ plane. In our experience, such a procedure has always given highly accurate determinations of the phase velocity and allows a considerable savings in time when compared with the standard root-bracketing and root-refining procedure.

Once the problem of an efficient determination of phase velocities is overcome, two other main problems must be solved at each frequency: (a) to correctly follow a mode and (b) to determine the minimum number of layers to be used.

The problem of correctly following a mode arises in the high-frequency domain ($f > 0.1$ Hz), where several higher modes are very close to each other. The determination of the minimum number of layers to be used—structure minimization—is critical in order to reach a high precision in phase velocity determination spending the minimum possible computer time.

In order to ensure high efficiency in the computation of synthetic seismograms, it is necessary to compute (in the frequency domain) the phase velocity, phase attenuation, group velocity, ellipticity, energy integral, and eigenfunctions and their maximum depth of penetration at constant frequency intervals. To reach a maximum frequency of 10 Hz, a satisfactory step is 0.05 Hz. To determine the total number of modes present in the frequency interval considered, we fix $c_0 = 0.98\,\beta_n$, where $\beta_n$ is the S-wave velocity in the half-space, and we increment $f$, using the Schwab and Knopoff (1972) algorithm, to find the values of $f$ corresponding to zeros of the dispersion function $F(f, c_0)$. Obviously, starting from $f = 0$, the first zero in $F(f, c_0)$ corresponds to the fundamental mode, the second to the first higher mode, and so on. The values of $f$ for which $F(f, c_0) = 0$ are used as starting frequencies (the lowest frequencies) for the computation of the different modes. Once the starting frequency for each mode is defined, it is possible to compute, beginning from the fundamental mode, all dispersion relations. This is accomplished by keeping $f$ fixed and varying $c$, the procedure being applied at all of the equally spaced frequency points of the chosen frequency interval.

2. The Mode Follower

The basic idea is to define an efficient method to follow a given mode $M$ in the phase velocity–frequency space, distinguishing it from the neighboring modes $M - 1$ and $M + 1$, a problem which is most severe near the osculation points, for example, those points characterizing the transition from crustal waves to channel waves (Panza *et al.*, 1972). For frequencies as high as 1 Hz, the fundamental mode is in general well separated from the remaining modes, while for higher frequencies this is no longer true. Thus, for the construction of strong motion synthetics, the mode follower must be applied to all modes, including the fundamental. On the basis of our experience, there are no other modes present in the proximity of the near osculations between the fundamental and the first higher mode. To follow the fundamental mode it is therefore sufficient to use these properties of $\partial F/\partial c$:

(a) for a given mode $M$, the sign of $\partial F/\partial c$ is constant with frequency, and

(b) going from a mode to the next $\partial F/\partial c$ changes sign with regularity.

In other words, once the sign of $\partial F/\partial c$ is computed at the initial frequency of the fundamental mode, in all subsequent points the simple check of this sign makes it possible to follow the mode correctly. In fact, with increasing frequency, as long as the sign of $\partial F/\partial c$ does not change, the obtained zero of $F(f, c)$ belongs to the fundamental mode. If the sign of $\partial F/\partial c$ changes, the zero of $F(f, c)$ does not belong to the fundamental mode and the search of the zero restarts from a lower value of $c$. In such a way it is possible to compute all of the dispersion curve for the fundamental mode quite rapidly.

For the higher modes, the above algorithm is not sufficient, because these modes are generally much closer to each other. However, the construction of an efficient mode follower is still possible.

In fact, from the work of Tolstoy (1956) for a given higher mode even if computations are made for structures containing very strong low-velocity layers, the phase velocity decreases with increasing frequency. Thus, for each higher mode $M$, the possible value of the phase velocity at a given frequency $f$ lies in the range $(c_1, c_2)$, where $c_1$ is the phase velocity of the mode $M - 1$ at the frequency $f$, and $c_2$ is the phase velocity of the mode $M$ at the frequency $f - \Delta f$. If the computations are carried to a maximum frequency of 1 Hz, we suggest a frequency step $\Delta f = 0.005$ Hz. This condition, combined with the property of the sign of $\partial F/\partial c$, recognizes an eventual jump from mode $M$ to modes $M \pm (2n + 1)$ $(n = 0, 1, \ldots)$. If in the domain $(c_1, c_2)$ and $(f - \Delta f, f)$ $2m + 1$ $(m = 1, 2, \ldots)$ modes are contained, the procedure just outlined is not sufficient to follow the mode. However, this very seldom happens; thus, we have not derived a very efficient algorithm for this problem. Our mode follower recognizes the mode jump only when the computation of modes $M + 1$ and $M + 2$ is completed. At this point the computation can be restarted, using the root-bracketing root-refining procedure, from the mode $M$ at the frequency $f$ using as the initial phase velocity a value just slightly greater than that of the mode $M - 1$ at the same frequency. We have carried out computations for a limited sample of continental and oceanic structural models (Panza, 1985; Chiaruttini *et al.*, 1985) and can state that this version of the mode follower is satisfactory to compute with high efficiency all of the frequency domain ingredients of synthetic seismograms.

3. Structure Minimization

The structure minimization is a critical point regarding the efficiency and the accuracy of the computation of eigenvalues, eigenfunctions, and related quantities. For each mode, in order to save computer time, it is necessary to determine for each frequency the minimum amount of structure to be used in the computation while retaining very high accuracy. In general for a structure made by $n$ layers, this can be done by computing the quantity

$$E_m = \bar{\rho}_m \left[ \left( \frac{u_m^*}{w_0} \right)^2 + \left( \frac{w_m}{w_0} \right)^2 \right], \qquad m = 1, \ldots, n - 1, \tag{7}$$

where

$$\bar{\rho}_m = \tfrac{1}{2}(\rho_{m-1} + \rho_m) \quad \text{if} \quad m \geq 2,$$

and $\rho_{m-1}$ and $\rho_m$ are the densities in layers $m - 1$ and $m$; if $m = 1$, then $\bar{\rho}_1 = \rho_1$.

Since we start from the lowest frequencies consistent with a value of $c = 0.98\beta_n$, the amount of structure to be used at the beginning of each mode coincides with the total number $n$ of layers in the structural model. Once the phase velocity is determined, $E_m$ can easily be computed, and starting from $m = n - 1$, it is easy to locate its deepest minimum value, and thus to determine the maximum depth of penetration of the considered mode at the considered frequency.

At this stage all the layers below the interface $j$ corresponding to the deepest minimum value of $E_m$ can be discarded and the parameters of the $j + 1$ layer used to define the terminating half-space. With the minimized structure, it is now possible to compute with the necessary accuracy (more than 8 figures) the final value of the phase velocity. In general, repeating this procedure for each frequency and mode gives very satisfactory results.

Particular care must be placed in the structure minimization when low velocity layers are present in the structural model. Let us consider here the case of only one low-velocity channel, the extension to many velocity inversions being quite obvious. For the waves propagating essentially in the low-velocity channel, the necessary accuracy is ensured by simply placing the terminating half-space just below the zone of velocity inversion. For the waves propagating above the low-velocity channel, i.e., for the waves with a phase velocity less than the minimum S-wave velocity in the channel, only the structure above the deepest minimum of $E_m$ located above the channel needs to be retained.

The situation is completely different when dealing with waves propagating with a phase velocity larger than the minimum S-wave velocity in the channel, i.e., for waves mainly propagating above the low-velocity channel but sampling also deeper. For these waves it is generally necessary to keep at least all the channel, assigning the properties of the layer immediately below it to the half-space. It must be observed that in many cases the penetration in the low-velocity channel is so small that the structure minimization can be performed without loss of precision by removing the whole channel, with evident time saving. The identification of the waves for which this reduction is possible, can be made by evaluating $E_m$, starting at $m = 0$. If in some of the layers just above the low-velocity layer $E_j \leq 10^{-4} E_0$, the structure can be terminated at the $j$th interface, using as half-space characteristics those of the $j$th layer. From the description just given, it is clear that the initial amount of structure used for the computation at a given frequency $f$ is determined by the result of the structure minimization at the frequency $f - \Delta f$. This is obviously not valid if at the frequency $f - \Delta f$ there was a wave sampling the channel very weakly ($E_j \leq 10^{-4} E_0$). In these cases the amount of structure initially used at the frequency $f$ always contains the low-velocity layer.

B. ANELASTIC MEDIA

1. General Background

The anelastic nature of the earth's interior manifests itself through the attenuation of elastic waves. To account for attenuation effects, Knopoff (1964b) introduced an additional term into the differential equation of motion. This term contains the nondimensional constant $Q$, which is related to the space ($e^{-\alpha x}$) and time ($e^{-\gamma t}$) attenuation coefficients as

$$\alpha = \omega/2Qc, \quad \gamma = \omega/2Q, \tag{8}$$

where $c$ is the phase velocity of the plane wave motion under consideration.

Brune (1962) and Knopoff et al. (1964) noted that there are some discrepancies for $Q$ obtained from propagating wave trains $Q_x$ and from free oscillations $Q_t$. The two values are connected by the relation $uQ_t = cQ_x$ where $c$ and $u$ are the phase and group velocity, respectively.

O'Connell and Budiansky (1978) derived the relation

$$Q = \tfrac{1}{2}((\omega/\alpha c) - (\alpha c/\omega)), \tag{9}$$

which is, however, relevant only for small values of $\omega$.

Attenuation also distorts dispersion properties. Futtermann (1962) pointed out that physical dispersion must accompany wave attenuation to preserve causality. In a medium with a constant $Q$, the correction to the dispersion of body waves can be expressed as

$$A_1(\omega) = A_1(\omega_0)/[1 + (2/\pi)A_1(\omega_0)A_2(\omega_0)\ln(\omega_0/\omega)], \tag{10}$$

$$B_1(\omega) = B_1(\omega_0)/[1 + (2/\pi)B_1(\omega_0)B_2(\omega_0)\ln(\omega_0/\omega)], \tag{11}$$

where $A_1(\omega)$ is the P-wave phase velocity, $A_2(\omega_0)$ the P-wave phase attenuation, $B_1(\omega)$ the S-wave phase velocity, and $B_2(\omega_0)$ the S-wave phase attenuation.

In the following computations, we have chosen $\omega_0 = 2\pi$ rad. The quantities $A_1$, $A_2$, $B_1$, and $B_2$ are related to the complex body wave velocities $\alpha$ and $\beta$, describing the properties of anelastic media, by

$$\alpha^{-1} = A_1^{-1} - iA_2, \quad \beta^{-1} = B_1^{-1} - iB_2 \tag{12}$$

(Schwab and Knopoff, 1972). In anelastic media the surface wave phase velocity $c$ must also be expressed as a complex quantity:

$$c^{-1} = C_1^{-1} - iC_2, \tag{13}$$

where $C_1$ is the attenuated phase velocity and $C_2$ the phase attenuation.

With these generalizations, the algorithms developed for perfectly elastic media can be readily applied to anelastic media. The attenuated phase velocity and the phase attenuation can thus be determined as roots of the dispersion function $F$, which now is a complex function. For the algorithm to be efficient, good estimates of $C_1$ and $C_2$ are necessary. As an initial estimate of $C_1$ it is convenient to use the phase velocity determined in the perfectly elastic case once body-wave dispersion is introduced [Eqs. (10) and (11)], while for $C_2$ it is necessary to make use of variational techniques (see, e.g., Takeuchi and Saito, 1972; Aki and Richards, 1980). For Rayleigh waves, as an intermediate step, it is necessary to compute the integrals

$$I_3 = \int_0^\infty \left\{ \left[ (\lambda + 2\mu) - \frac{\lambda^2}{(\lambda + 2\mu)} \right] y_3^2 + \frac{1}{k} \left( y_1 y_4 - \frac{\lambda}{(\lambda + 2\mu)} y_2 y_3 \right) \right\} dz, \tag{14}$$

$$I_4 = \int_0^\infty \left\{ \delta(\lambda + 2\mu) \left[ \frac{1}{(\lambda + 2\mu)^2} (y_2^2 + 2k\lambda y_2 y_3) \right. \right.$$
$$\left. + k^2 \left( 1 + \frac{\lambda^2}{(\lambda + 2\mu)^2} \right) y_3^2 \right]$$
$$\left. + \delta\mu \frac{1}{\mu^2} y_4^2 + \delta\lambda \left[ \frac{2k}{(\lambda + 2\mu)} (y_2 y_3 + k\lambda y_3^2) \right] \right\} dz, \tag{15}$$

where

$$y_1 = \frac{w(z)}{w_0}, \quad iy_3 = \frac{u(z)}{w_0}, \quad y_2 = \frac{\sigma(z)}{w_0}, \quad iy_4 = \frac{\tau(z)}{w_0}; \tag{16}$$

$w_0$ is the vertical component of displacement at the free surface, $z$ is the downward increasing depth, and

$$\delta\mu = \rho(\beta_1^2 - \beta_2^2 - \bar{\beta}^2) + i2\rho\beta_1\beta_2, \tag{17}$$

$$\delta\lambda = \rho[(\alpha_1^2 - \alpha_2^2 - \bar{\alpha}^2) - 2(\beta_1^2 - \beta_2^2 - \bar{\beta}^2)]$$
$$+ i\rho 2(\alpha_1\alpha_2 - 2\beta_1\beta_2), \tag{18}$$

$$\delta(\lambda + 2\mu) = \rho(\alpha_1^2 - \alpha_2^2 - \bar{\alpha}^2) + i2\rho\alpha_1\alpha_2. \tag{19}$$

In these expressions $\bar{\alpha}$ and $\bar{\beta}$ are the compressional and shear-wave velocities in the perfectly elastic case; in other words

$$\rho(\beta_1 + i\beta_2)^2 = \mu + \delta\mu, \quad \rho(\alpha_1 + i\alpha_2)^2 = (\lambda + 2\mu) + \delta(\lambda + 2\mu), \tag{20}$$

with $\lambda$ and $\mu$ indicating Lamé constants.

The integrals $I_3$ and $I_4$ can be computed analytically from the layer constants (Schwab et al., 1987) yielding the phase attenuation

$$C_2 = (2\omega \bar{k} I_3)^{-1} \mathrm{Im}(I_4), \qquad (21)$$

where $\bar{k}$ is the wave number in the perfectly elastic case.

For Love waves $C_2$ is given by

$$C_2 = \frac{1}{\bar{c}} \frac{\int_0^\infty \rho \beta_1 \beta_2 \left( \frac{1}{\mu^2 k^2} y_2^2 + y_1^2 \right) dz}{\int_0^\infty \mu y_1^2 \, dz}. \qquad (21a)$$

Thus, they simply imply integration versus depth of squared displacement $y_1$ and stress $y_2$.

Having estimated the initial values for $C_1$ and $C_2$, the exact mathematical treatment of the attenuation due to anelasticity as described by Schwab and Knopoff (1971, 1972, 1973) can be applied.

2. The Mode Follower

As we have seen in the Section III.B.1 body waves are dispersed in anelastic media. The frequency dependence of body waves requires the introduction of a small but essential variation in the mode follower.

We have seen in Section II.A.2 that in the perfectly elastic case for each higher mode $M$ the possible value of the phase velocity at a given frequency lies in the range $(c_1, c_2)$. When body wave dispersion is present, the upper limit $c_2$ has to be redefined. The phase velocity of body waves, in fact, increases with increasing frequency [see Eqs. (40) and (41)] and this may cause an increase of the phase velocity of higher modes with frequency. This effect is evident in those parts of the mode curves that are almost undispersed in the perfectly elastic case.

One has therefore to estimate the increase in phase velocity of a given mode at a given frequency $f$ with respect to the frequency $f - \Delta f$ due to the dispersion of body waves. Let us denote by $\Delta c_2$ the maximum possible increment of $c_2$.

When using Eqs. (10) and (11), it is convenient to express the difference $\Delta c$ in the phase velocity between the frequencies $f - \Delta f$ and $f$ due to the effect of the body wave dispersion by

$$\Delta c = \frac{c(f)}{1 + x \ln[f_0/(f - \Delta f)]} \frac{x \ln[f/(f - \Delta f)]}{1 + x \ln(f_0/f)}, \qquad (22)$$

with

$$x = (2/\pi) c(f) C_2(f).$$

The use of Eq. (22) is not straightforward, since the value of $\Delta c$ depends upon $c(f)$ and $C_2(f)$, quantities that are obviously unknown at this stage of the computation.

In order to estimate $\Delta c$, one can substitute into $c(f)$ and $C_2(f)$ a weighted average of the S-wave velocities $B_1$ and of the S-wave phase attenuations $B_2$. As weights, we use the eigenfunctions at the frequency $f - \Delta f$, in particular the sum of the squared displacements. Therefore,

$$\Delta c_2 = \frac{\overline{B}_1(f - \Delta f)}{1 + \bar{x} \ln[f_0/(f - \Delta f)]} \frac{\bar{x} \ln[f/(f - \Delta f)]}{1 + \bar{x} \ln(f_0/f)}, \qquad (23)$$

with

$$\bar{x} = (2/\pi)\overline{B}_1(f - \Delta f)\overline{B}_2(f - \Delta f).$$

It has been found with extensive numerical testing that the above relations yield a very satisfactory definition of the upper limit $c_2$. In case the wave at the frequency $f - \Delta f$ penetrates to a much smaller depth than that at the frequency $f$, as, for instance, in the channel wave–crustal wave sequence, the weighted averages $\overline{B}_1$ and $\overline{B}_2$ are computed at the last frequency $f - N\Delta f$, where the wave reaches about the same penetration depth as that at the frequency $f$. The main advantage of this modification is in keeping the general scheme of the perfectly elastic mode follower the same.

Due to body wave dispersion, care must also be taken in computing group velocities using implicit function theory [Eqs. (5) and (6)]. When computing $\partial F/\partial \omega$, one has to remember that body wave velocities are functions of frequency. In this case Eqs. (22) and (87) of Schwab and Knopoff (1972), respectively, for Love and Rayleigh modes, contain terms associated with the derivative with respect to the angular frequency of the compressional and shear wave velocities.

The effects of body wave dispersion are not very relevant in practice, at least in the frequency band that characterizes strong motion records; however, we want to stress that the introduction of body wave dispersion in anelastic media is a physical necessity.

## C. Response to Buried Sources

### 1. Perfectly Elastic Media

A detailed description of the fault model of an earthquake used in the following computations is given by Panza, Schwab, and Knopoff (1973). Accordingly, in the reference system shown in Fig. 1, the asymptotic expression of the Fourier time transform of the $j$th Rayleigh-mode displacement at the free surface of a perfectly elastic earth model at a distance

# 4. COMPLETE STRONG MOTION SYNTHETICS

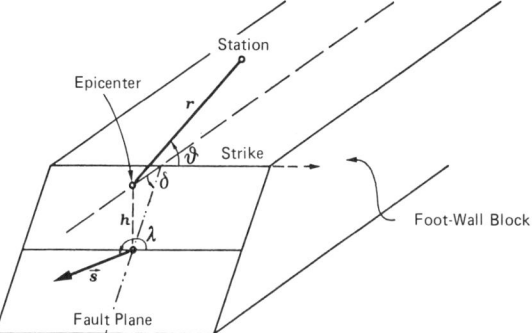

**FIG. 1.** Source geometry and coordinate system associated with free surface; $\theta$ is the angle between the strike of the fault and the epicenter–station direction, $\delta$ is the dip angle, $\lambda$ is the rake angle, and $h$ is the source depth.

$r$ from the source can be written as

$$U_r^{DC} = \{|R(\omega)|\exp(i\phi_0)\}|\mathbf{n}|k^{1/2}$$
$$\times \exp(-i3\pi/4)\chi(\theta, h)\varepsilon_0 G \exp(-ikr)/\sqrt{2\pi r}$$

and $\qquad(24)$

$$U_z^{DC} = (\varepsilon_0 \exp(i\pi/2))^{-1} U_r^{DC}, \qquad U_\theta^{DC} = 0,$$

where $R(\omega)$ is the Fourier transform of the equivalent point-force time function, the quantity $\mathbf{n}$ is the unit vector perpendicular to the fault and has units of length,

$$\phi_0 = \arg R(\omega) \tag{25}$$

is the initial phase, $k$ is the wave number, and

$$\varepsilon_0 = -u_0^*/w_0. \tag{26}$$

The factor $G$ is given by

$$G^{-1} = 2cuI_1, \tag{27}$$

where

$$I_1 = \int_0^\infty \rho(z)[y_1^2(z) + y_3^2(z)]\, dz, \tag{28}$$

and $\rho(z)$ is the density. The azimuthal dependence of the response is given by

$$\chi(\theta, h) = d_0 + i(d_1 \sin\theta + d_2 \cos\theta) + d_3 \sin 2\theta + d_4 \cos 2\theta, \tag{29}$$

where $\theta$ is the angle between the strike of the fault and the epicenter–station direction, and $h$ is the hypocentral depth. The quantities $d_i$ ($i = 0, \ldots, 4$) are

$$d_0 = \tfrac{1}{2} \sin \lambda \sin 2\delta B(h), \qquad d_3 = \cos \lambda \sin \delta A(h),$$
$$d_1 = -\sin \lambda \cos 2\delta C(h), \qquad d_4 = -\tfrac{1}{2} \sin \lambda \sin 2\delta A(h), \qquad (30)$$
$$d_2 = -\cos \lambda \cos \delta C(h),$$

where $\lambda$ is the rake angle, and $\delta$ is the dip angle. Furthermore,

$$A(h) = -u^*(h)/w_0,$$
$$B(h) = -\left(3 - 4\frac{\beta(h)^2}{\alpha(h)^2}\right)\frac{u^*(h)}{w_0} - \frac{2}{\rho(h)\alpha(h)^2}\frac{\sigma^*(h)}{\dot{w}_0/c}, \qquad (31)$$
$$C(h) = -\frac{1}{\mu(h)}\frac{\tau(h)}{\dot{w}_0/c},$$

where $\alpha(h)$ is the P-wave velocity at the source depth, $\beta(h) = [\mu(h)/\rho(h)]^{1/2}$ the S-wave velocity at the source depth, $u^*(h)$, $\sigma^*(h)$, $\tau(h)$, and $w_0$ are the eigenfunctions at the source depth, and at the free surface, respectively.

If one adopts the far-field relation given by Ben-Menahem and Harkrider (1964),

$$U_r/U_z = \varepsilon_0 e^{i\pi/2}, \qquad (32)$$

then for a wave propagating in the positive $r$ direction with retrograde elliptical particle motion, $U_r$ leads $U_z$ by $\pi/2$ radians and $\varepsilon_0$ is positive only if $z$ is chosen to increase upward. If, however, as in Haskell (1953), Panza et al. (1972), and the first part of Harkrider (1964), $z$ is chosen positive downward, $U_r$ leads $U_z$ by $3\pi/2$ radians. If Eq. (32) is used to define $\varepsilon_0$, retrograde particle motion will be defined by negative values of the ellipticity. In relation to the formalism given by Ben-Menahem and Harkrider (1964), the following observation is relevant for programming. Since the depth-dependent quantities $u^*(h)/w_0$, $\sigma^*(h)/\dot{w}_0/c)$, and $\tau(h)/\dot{w}_0/c)$ are to be computed from the usual Haskell (1953) formalism, in which $z$ is positive in the downward direction, $U_r$ must lead $U_z$ by $3\pi/2$ radians.

The asymptotic expression just described allows the computation of synthetic seismograms with at least 3 significant figures as long as $kr \geq 10$ (Panza et al., 1973) and is equivalent to the expression in terms of the seismic moment [e.g., see Eqs. (7.149) and (7.150) of Aki and Richards (1980)].

## 2. Anelastic Media

When considering anelastic media, the wave number $k$ is complex:

$$k = (\omega/C_1) - i\omega C_2; \tag{33}$$

thus the term $\exp(-ikr)$ in Eq. (24) can be written as

$$\exp(-i\omega r/C_1)\exp(-\omega C_2 r). \tag{34}$$

The term $e^{-\omega C_2 r}$ represents the amplitude damping and is the main effect introduced by anelasticity. Smaller effects, like the ones arising from complex group velocities and eigenfunctions, are not included in the present calculations. They may become important when lateral variations are treated.

The extension of these results to the available formalism for sources with finite dimensions and durations is quite straightforward. In case the source is not instantaneous but has a finite rise time, the derivative of the time source function changes from a delta function to a triangularlike function of duration $T$, with the effect of filtering out periods smaller than $T$.

Finite-length sources can be dealt with in two ways. In case the source receiver distance is much bigger (at least a factor of 10) than the source dimensions, the Ben-Menahem (1961) factor

$$[(\sin X)/X]e^{-iX}, \tag{35}$$

with

$$X = \frac{\omega L}{2}\left[\frac{1}{v} - \frac{\cos\psi}{c}\right], \tag{36}$$

may be used, as described in detail by Båth (1974). In Eq. (36) $\omega$ is the angular frequency, $L$ the source length, $v$ the rupture velocity, $c$ the phase velocity, and $\psi$ the azimuth of the station measured from the rupture direction. A second possibility is to compute the seismogram as a sum of point sources shifted in space and time. This last method has been extensively used in connection with empirical Green's functions (see, e.g., Hartzell, 1978; Kanamori, 1979).

The seismogram $S(t, \bar{r})$ at the receiver $R$ due to the extended fault $\Sigma$ can be expressed as

$$S(t, \bar{r}) = \int\int_\Sigma s(t + \tau(\bar{r}'); \bar{r} - \bar{r}')w(\bar{r}')\,d\bar{r}', \tag{37}$$

where $s$ is the response at the receiver, $R$, due to a point source at the point $\bar{r}'$ on the fault $\Sigma$, and $w$ is a weighting or scaling factor with inverse area units; $\tau$ is a time expressing the delay at the point $\bar{r}'$ due to the rupture propagation and is equal to $|\bar{r}'|/v$ for a rupture propagating with a constant velocity $v$.

Since we sum discrete point sources, the integral in Eq. (37) will be replaced by a sum

$$S(t, \bar{r}) = \sum_j s_j(t + \tau_j; \bar{r}_j) w'_j, \tag{38}$$

where $j$ runs over the point sources on the fault plane $\Sigma$ and $w'_j$ is now nondimensional. This equation is equivalent to Eq. (1) in Hartzell (1978):

$$S(t, \bar{r}) = \sum_j [s_j(t) * Q_j(t)] H(t - \tau_j), \tag{39}$$

where the operation $*$ indicates convolution, $Q_j(t)$ is a generalized scaling factor (a delta function with a given amplitude in the simplest case), and $H$ indicates the Heaviside step function.

When the successive excitation of regularly spaced point sources is used to approximate a rupture propagation, the condition that the apparent time separation between point sources, as seen at the receiver, must be smaller than the Nyquist period of the synthetic signals has to be met. In the case when just a few point sources that model irregularly spaced asperities are considered, this condition obviously does not have to be satisfied.

It is also important to note that the expressions for sources of finite dimensions are valid in the far-field approximation, which can be roughly expressed by the condition that the source–receiver distance must be an order of magnitude greater than the source dimensions. If this condition is not satisfied while the condition $kr \geq 10$ still holds, the synthetic signal can be constructed as a proper sum of seismograms given by point sources separated in time and space. With the modal approach this is easily done. For a given earth model, different seismograms corresponding to different sources can be computed with very little computer time (essentially the time required for a fast Fourier transform), since all the time consuming computations (eigenvalues and eigenfunctions) are independent of source specifications.

## III. Examples of Computation of Synthetic Signals

A. Frequency Domain

The construction of realistic strong motion records requires handling earth models formed by a large number of layers including low-velocity zones (LVZ). According to the more recent models of the crust and upper mantle, these layers correspond to sedimentary layers, the laccolithic zone of granitic intrusion (sialic low-velocity zone), granulitic layers (lower crustal layer), and the asthenospheric low-velocity layer (see, e.g., Mueller, 1977; Panza, 1980). The presence of several low velocity zones removes from the dispersion diagrams (multimode phase velocities) the regular patterns proper, for instance, of channel and crustal waves. This in turn prevents the procedure of joining points on the dispersion curves by straight line segments [see e.g., Kerry (1981)] to handle the multimode summation in an approximate way.

For the construction of complete synthetic accelerograms, using mode summation, it is convenient to consider models of the earth with elastic and anelastic properties specified to depths on the order of 100 km, which is a representative thickness of the lithosphere.

We present examples of exact computations for a continental structure containing both low-velocity layers in the crust and a sedimentary cover (see Fig. 2). This structure is meant to simulate the elastic and anelastic properties of the southern pre-Alps close to the May 6, 1976, Friuli earthquake, and is obtained from the interpretation of deep seismic sounding experiments [see, e.g., Italian Explosion Seismology Group (1981) and Angenheister et al. (1972)] and surface wave dispersion measurements [Calcagnile and Panza (1981)].

The structural properties are specified to depths of ~ 50 km, where the S-wave velocity reaches 4.65 km/sec. The handling of structural models extending to these depths in an efficient way makes it also possible to synthesize early P-wave arrivals without the necessity of introducing any unrealistic high-velocity half-space with the consequent generation of spurious S-wave arrivals as in the case of the locked mode approximation (Harvey, 1981). In fact all those body waves, including P waves, will be synthesized, for which the propagation direction—defined as the angle, with respect to the vertical axis, $\theta = \cot^{-1}\left(\dfrac{r_{\alpha m}}{r_{\beta m}}\right)$—is real.

It is important that the algorithm described in this paper does not limit computations to layered models, because it makes feasible the simulation of gradients by means of a sequence of thin layers.

1. Phase Velocities

The Rayleigh wave dispersion curves for the first 154 modes are shown in Fig. 3a. The modes are well separated for phase velocities less than ~ 3.35 km/sec, which corresponds to the S-wave velocity in the upper part of the upper crustal low-velocity channel. For higher phase velocities, the dispersion curves are closely packed together. The first curves correspond to waves that sample only the first few kilometers of the crust, while the latter

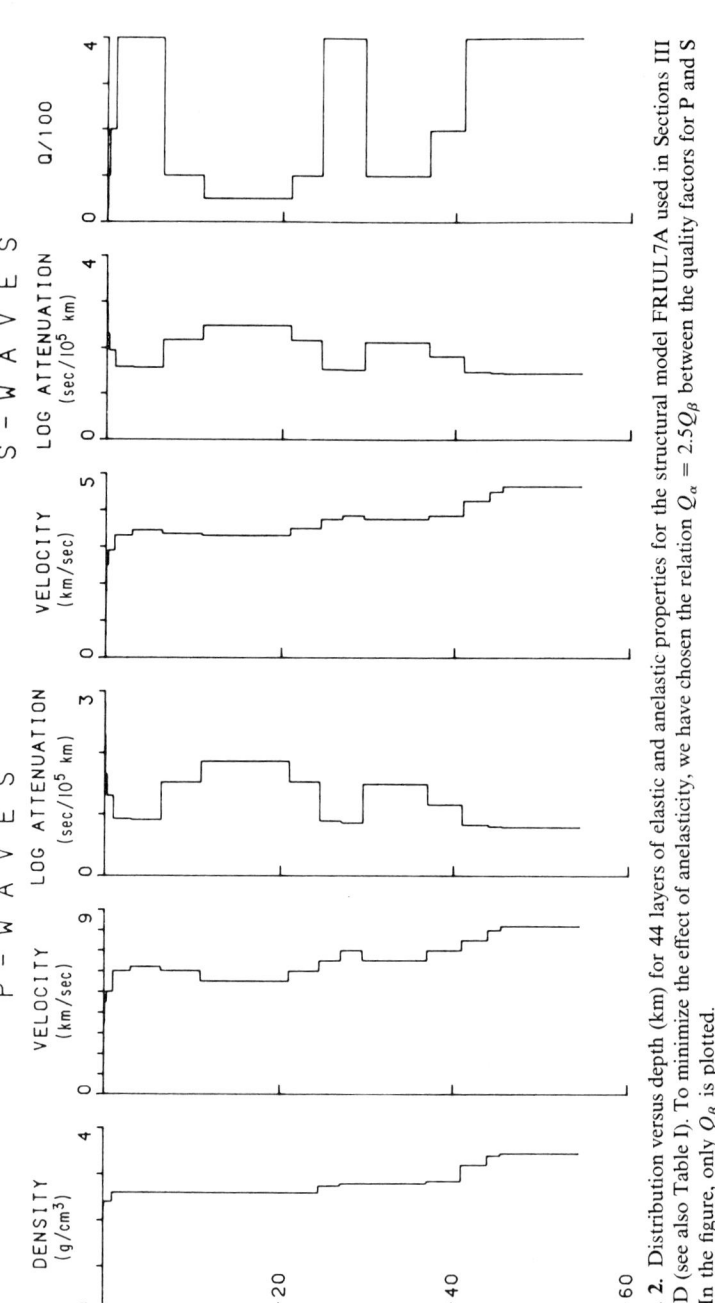

**FIG. 2.** Distribution versus depth (km) for 44 layers of elastic and anelastic properties for the structural model FRIUL7A used in Sections III and IV.D (see also Table I). To minimize the effect of anelasticity, we have chosen the relation $Q_\alpha = 2.5 Q_\beta$ between the quality factors for P and S waves. In the figure, only $Q_\beta$ is plotted.

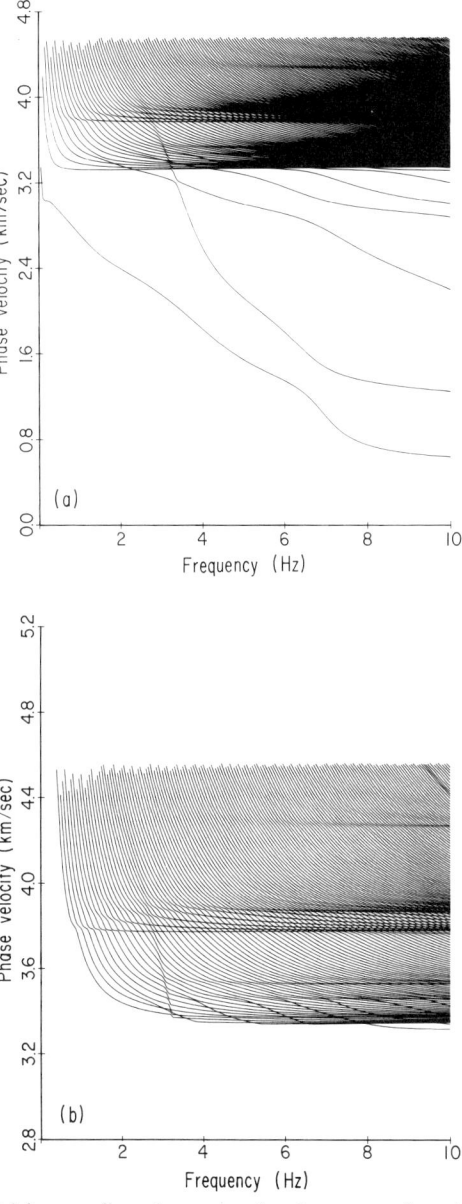

**FIG. 3.** (a) Rayleigh-wave dispersion curves for the structural model FRIUL7A. Mode numbering: 0 for the fundamental mode, 1 for the first higher mode, 2 for the second higher mode, and so on up to 154. (b) Enlarged portion (mode 6–154) of part (a) showing in detail the effect of the presence of low-velocity waveguides.

TABLE I. STRUCTURE FRIUL7A USED IN ALL THEORETICAL COMPUTATIONS OF SECTION III[a]

| Thickness (km) | Density (g/cm$^3$) | P-wave velocity (km/sec) | S-wave velocity (km/sec) | $Q_\beta$ |
|---|---|---|---|---|
| 0.04 | 2.00 | 1.5 | 0.60 | 20 |
| 0.06 | 2.30 | 3.5 | 1.80 | 30 |
| 0.20 | 2.40 | 4.5 | 2.50 | 100 |
| 0.70 | 2.40 | 5.0 | 2.90 | 200 |
| 2.00 | 2.60 | 6.0 | 3.30 | 400 |
| 3.50 | 2.60 | 6.2 | 3.45 | 400 |
| 4.50 | 2.60 | 6.0 | 3.35 | 100 |
| 10.00 | 2.60 | 5.5 | 3.30 | 50 |
| 3.50 | 2.60 | 6.0 | 3.50 | 100 |
| 2.50 | 2.75 | 6.5 | 3.75 | 400 |
| 2.50 | 2.80 | 7.0 | 3.85 | 400 |
| 7.50 | 2.80 | 6.5 | 3.75 | 100 |
| 4.00 | 2.85 | 7.0 | 3.85 | 200 |
| 3.00 | 3.20 | 7.5 | 4.25 | 400 |
| 1.50 | 3.40 | 8.0 | 4.50 | 400 |
| 9.00 | 3.45 | 8.2 | 4.65 | 400 |

[a] Used also for the Friuli 1976 aftershock modeling (Section IV.D). It has been assumed that $Q_\alpha = 2.5 Q_\beta$. See also Fig. 2.

ones correspond to waves that sample the entire structure. Essentially three "quasi-osculation" types can be recognized (see Fig. 3b). The first corresponds to horizontal "quasi-oscultations" around a constant phase velocity (for example 4.25 and 3.50 km/sec) and are related to the structural layering. The second is the standard sequence of channel and crustal waves (Panza et al., 1972) due to the presence of a low-velocity layer. The presence of two low-velocity layers is clearly recognizable in the range of phase velocities between 3.35–3.45 km/sec and 3.75–3.85 km/sec (Fig. 3b). The third is made up of very steep "quasi-osculations" and corresponds to a family of waves mainly sampling the wave guide formed by the sedimentary layers (Chiaruttini et al., 1985). Sedimentary layering begins to be barely visible in the phase velocity curves for frequencies greater than about 8 Hz. Thus, to get significant information on the physical properties of the sedimentary layering as detailed as that in Table I, it is necessary to use data with maximum frequencies much larger than 10 Hz.

From Fig. 3 the progressive reduction of the spacing between modes as the frequency increases is clearly visible. This requires a very high accuracy in the computation of phase velocities. This is not a difficult task if use is made of the algorithms previously mentioned, but it is impossible to reach the required accuracy if approximate methods are used in the computation of phase velocities (Panza, 1985).

## 2. Group Velocities

The group velocities are shown in Fig. 4. The group velocity diagram has been subdivided into three parts due to the complexity of the pattern; Fig. 4a gives the first 31 modes. Although it is difficult to follow an individual mode, it is relatively easy to follow the behavior of channel and crustal waves as well as that of the sedimentary waves. The stationary phases with group velocity between 0.3 and 1.6 km/sec, visible for frequencies greater than ~ 3 Hz, correspond to waves essentially propagating in the low-velocity sediments. For group velocities around 2.4–2.8 km/sec, stationary phases are visible at frequencies greater than 5 Hz; these phases can be associated with waves propagating near the bottom of the sediments. The stationary phases, formed by the combination of several higher modes visible in the group velocity interval of 3.0–3.3 km/sec, starting from frequencies on the order of 3 Hz, can be considered the high-frequency equivalent of Li and Lg phases (see, e.g., Panza and Calcagnile, 1974).

For frequencies larger than 1 Hz, two flat envelopes corresponding to the two-channel velocities can be easily seen at about 3.75 and 3.35 km/sec. Near 3 Hz an envelope of group velocities as low as 1.6 km/sec is clearly seen. This corresponds to the waves sampling the sedimentary layers wave guide.

## 3. Energy Integral

As for group velocities, a single plot of the energy integral $I_1$ is not suitable for interpretation. Due to the large variations of the energy integral, it is convenient to plot the log $I_1$ (Figs. 5a and b). From Fig. 5a the effect of trapping in the low-velocity layers is clearly visible; in fact, in general, large values of $I_1$ correspond to practically no motion at the free surface, while small values of $I_1$ correspond to significant surface displacement. It is quite interesting to observe that for frequencies smaller than 3 Hz the fundamental mode dominates, while around 3 Hz several higher modes are characterized by small values of $I_1$. These modes mainly sample the sediments. In the frequency range between 3–6 Hz, the first higher mode is dominant, while for larger frequencies the fundamental mode again dominates the surface displacement. The exact prediction of the surface displacement from $I_1$ is not straightforward since it depends strongly upon many factors, as can be seen from Eq. (24). In Fig. 5b the trapping in the low-velocity layers is not as dramatic as for the first higher modes. This indicates, in general, that modes with high-order number are homogeneously sampling the whole structure. A common feature of Figs. 5a and b are the very narrow peaks around 3 Hz associated with the presence of sediments. If the source is located near the sedimentary layers, one may

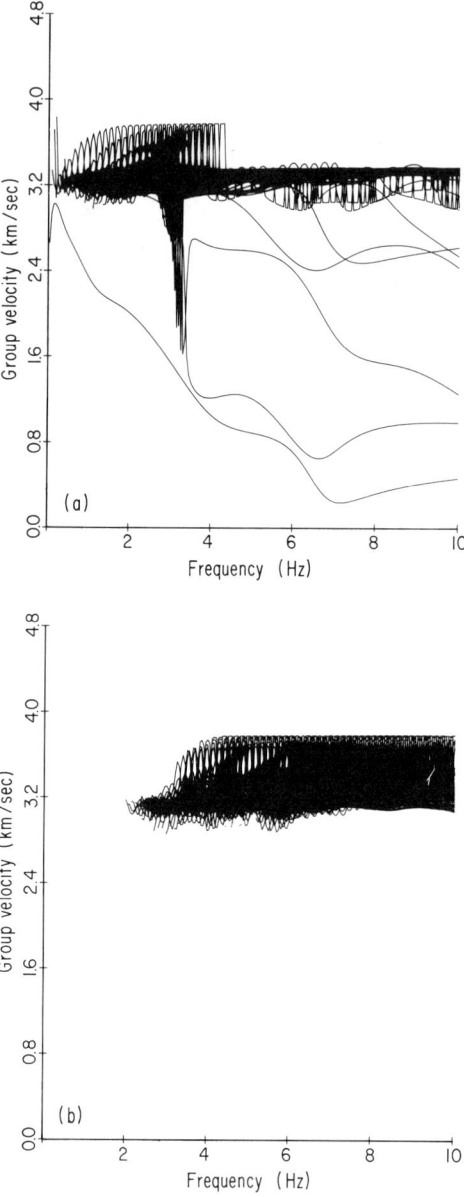

FIG. 4. (a)–(c): Rayleigh-wave group velocities for the structural model FRIUL7A; Rayleigh modes: (a) 0–30, (b) 31–90, and (c) 91–154.

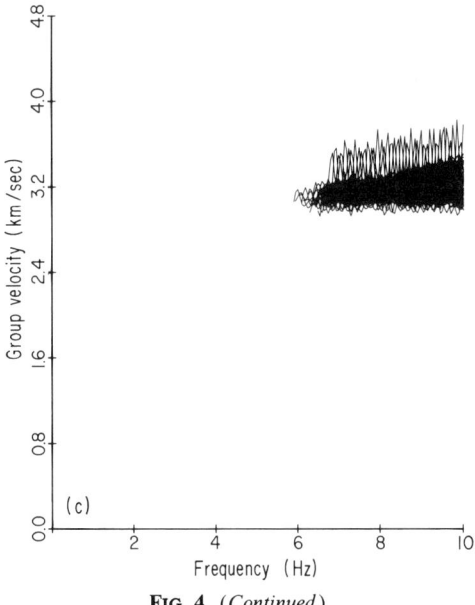

**FIG. 4.** (*Continued*)

expect significant surface motion mainly in the horizontal component even if $I_1$ is quite large. In these portions of the spectrum, the eigenfunctions are characterized by large lobes concentrated in the sedimentary layers, and the ellipticity (see Section III.A.4) becomes very large. Also, the fundamental mode is not dominant, i.e., it does not have the smallest $I_1$, over the entire spectrum due to the sedimentary layers.

4. Ellipticity

Another important quantity describing Rayleigh-mode particle motion is the ellipticity $\varepsilon_0 = -u_0^*/w_0$, i.e., the ratio between the horizontal and vertical components of motion at the free surface. It is very important to observe that $\varepsilon_0$ has abrupt discontinuities. More precisely, at some frequencies $\varepsilon_0 \to \pm \infty$ as a consequence of the fact that $w_0$ passes through zero. This is not an obvious property, and it strongly depends upon the elastic properties of the layers closest to the free surface, as was suggested by Mooney and Bolt (1966) for the first few modes of relatively simple structural models. For frequencies not exceeding 1 Hz and for earth models without sedimentary layers, these discontinuities are present only once in a given mode and only for modes with large-order number (Panza, 1985). However, when there are sediments at the top of the earth model, several

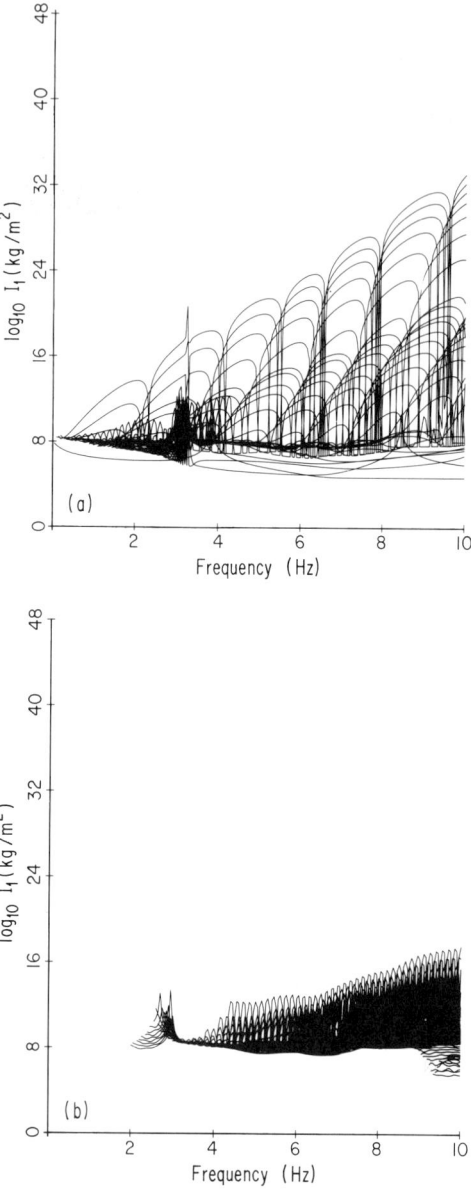

FIG. 5. (a) and (b): Rayleigh-wave energy integral $I_1$ for the structural model FRIUL7A; Rayleigh modes: (a) 0–30 and (b) 31–154.

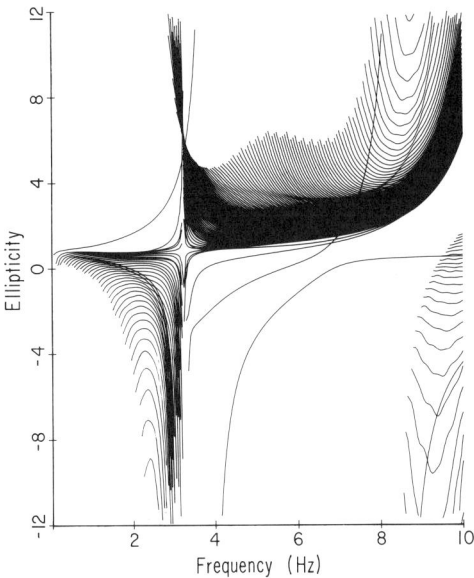

**FIG. 6.** Ellipticity, $\varepsilon_0$ for the structural model FRIUL7A; Rayleigh modes 0–154. The clearly visible discontinuities (e.g., at about 3 Hz) are due to the crossing of $w_0$ through zero.

discontinuities are also present in each of the first higher modes. For higher frequencies in the presence of sediments, as in our example (Fig. 6), even the fundamental mode has a discontinuity around 3 Hz. Thus, due to the presence of sediments, the particle motion of several modes is essentially horizontal, i.e., $\varepsilon_0 > 10$, over a quite wide frequency range. This is an extremely important observation that has several practical implications in engineering seismology. The concentration of Rayleigh motion in the horizontal direction may play on important role in the "amplification effect" introduced by sediments. Thus, sedimentary layers significantly increase the seismic hazard of a region not only due to energy trapping, but also because the layers tend to make the horizontal component of motion of Rayleigh modes dominant (Chiaruttini *et al.*, 1985).

## 5. Phase Attenuation

For large frequencies, the phase attenuation of surface waves $C_2$ can be related to the quality factor $(Q_x)$ by the relation

$$1/(Q_x) = 2C_1C_2. \tag{40}$$

Also for Rayleigh waves, the difference between the anelastic phase velocity $C_1$ and the perfectly elastic phase velocity $c$ can be either positive

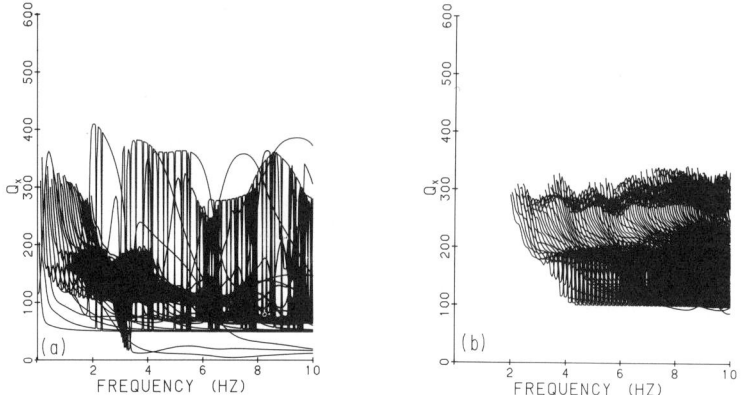

FIG. 7. The $Q_x$ curves showing the effect of the layering in $Q_\alpha$ and $Q_\beta$ for the structural model FRIUL7A; Rayleigh modes: (a) 0–30 and (b) 31–154.

or negative as has been shown by Schwab and Knopoff (1971) for the first few modes of Love waves.

In Fig. 7a, it is practically impossible to follow individual modes in their entirety, yet it is relatively easy to see the effect of the layering in $Q_\alpha$ and $Q_\beta$. For instance, the fundamental mode shows a large peak around 0.5 Hz and this corresponds to wave propagation in the crust outside the low-velocity layers where $Q_\beta = 400$. For frequencies larger than 3 Hz, several modes are characterized by very low values and this indicates wave propagation in the upper sedimentary layers where $Q_\beta$ does not exceed 100.

The trapping in the low-velocity layers, characterized by $Q_\beta = 100$ and $Q_\beta = 50$, is clearly visible for several nearby modes that have almost constant $Q_x$ close to 50, 65, and 100. The value 65 represents the weighted average by layer thickness of the values 50 and 100, both present in the upper crustal LVZ. In Fig. 7b the values of $Q_x$ in the range 180–200 correspond to sedimentary waves, while the values of $Q_x$ around 300 correspond to crustal waves propagating above the upper crustal LVZ.

In order to have interpretable measurements of $Q_x$, it is necessary to apply very accurate time or group velocity windows to the records before any further processing. In fact, an indiscriminate use of amplitude spectra may lead to reasonable $Q_x$ values, which cannot be easily related to the anelastic properties of the area under investigation.

## B. Time Domain

### 1. Point Sources

By summing the modes of oscillation of a given structure approximated with flat parallel layers and using realistic models for the seismic source, it

## 4. COMPLETE STRONG MOTION SYNTHETICS

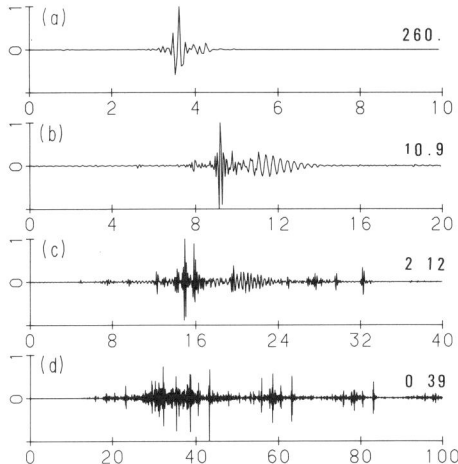

**FIG. 8.** Radial component of acceleration for an instantaneous point source ($h = 6$ km, $\theta = 68°$, $\delta = 15°$, and $\lambda = 75°$) with seismic moment $|M_0| = 1$ dyn cm as a function of the epicentral distance $r$; from top to bottom $r =$ (a) 10, (b) 30, (c) 50, and (d) 100 km. The maximum zero-to-peak amplitudes are normalized to one. In this and all subsequent figures, the number above each seismogram gives the peak acceleration (here in units of $10^{-22}$ cm/sec$^2$), and on the horizontal axis the time is given in seconds. The structure used in all theoretical examples is FRIUL7A (Fig. 2).

is easy to construct complete synthetic accelerograms. The use of Rayleigh modes models the vertical and radial components; the use of Love modes models the transverse component.

Some examples of synthetic seismograms using Eq. (24) with $R(\omega)$ equal to a unit-step function and for the model shown in Fig. 2, are given in Figs. 8 and 9 for the radial and vertical component of acceleration, respectively, for different epicentral distances. The source at the 6-km depth is a dip-slip with a small strike-slip component ($\delta = 15°$, $\lambda = 75°$). The strike angle is 68°. The amplitude of the radial component is about twice the vertical, and this is in quite good agreement with the ellipticity discussed in Section III.A.4. In Figs. 10 and 11, synthetic strong motions computed for point sources with a finite duration of 1 sec are shown; more precisely $R(\omega) = (i/\omega)S(\omega)$, where $S(\omega)$ is a symmetric triangular function and is the Fourier transform of the derivative of the source time function. A general feature common to all these synthetic signals is the large increase in duration with increasing distance mainly due to the dispersion of the fundamental and first few higher modes, which are trapped in the uppermost low-velocity layers.

The difference in frequency content between the signals generated by instantaneous and finite time rise point sources is quite striking. The

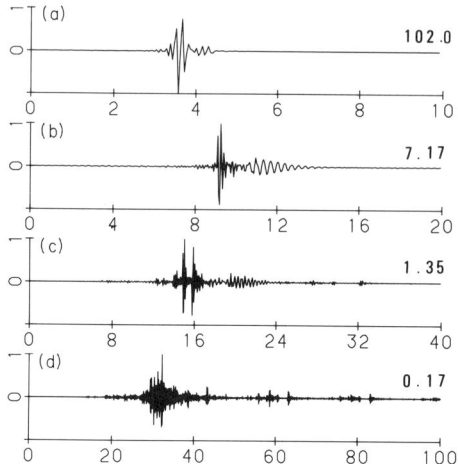

FIG. 9. Vertical component of acceleration for the cases shown in Fig. 8.

dominant waves radiated by sources with a finite time rise $t_0$ are in general characterized by lower and lower frequencies as $t_0$ increases. Thus, a large fault radiating high frequencies cannot be modeled as a continuous process in time but rather as a sequence of sources with small durations. Also, the maximum amplitudes for the finite rise time accelerograms are much smaller than the corresponding ones computed with instantaneous sources. In our examples the difference is about two orders of magnitude.

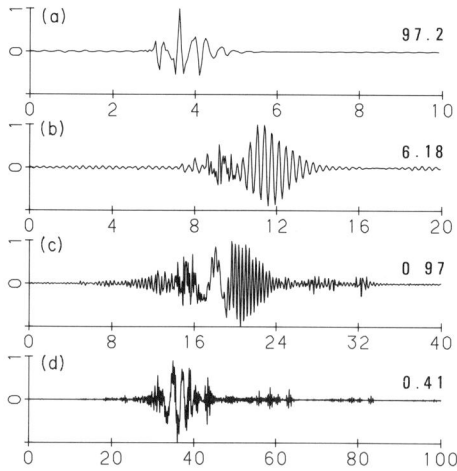

FIG. 10. The same as Fig. 8, but for a point source of 1-sec duration and $|M_0| = 0.5$ dyn cm. The peak acceleration is in units of $10^{-24}$ cm/sec$^2$.

## 4. COMPLETE STRONG MOTION SYNTHETICS 183

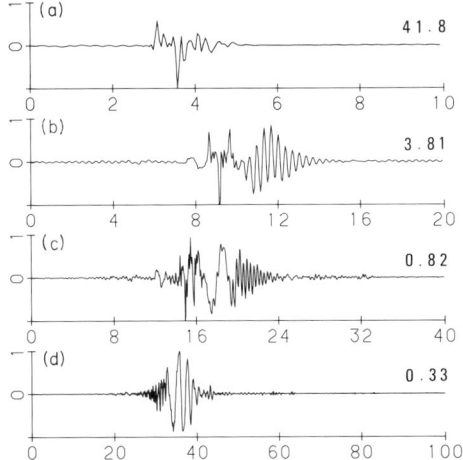

**FIG. 11.** The same as Fig. 9 but for a point source of 1-sec duration and $|M_0| = 0.5$ dyn cm. The peak acceleration is in units of $10^{-24}$ cm/sec$^2$.

Other examples of synthetic accelerograms are given by Suhadolc and Chiaruttini (1986). In their work the synthetics are used to compute theoretical peak ground accelerations. These are found to be strongly dependent upon the crustal model and the source mechanism, but wave propagation effects also play an important role. The approach of using complete synthetic signals in future developments will certainly lead to more appropriate estimates of the characteristics of strong ground shaking than will the empirical methods now used in seismic hazard studies.

2. Extended Sources

*a. Finiteness Factor.* A realistic description of a seismic source requires the capability of modeling sources with finite dimensions. For large epicentral distances $r \geq 10\ L$, where $L$ is the fault length, the use of the finiteness factor (Ben-Menahem, 1961) allows an easy extension of the results obtained for point sources to line sources.

We show some results for 10-km long ($L = 10$ km) line sources. Since the finiteness factor cannot be used for epicentral distances of 10 and 30 km, only examples of computations for epicentral distances of 100 km are given in Fig. 12 for the radial and Fig. 13 for the vertical component. The point-source parameters used in this case are the same as those given in the previous section. The accelerograms refer, from top to bottom, to a unilateral along-strike rupture, to a unilateral antistrike rupture, and to a bilateral symmetric rupture. In the last case, the fracture starts at the time

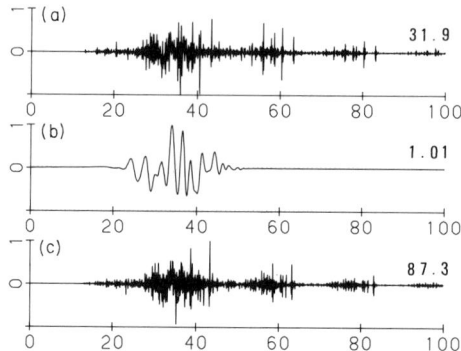

**FIG. 12.** Radial component of acceleration for a line source. The fracture, 10 km long and 6 km deep, propagates with a velocity of 2.76 km/sec, i.e., 80% of the S-wave velocity of the layer containing the source. The three seismograms refer to (a) a unilateral along-strike, (b) a unilateral antistrike, and (c) a symmetric bilateral rupture. The epicentral distance is 100 km. Point-source parameters are the same as in Fig. 8. The peak acceleration is in units of $10^{-24}$ cm/sec$^2$.

$t = 0$ in the center of the 10-km long line source and propagates with the same constant velocity (2.76 km/sec) for about 1.8 sec both toward the receiver and away from it. It will be shown in Section IV.C how the method we use to generate synthetic signals turns out to be particularly suitable to handle the problem of generating strong motions for finite sources in space when the finiteness factor cannot be applied.

The difference between the records of Fig. 12 with the corresponding records shown in Figs. 8 and 9 is quite evident. The frequency content is very different, especially when we consider the antistrike rupture and the

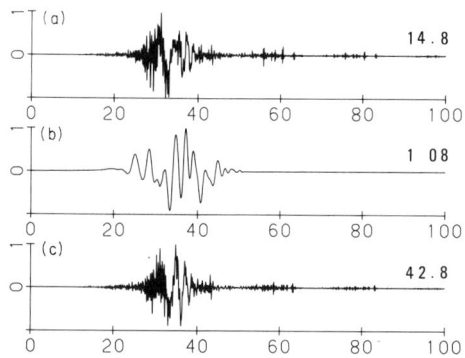

**FIG. 13.** Vertical component of acceleration for a line source. The parameterization is the same as in Fig. 12.

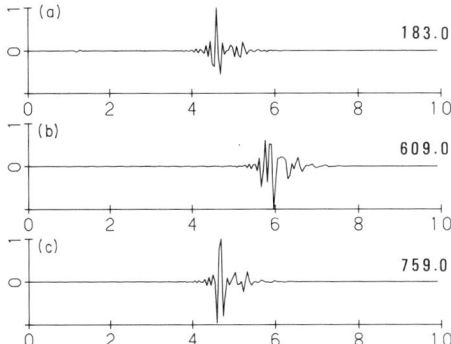

FIG. 14. Radial component of acceleration at an epicentral distance of 13 km: (a) instantaneous single point source, (b) unilateral along-strike line source, and (c) symmetric bilateral line source. The source mechanism of the single point source and of the instantaneous point sources, with $|M_0| = 1$ dyn cm, modeling the line source is a thrust: rake 75°, dip 23.5°, and strike 78°. The single point source depth is 7.5 km. The line source, oriented in a N–S direction, is 7.5 km long, dips 23.5° toward N, and lies between 6 and 9 km deep. The receiver is placed at an angle of 192° from N. The rupture velocity is 2.76 km/sec. The peak acceleration is in units of $10^{-22}$ cm/sec$^2$.

maximum acceleration changes. The bilateral rupture gives rise to accelerations that are about 3 times larger than the unilateral rupture propagating toward the receiver. These examples clearly show the strong dependence of the strong ground motion on directivity.

*b. Point Sources Summation.* For distances on the order of the fault length, the finiteness factor cannot be applied. A fault of finite length can be modeled in this case as a series of point sources on a defined grid placed along the fault line with an appropriate spacing $\Delta s$. In order not to artificially enhance certain frequencies, $\Delta s$ must be such that the apparent time separation between individual sources, as seen at the receiver, is smaller than the Nyquist period. Also, $\Delta s$ depends on the rupture velocity, since the time interval between the "ruptures" of two nearby points has to be smaller or equal to the time sampling interval of the seismograms. Another "equivalent" condition is that the smallest wavelength considered has to be sampled by at least three grid points. The same considerations apply if the grid is two- or three-dimensional.

Examples of computations for a one-dimensional grid with a spacing $\Delta s = 150$ m approximating a 7.5-km long rupture with a dip of 23.5° and the grid center 7.5-km deep are given in Fig. 14; the epicentral distance from the center of the rupture is 13 km and the angle between the rupture line and the line joining the center of the rupture and the observer is 22°.

The two examples refer to the radial component and unilateral along-strike and symmetric bilateral rupture. An accelerogram due to an instantaneous point source located at the center of the fault is also shown for comparison at the top of the figure. To simulate a smooth stopping of the rupture, the sources have been summed by applying a Parzen taper.

The rupture velocity is 80% of the shear wave velocity at the source. In this case the differences in frequency content are not striking; however apart from the obvious difference in the first arrival time the overall wave shape differs considerably in the two cases.

Finally, let us consider a two-dimensional vertical fault plane. The fault is 3 km long and 1 km wide, is between 5 and 6 km deep, and has been modeled with a rectangular grid with $\Delta s = 100$ m. This geometry is taken from Spudich and Frazer (1984), and the observer is 70° off strike at an

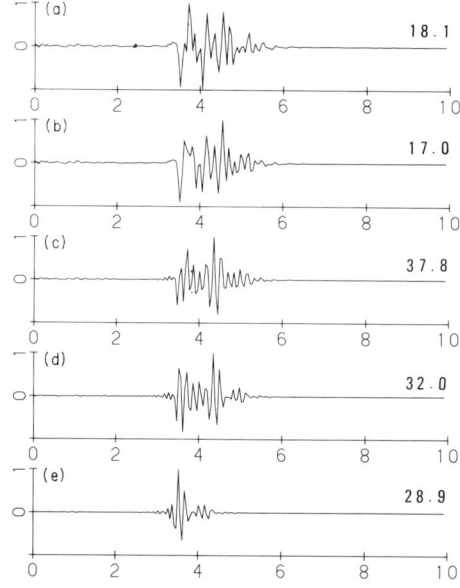

**FIG. 15.** Examples of vertical component accelerations due to the two-dimensional fault of Fig. 16 modeled by a grid of instantaneous point sources with $|M_0| = 1$ dyn cm. The receiver is placed 70° off strike at an epicentral distance of 10 km. (a) All grid points summed with equal weight. (b) All grid points summed with equal weight, cosine tapering applied at the edges. (c) Nonequal weights applied: nonunitary weights are shown in Fig. 16, cosine tapered at the edges. (d) Only the 10 sources with nonunitary weights are summed. (e) Only the point source signal corresponding to the hypocenter marked by a star in Fig. 16 is shown. The peak acceleration is in units of $10^{-23}$ cm/sec$^2$.

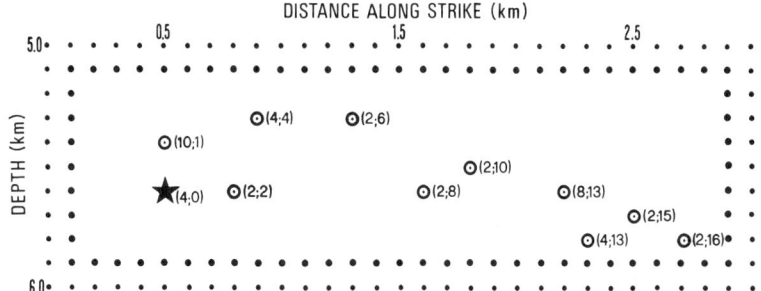

FIG. 16. Grid modeling of a two-dimensional source. This example is adjusted after Spudich and Frazer (1984). Only the edge lines and some interior points are shown. The small dot points are given a weight of 0.25, and the big dot points are given a weight of 0.50 when cosine tapering is applied. The star represents the point of rupture initiation—the hypocenter; the open circles are points with nonunitary weights. The numbers in parenthesis near the interior points are, respectively, the weight, and the rupture time in units of the time series sampling interval (0.04883 sec). The rupture spreads out uniformly at a velocity of $0.9\beta$.

epicentral distance of 10 km and the hypocenter is placed 5.6 km deep, at 0.5 km from one end of the fault.

Several examples have been considered, all having a constant rupture velocity, $v_r = 0.9\beta$, the rupture spreading circularly from the hypocenter. In Fig. 15a all grid points have been summed with an equal weight. In Fig. 15b, where the accelerogram is characterized by a relatively long-period initial swing, all grid points have been summed with an equal weight, but cosine tapering has been applied to the points at the edges. In the third example (Fig. 15c), not all the central points have been given equal weights in order to model asperities. The locations and weights of the asperities, adjusted after Spudich and Frazer (1984), are shown in Fig. 16. In this case the initial long-period swing disappears, and the accelerogram is seen to contain more high frequencies. The result shown in Fig. 15d has been obtained by summing just the 10 seismograms relative to the asperities of Fig. 16, without considering the other grid points simulating the background fault rupture. As expected, this example is very similar to the previous one, but there are some differences, especially in the amplitudes of the initial peaks. A reference accelerogram for an instantaneous point source located at the hypocenter is shown in Fig. 15e.

In conclusion, the illustrated summation method is also well suited to investigate volume sources that have been largely neglected. Also, the computer time used to generate the first two-dimensional uniform rupture was about 200 min on our IBM 370/168 computer, which is equivalent to about 20 min on a CRAY-XMP. All subsequent two-dimensional examples took only 200 sec on the IBM machine.

## IV. Comparison with Real Data

To show how suitable our method for the summation of Rayleigh-wave modes is in modeling observational data, we present in this chapter some examples.

### A. Borrego Mountain, California, 1968 Event

First, we try to synthetize the recordings of the Carder displacement meter of the El Centro (ELC) station for the Borrego Mountain, California, ($M = 6.4$), earthquake of April 9, 1968, as given by Heaton and Helmberger (1977). An attempt to model this earthquake in terms of addition of surface wave modes has been done by Swanger and Boore (1978). These two authors were able to reproduce the transversal component of motion quite well, but failed to obtain a good fit for the radial and vertical component.

The initial crustal structure BORN (Table II), used in the computations (Suhadolc and Panza, 1985), is taken from the structure used by Swanger and Boore (1978). The mantle portion is taken from the models proposed by Biswas and Knopoff (1974) for the western United States.

In the initial step of fitting the radial component, we held the structure fixed and varied some of the source parameters. At the beginning we used a single point source model, while later a two-point model was shown to provide a better fit.

1. Single Point Source

The radial and vertical components of the observed displacements are shown in Figs. 17a and b, respectively. The source parameters are given by

TABLE II. Structure BORN Used in Borrego Mountain Earthquake Modeling[a]

| Thickness (km) | Density (g/cm$^3$) | P-wave velocity (km/sec) | S-wave velocity (km/sec) | $Q_\beta$ |
|---|---|---|---|---|
| 0.25 | 2.0 | 1.7 | 1.0 | 20 |
| 0.30 | 2.2 | 2.1 | 1.2 | 30 |
| 1.35 | 2.2 | 2.4 | 1.4 | 40 |
| 0.95 | 2.4 | 3.3 | 1.9 | 100 |
| 1.65 | 2.5 | 4.3 | 2.5 | 200 |

[a] Only the uppermost layers are shown. For the remaining structure see Suhadolc and Panza (1985). It has been assumed that $Q_\alpha = 2.5 Q_\beta$.

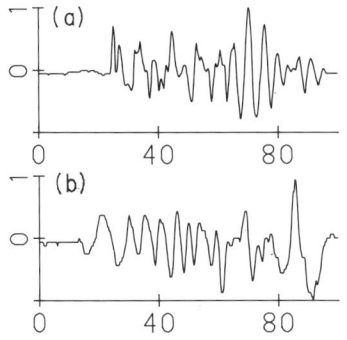

FIG. 17. Observed ground motion due to the 1968 Borrego Mountain, California, earthquake recorded at El Centro. (a) Radial component, maximum zero-to-peak amplitude about 7.3 cm. (b) Vertical component, maximum zero-to-peak amplitude about 3.1 cm. (After Swanger and Boore, 1978.)

Burdick and Mellman (1976). These authors modeled the teleseismic P pulse by adding the contribution of three point sources occurring in the time span of about 15 sec. We used the parameters of the first of these three events ($\lambda = 178°$, $\delta = 81°$), which are also in good accord with the values proposed by Allen and Nordquist (1972). Initially, a Heaviside step function was used for the source time function.

The synthetic seismogram, sum of 218 modes, obtained with these parameters is shown in Fig. 18a; its peak half-amplitude is $6.1 \times 10^{-26}$ cm, thus to obtain the observed maximum displacement of $\sim 3$ cm, we need a seismic moment of $\sim 12 \times 10^{25}$ dyn cm.

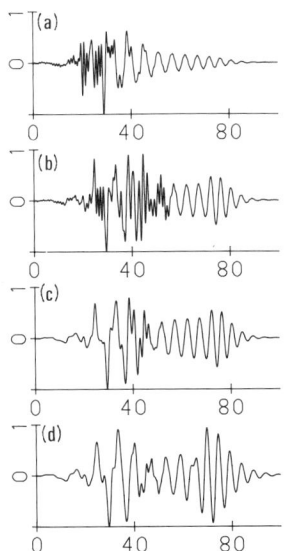

FIG. 18. Single-point-source radial component synthetic seismograms referring to the 1968 Borrego Mountain earthquake: parts (a), (b), and (c) refer to the structure BORN (see Table II and text), while part (d) refers to the structure BORY (see Table III). The parameters of the sources are described in the text. The amplitude scale factors (in units of $10^{-26}$ cm) are, from top to bottom, 6.1, 8.1, 5.6, and 4.2.

It is evident that the amplitude of the observed coda is much larger than the synthetic coda amplitude. This problem may be removed by a variation in the depth of the source, which greatly affects the relative amplitudes between the early and later parts of the recording. The synthetics obtained with a source depth of 4.5 km (and leaving the other parameters the same) is shown in Fig. 18b. The improvement is quite evident. The larger excitation of Rayleigh waves in the sedimentary surficial layers, which is probably responsible for the high-amplitude late arrivals, is accomplished by a shallower source. The high frequencies present in the early arrivals of the synthetic seismogram of Fig. 18b can be removed by using a finite rise time source function. We adopted a symmetric triangular function for the derivative of the source time history. The resulting synthetic for a source duration of 2 sec is shown in Fig. 18c.

The principal difference seen between the synthetic and the observed recording is still in the amplitudes. The amplitudes of the later arrivals dominate in the observed recording, while in the synthetic, the early arrivals are the bigger ones.

To reduce further the ratio between the early and later arrivals, we tried to increase the $Q$ factor in the sediments and to diminish it in the crust, obtaining a very small improvement.

Thus second-order variations, like the ones in $Q$, are not sufficient to improve the fit significantly. Therefore some structural parameters, describing the elastic properties, have to be changed. In particular we note that the frequency of the fundamental mode, which produces the large amplitude later arrivals, is lower in the observed seismogram than in the synthetic one. The frequency can be changed by increasing the thicknesses of the uppermost layers. After some adjustments in the corresponding velocities in order to maintain the correct arrival times, the structural model, BORY, shown in Table III, was found to give satisfactory results. The synthetic displacement

TABLE III. STRUCTURE BORY USED IN BORREGO MOUNTAIN EARTHQUAKE MODELING[a]

| Thickness (km) | Density (g/cm$^3$) | P-wave velocity (km/sec) | S-wave velocity (km/sec) | $Q_\beta$ |
|---|---|---|---|---|
| 0.50 | 2.0 | 1.70 | 1.00 | 150 |
| 0.25 | 2.2 | 2.30 | 1.35 | 150 |
| 0.25 | 2.2 | 2.45 | 1.45 | 150 |
| 1.60 | 2.2 | 2.70 | 1.60 | 150 |
| 0.55 | 2.4 | 3.30 | 1.90 | 80 |
| 1.35 | 2.5 | 4.30 | 2.50 | 80 |

[a] Only the uppermost layers are shown. For the remaining structure see Suhadolc and Panza (1985). It has been assumed that $Q_\alpha = 2.5 Q_\beta$.

corresponding to this structure for a source duration of 3 sec, is given in Fig. 18d. The fit on the later arrivals is now quite good. The corresponding seismic moment is about $17 \times 10^{25}$ dyn cm, in accord with the values given by Swanger and Boore (1978) and Heaton and Helmberger (1977). We note incidentally that an idea of the resolving power, with respect to structural parameters, connected with the use of complete signals may be deduced from the differences between the models BORN and BORY.

The discrepancies that persist between the observed and synthetic data are (1) too large early arrivals, (2) the initial double peak, and (3) the last arrival, which has not been modeled. These problems can be resolved by considering more than one point source.

2. Two-Point Source

An easy way to reduce the amplitudes of the first arrivals is to consider more than one point source. Heaton and Helmberger (1977), Ebel and Helmberger (1982), and Burdick and Mellman (1976) admit that the event cannot be modeled with a single point source. In order to preserve the fit of the later arrivals and change only the early part of the synthetic shown in Fig. 18d, a second deeper point source, having the same mechanism as the first one, is considered. Since the early arrivals show more high frequency content, a source 8 km deep with a time duration of 1 sec was chosen. The seismogram is shown in Fig. 19a. The effect of the sum of the two point sources, the deeper one being 6 sec later with respect to the shallower one (Fig. 19b), is shown in Fig. 19c. The weights of the two point sources were chosen to be identical, thus the seismic moment of the earlier source can be estimated around $2.7 \times 10^{26}$ dyn cm, while that of the later one around $0.9 \times 10^{26}$ dyn cm.

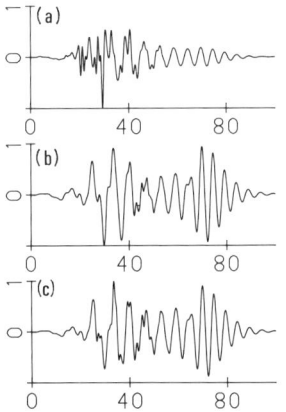

FIG. 19. Superposition of two point sources: radial component seismograms. (a) 8-km deep point source, duration 1 sec. (b) 4.5-km deep point source, duration 3 sec. (c) Superposition of the two previous point sources. The amplitude scale factors (in units of $10^{-26}$ cm) are from top to bottom, 6.6, 4.2, and 3.6.

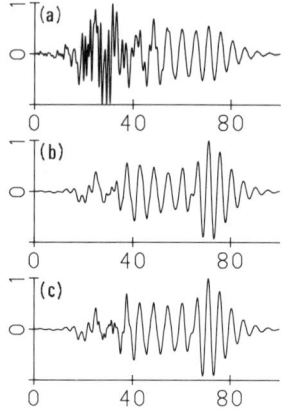

FIG. 20. Vertical component for the same cases shown in Fig. 19. Scale factors are (a) $2 \times 10^{-26}$ cm, (b) $3.9 \times 10^{-26}$ cm, and (c) $3.1 \times 10^{-26}$ cm.

Other features could be modeled as well by assuming more point sources not necessarily having the same mechanism and located at different hypocentres, but the problem will not be dealt with in this paper, since it is not our aim to do a particular study of the Borrego Mountain event.

Let us now consider the vertical component: the observed is shown in Fig. 17b, and the synthetic, corresponding to the two-point source model, is shown in Fig. 20. The overall fit is not bad, but the late arrivals of the observed recording (Fig. 17b) show sharp phase changes and long periods and amplitudes, features missing in the synthetics of Fig. 20. These are very probably due to the interference of more point sources and to the effect of lateral variations. Further attempts to model this component seem useless, since the size of the first part of the signal is just above the noise level, and thus any later phase arising from lateral variations can become a dominant feature in the record.

## B. Brawley, California, 1976 Event

Strong motion displacements due to the November 4, 1976, Brawley, California, earthquake (magnitude 4.9), recorded at the ELC and IVC stations, as given by Heaton and Helmberger (1978), are modeled in this section. The initial part of the tangential components of the recordings were modeled by Heaton and Helmberger (1978) using the Cagniard–de Hoop technique. In this section we will concentrate on the radial components (Fig. 21).

The recorded horizontal ground motion is much larger than the vertical one, and the radial displacement at ELC is greater by about a factor of two than at IVC, which lies near a P–SV node. The radial components at the

## 4. COMPLETE STRONG MOTION SYNTHETICS

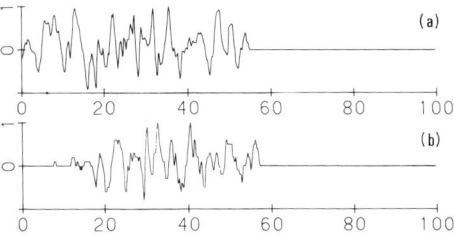

FIG. 21. Observed ground motions, from radial component seismograms, due to the 1976 Brawley, California, earthquake. (a) El Centro (ELC) recording, maximum zero-to-peak amplitude about 1.4 mm. (b) Imperial Valley College (IVC) recording, maximum zero-to-peak amplitude about 0.7 mm.

two stations show a high degree of overall coherence, only the relative amplitudes of the single peaks are different. This is to be expected, since, from the USGS determined epicenter, the stations have similar azimuths (160° and 174° for IVC and ELC, respectively) and distances (33 km and 36 km for IVC and ELC, respectively) from the source. The hypocentral depth, according to the USGS, is 4.5 km. The mechanism determined from the local stations' first arrivals is a right-lateral strike-slip on a plane dipping almost vertically.

The five uppermost low-velocity layers used in our structural model BRAW (Table IV) are those given by Biehler et al. (1964), while other crust and upper mantle specifications are practically the same as those used in the Borrego Mountain earthquake modeling.

Several hypocentral depths, around the USGS estimate, have been tried; a relatively good match has been obtained with a single right lateral strike-slip point source on a vertical fault plane, 3.5 km deep and with a source duration of 1 sec (Fig. 22a).

TABLE IV. STRUCTURE BRAW USED IN BRAWLEY EARTHQUAKE MODELING[a]

| Thickness (km) | Density (g/cm$^3$) | P-wave velocity (km/sec) | S-wave velocity (km/sec) | $Q_\beta$ |
|---|---|---|---|---|
| 0.45 | 2.0 | 1.7 | 0.75 | 20 |
| 0.50 | 2.2 | 2.1 | 0.92 | 20 |
| 1.15 | 2.3 | 2.6 | 1.50 | 30 |
| 1.30 | 2.4 | 3.7 | 2.13 | 100 |
| 2.50 | 2.6 | 4.7 | 2.71 | 200 |

[a] Only the uppermost layers are shown, the remaining structure being practically identical to that in BORY. It has been assumed that $Q_\alpha = 2.5 Q_\beta$.

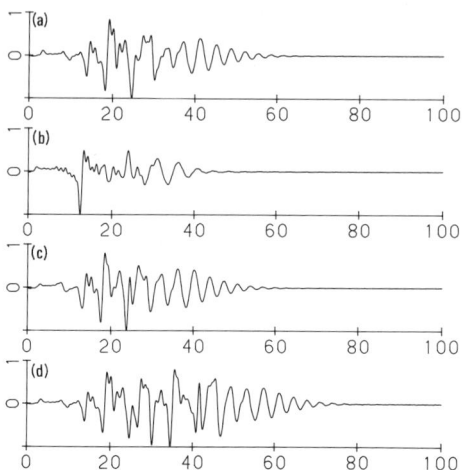

FIG. 22. Synthetic seismograms referring to the 1976 Brawley event and to the ELC station: (a) 3.5-km deep point source, epicentral distance of 38 km, source duration 1 sec, $\lambda = 180°$, $\delta = 90°$, amplitude scale factor $3.4 \times 10^{-25}$ cm; (b) 7-km deep point source, epicentral distance of 36 km, source duration 1 sec, $\lambda = 180°$, $\delta = 80°$, amplitude scale factor $2 \times 10^{-25}$ cm; (c) 3.5-km deep point source, epicentral distance of 38 km, source duration 1 sec, $\lambda = 180°$, $\delta = 80°$, amplitude scale factor $4.2 \times 10^{-25}$ cm; and (d) superposition of seismograms (a), (b), and (c), amplitude scale factor $2.1 \times 10^{-25}$ cm. All amplitude scale factors refer to $|M_0| = 1$ dyn cm. See text for details.

However, the duration of the synthetic signal is still too short. Since a certain degree of coherence may be observed between the initial and the final part of the seismogram, another shallow source (Fig. 22c) has been added 17 sec after the first source. In order to match the central portion, a third point source (7 km deep and 14.3 sec after the first source) has also been added (Fig. 22b). The three sources and their sum—all have been given equal weight—are shown in Fig. 22d for ELC and Fig. 23d for IVC. In order to preserve the relative amplitudes of the initial and final part of the seismogram at both stations, the fault dip of the second shallow source and that of the deep one have been set to 80°. The deep event is placed at the USGS epicenter, while the shallow ones are placed about 3 km to the northwest, along the fault line. This gave a better fit on the initial part of the record than records (see Figs. 24 and 25) obtained by summing the three point sources—3.5 km, 7 km, and 3.5 km deep—with the USGS epicenter and focal mechanism.

It is interesting to note that the location of the epicenter 3 km to the northwest is fairly well determined to lie along the fault passing through the USGS epicenter (33° 05′ N, 115° 36′ W) by the ratio of the amplitudes at

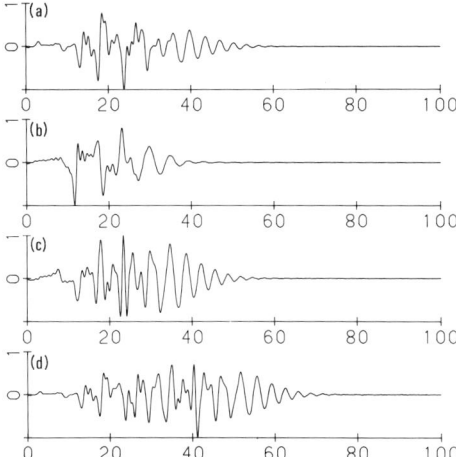

**FIG. 23.** Same as Fig. 22 for the Imperial Valley College (IVC) station. The amplitude scale factors (in units of $10^{-25}$ cm) are (a) 1.5, (b) 1.2, (c) 1.7, and (d) 1.1.

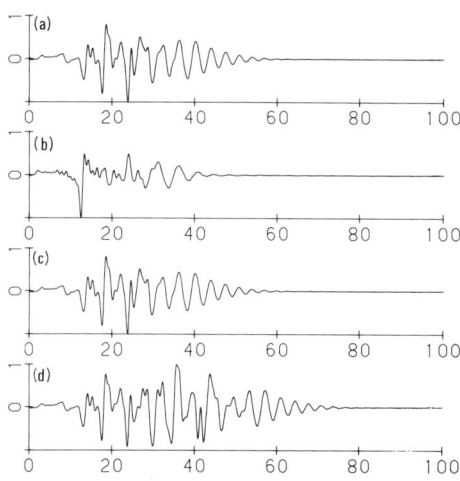

**FIG. 24.** Same as Fig. 22, the epicentral distance of all three single point sources now being 36 km. The amplitude scale factors (in units of $10^{-25}$ cm) are (a) 4.2, (b) 2, (c) 4.2, and (d) 4.4.

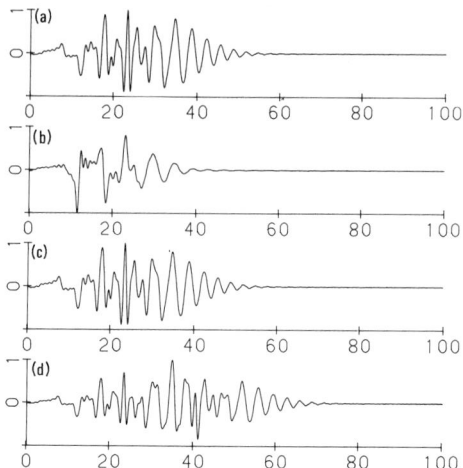

**FIG. 25.** Same as Fig. 24 for the IVC station. The amplitude scale factor (in units of $10^{-25}$ cm) are (a) 1.7, (b) 1.2, (c) 1.7, and (d) 1.3.

the two stations. This ratio was found to be maximized for a vertical fault and for shallow sources.

In the first part of the seismogram, the fit of the IVC record is a little bit worse than the ELC one, but the overall features are still reproduced fairly well. Also, in this case the small amplitude of the signal can make the effect of lateral variations relevant, which may be responsible for the main observed discrepancies. The seismic moment of each point source is found to be about $3 \times 10^{23}$ dyn cm, the same value proposed by Heaton and Helmberger (1978).

### C. Irpinia, Italy, 1980 Event

The earthquake source behavior is relatively well understood by waveform matching of synthetic and observed ground motions for frequencies up to 1 Hz, with some examples shown in the previous sections. At higher frequencies, smaller scale details of the earthquake source process and the structure surrounding the source volume become essential for a deterministic prediction of the strong ground motion. Since these details are not known at present, a statistical approach has been taken up to now to predict ground motion above 1 Hz (see, e.g., Boore and Joyner, 1978; Boatwright, 1982; Koyama, 1985).

In the following we present a first attempt to model deterministically some of the Ente Nazionale Energie Altenative–Ente Nazionale Energia

## 4. COMPLETE STRONG MOTION SYNTHETICS

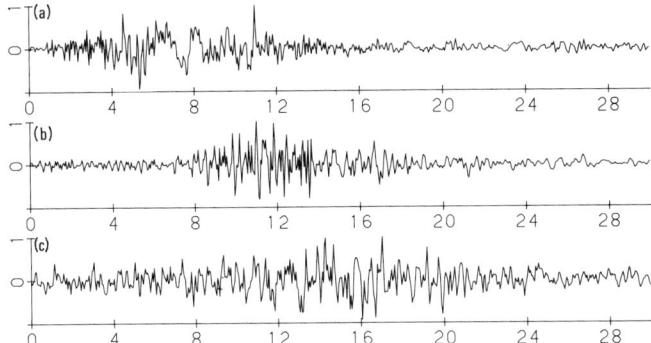

FIG. 26. Observed accelerations, after Gaussian filtering with a cutoff frequency at 10 Hz, due to the 1980 Irpinia, Italy, earthquake. (a) Sturno station, maximum peak ground acceleration about 132 cm/sec². (b) Brienza station, maximum peak ground acceleration about 83 cm/sec². (c) Auletta station, maximum peak ground acceleration about 25 cm/sec². For station locations see Fig. 27. The zero of the time axis does not coincide with the earthquake origin time.

Elettrica (ENEA–ENEL) strong ground motion recordings (Berardi et al., 1981) of the 1980 Irpinia, Italy, earthquake ($M_s = 6.9$) [see Deschamps and King (1983) and Del Pezzo et al. (1983) for a review of the source parameters of this event].

By comparing the durations of the observed ground motions (Fig. 26) with those of the synthetic examples (Fig. 9), computed at comparable distances—few tens of kilometers, one can immediately see that in the frequency domain under consideration a single point source is not a realistic representation of the source process. On the basis of our experience most of the observed signal can be reproduced by a superposition of point sources that model asperities. When the location and rupture time of these asperities is determined by rough waveform fitting, it is then possible to model a more detailed finite dimension fault rupture along the lines outlined in Section III.B.2.b. The lithospheric model IRPI used in the following computations is given in Table V.

In order to obtain a gross agreement of both the waveforms and relative amplitudes at different stations, it was found that a 17.5-km deep source located at A (see Fig. 27) with a rake of 230° on a fault dipping 70° toward the NE had to be assumed. A source duration of 0.6 sec reproduces quite well the frequency content of most of the signal, provided the shock is repeated after 0.3 sec. To model the longer duration of the observed recordings, more point sources were added some seconds later. A clear example of the need of another point source is represented by the large acceleration peak in the Sturno recording arriving at about 11 sec (Fig. 26).

**TABLE V.** STRUCTURE IRPI USED IN IRPINIA EARTHQUAKE MODELING[a]

| Thickness (km) | Density (g/cm³) | P-wave velocity (km/sec) | S-wave velocity (km/sec) | $Q_\beta$ |
|---|---|---|---|---|
| 0.05 | 2.3 | 1.55 | 0.90 | 20 |
| 0.20 | 2.3 | 1.90 | 1.10 | 20 |
| 0.25 | 2.3 | 2.25 | 1.30 | 30 |
| 0.25 | 2.3 | 2.60 | 1.50 | 30 |
| 0.25 | 2.3 | 2.77 | 1.60 | 50 |
| 1.00 | 2.3 | 3.00 | 1.70 | 100 |
| 8.00 | 2.8 | 5.60 | 3.20 | 400 |
| 5.00 | 2.8 | 5.70 | 3.30 | 400 |
| 5.00 | 2.8 | 4.80 | 3.10 | 400 |
| 17.00 | 2.9 | 6.80 | 3.90 | 400 |
| 47.50 | 3.4 | 8.10 | 4.65 | 400 |

[a] It has been assumed that $Q_\alpha = 2.5 Q_\beta$.

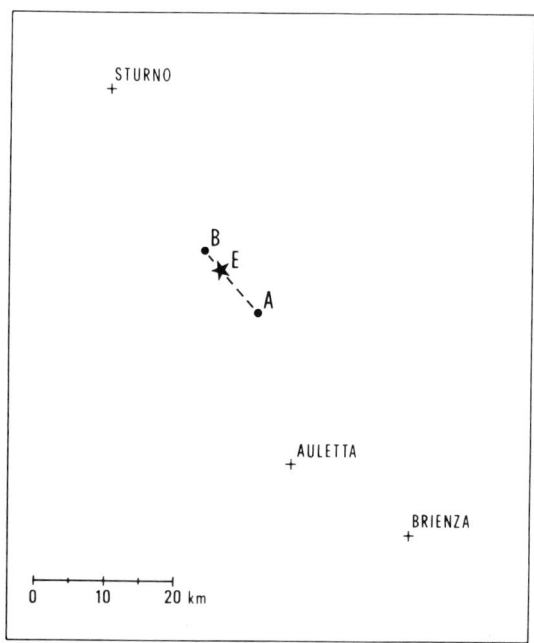

**FIG. 27.** Map showing the positions of the epicenter E (40°46'N, 15°18'E) and the point sources A and B, which were used in the construction of the synthetic signals shown in Fig. 28 and the considered strong motion stations. The coordinates are Auletta (40°33'37"N 15°23'30"E), Brienza (40°28'27"N 15°38'06"E), and Sturno (41°01'21"N 15°07'02"E).

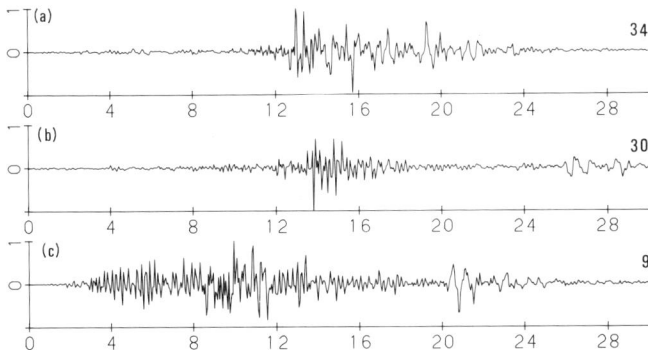

FIG. 28. Synthetic accelerations: (a) Sturno, (b) Brienza, and (c) Auletta stations. The peak accelerations, in units of $10^{-25}$ cm/sec$^2$, correspond to $|M_0| = 0.3$ dyn cm.

This peak can be modeled by assuming a point source with the same focal mechanism but located at B (see Fig. 27) with origin time shifted by 9.4 sec.

The synthetic accelerograms obtained in this way are presented in Fig. 28 for the three stations Sturno, Brienza, and Auletta. The amplitudes and the overall appearance of these signals match quite well the observed ones and allow an estimate of the seismic moment of the first two sources to be $\sim 3 \times 10^{25}$ dyn cm, while the moment of the third source is $\sim 5 \times 10^{25}$ dyn cm.

These values are obviously smaller than the one obtained with body waves by Deschamps and King (1983) and are one order of magnitude smaller than those obtained with long period surface waves [see, e.g., Kanamori and Given (1982)]. This shows that the average seismic moment should be estimated from teleseismic waves, for which the point source approximation is valid. The high frequency data on the other end give information on local fault heterogeneities.

A closer waveform matching depends on a better understanding of the high frequency behavior of earthquake rupturing. For instance, the start of the big amplitudes in the theoretical records, relatively abrupt when compared with the observed recordings, can be made more gradual by considering the rupture nucleation (modeled with increasingly stronger point sources delayed in time) to occur before the main energy release.

Furthermore this high frequencies form matching implies the structural parameters to be known on the scale of the involved wavelengths. This requires the development of codes for the treatment of detailed two- and three-dimensional laterally heterogeneous structures, presently in progress.

## D. Friuli, Italy, 1976 Aftershock

On September 11, 1976, at 16.35, an $M_L = 5.6$ aftershock of the May 6, 1976, event occurred in Friuli. The fault-plane solution of this event is not well constrained (Lyon-Caen, 1980), and it is only possible to infer that it is consistent with the one of the main shock: a thrust event on a very shallow NW dipping plane, as confirmed in a recent study by Slejko and Renner (1984). The depth of the September event according to these authors is around 3.7 km.

This earthquake has been recorded by various accelerograph stations (CNEN-ENEL, 1977). The vertical component recorded at the station of Buia has been chosen in order to make a first analysis of the possible source

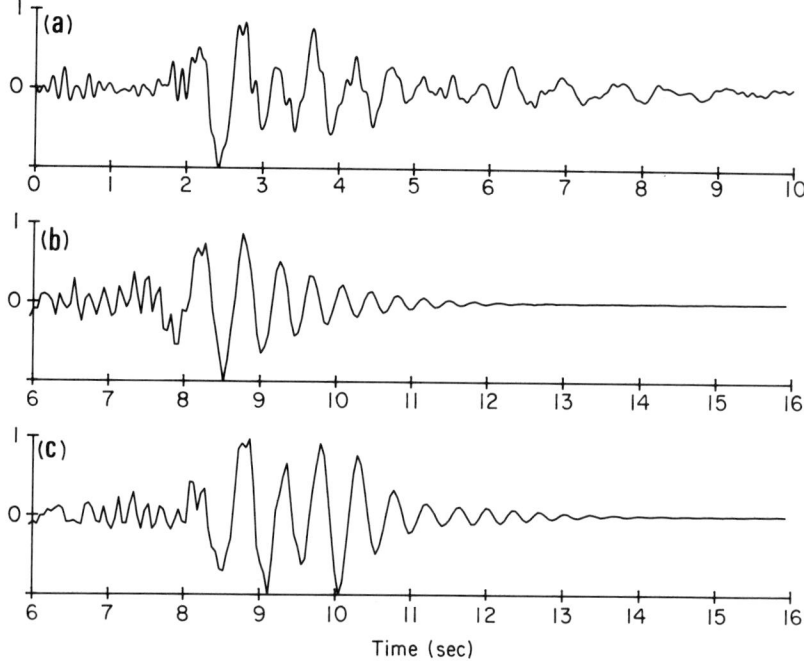

**FIG. 29.** (a) Vertical component accelerogram, after Gaussian filtering with a cutoff frequency of 10 Hz, for the September 11, 1976, Friuli $M_L = 5.6$ aftershock. The zero of the time axis does not coincide with the earthquake origin time, the amplitude scale factor is 91 cm/sec². (b) Synthetic accelerogram due to a 0.5-km deep point source, epicentral distance 17 km, source duration 0.5 sec, $\delta = 24°$, $\lambda = 75°$, amplitude scale factor $5.4 \times 10^{-22}$ cm/sec². (c) Synthetic accelerogram for three point sources (see text for more details), amplitude scale factor $5.8 \times 10^{-22}$ cm/sec². For the synthetic signals the zero of the time axis coincides with the earthquake origin time and $|M_0| = 1$ dyn cm.

solutions responsible for this accelerogram trace. This strong motion record, Gaussian filtered with a cut off frequency of 10 Hz is shown in Fig. 29a.

Several tries have been made starting from the source parameters given in the literature to model the observed waveform. The structure used in the computation is the model FRIUL7A (Table I). Good results have been obtained only when considering a very shallow source, as can be seen from Fig. 29b, where the synthetic signal due to a point source located at a depth of 0.5 km at a distance of 17 km and with the strike, dip, and rake being, respectively, 270°, 24°, and 75°, is shown. The agreement between experimental and theoretical data is greatly enhanced if more than one point source is considered. The result of summing three point sources all having the same focal depth and mechanism as the single point source of Fig. 29b, but with different weights and time shifts (0.4, 1.0, 0.8, and 0 sec, 0.6 sec, 1.5 sec, respectively) is shown in Fig. 29c. The seismic moment of the three sources turns out to be about $0.5 \times 10^{23}$ dyn cm, $1.3 \times 10^{23}$ dyn cm, and $1 \times 10^{23}$ dyn cm, respectively.

The conclusions which can be drawn from this modeling are the following. The accelerogram seems to be either due to a very shallow complex source or the rupture(s) did initiate at greater depths but propagated almost to the surface. The second hypothesis seems to be more sound, if we think of the relatively big magnitude—5.6—of this event. At the surface the rupture might have generated several stopping phases, giving rise, when propagating in the sedimentary layers, to the dominant part of the signal.

The implication we can draw for engineering strong motion analyses is that in the case of stations located on sediments the dominant part of the motion seems to be due to waves in sedimentary layers excited by either very shallow asperities or stopping phases.

## V. Conclusions

The computation of eigenvalues and eigenfunctions of Rayleigh waves for flat layered anelastic models of the earth allows one to construct, with satisfactory efficiency, "complete" synthetic strong motions up to frequencies of 10 Hz. Routinely, it is possible to consider earth models made up of 70 layers or more, and thus it is feasible to model through fine layering any sort of gradient in the distribution of elastic and anelastic properties versus depth.

Therefore, computer programs based on the addition of surface wave modes are quite versatile and can be used to model observed recordings in a realistic way provided the maximum frequency does not exceed ~ 1 Hz.

Nevertheless, our first attempts of waveform fitting at higher frequencies (~ 10 Hz) show that it is possible to unravel the main features of the rupture process. A full exploitation of the algorithms we have developed requires the capability of treating lateral heterogeneities. Once this is achieved it will be possible to attack the basic problem of understanding the high-frequency radiation associated with earthquake rupture.

#### Acknowledgments

We are indebted to Professor B. Mitchell and Dr. F. Schwab for several useful discussions. Special thanks go to Mr. Franco Vaccari, who supplied the synthetic signals of Fig. 28, which were part of his thesis. The help of Mrs. I. Galante for the patient and accurate typing of the manuscript and that of Mr. G. Cavicchi, Mr. M. Gergolet, and Mr. S. Zidarich in elaborating the figures, most of them produced at the Computer Center of the University of Trieste, is gratefully acknowledged. This research was supported by Ente Nazionale Energie Alternative (ENEA) (contract n. 16094), Consiglio Nazionale delle Ricerche (CNR) (contribution n. 85.00933.05), and Ministero Pubblica Istruzione (MPI) (40% and 60%) funds.

#### References

Aki, K., and Richards, P. G. (1980). "Quantitative Seismology." Freeman, San Francisco, California.
Allen, C. R., and Nordquist, J. M. (1972). *Geol. Surv. Prof. Pap. (U.S.)* No. 787, 16–23.
Angenheister, G., Bogel, H., Gebrande, P., Giese, P., Schmidt-Thome, P., and Zeil, W. (1972). *Geol. Runds.* **61**, 349–395.
Båth, M. (1974). "Spectral Analysis in Geophysics." Elsevier, Amsterdam.
Ben-Menahem, A. (1961). *Bull. Seismol. Soc. Am.* **51**, 401–435.
Ben-Menahem, A., and Harkrider, D. G. (1964). *J. Geophys. Res.* **69**, 2605–2620.
Berardi, R., Berenzi, A., and Capozza, F. (1981). *Tech. Rep. CNEN-ENEL, Rome.*
Biehler, S., Kovach, R. L., and Allen, C. R. (1964). *In* "Marine Geology of Gulf of California" (T. Van Andel and G. Shor, eds.), *Mem. Am. Assoc. Pet. Geol.* **3**, 126–196.
Biswas, N. N., and Knopoff, L. (1974). *Geophys. J. R. Astron. Soc.* **36**, 515–539.
Boatwright, J. (1982). *Bull. Seismol. Soc. Am.* **72**, 1049–1068.
Boore, D. M., and Joyner, W. B. (1978). *Bull. Seismol. Soc. Am.* **68**, 283–300.
Brune, J. N. (1962). *Bull. Seismol. Soc. Am.* **52**, 109–112.
Burdick, L. J., and Mellman, G. R. (1976). *Bull. Seismol. Soc. Am.* **66**, 1485–1499.
Burridge, R., and Knopoff, L. (1964). *Bull. Seismol. Soc. Am.* **54**, 1875–1888.
Calcagnile, G., and Panza, G. F. (1981). *PAGEOPH* **119**, 865–879.
Červený, V. (1985). *J. Geophys.* **58**, 44–72.
Chapman, C. H. (1985). *J. Geophys.* **58**, 27–43.
Chiaruttini, C., Costa, G., and Panza, G. F. (1985). *J. Geophys.* **58**, 189–196.
CNEN-ENEL (1977). Uncorrected accelerograms. Accelerograms from the Friuli, Italy, earthquake of May 6, 1976 and aftershocks: part 3. Rome, Italy, November 1977.
Del Pezzo, E., Iannaccone, G., Martini, M., and Scarpa, R. (1983). *Bull. Seismol. Soc. Am.* **73**, 187–200.
Deschamps, A., and King, G. C. P. (1983). *Earth Planet. Sci. Lett.* **62**, 296–304.

Dunkin, J. W. (1965). *Bull. Seismol. Soc. Am.* **55**, 335-358.
Ebel, J. E., and Helmberger, D. V. (1982). *Bull. Seismol. Soc. Am.* **72**, 413-437.
Futtermann, W. I. (1962). *J. Geophys. Res.* **67**, 5279-5291.
Hanks, T. C. (1982). *Bull. Seismol. Soc. Am.* **72**, 1867-1879.
Harkrider, D. G. (1964). *Bull. Seismol. Soc. Am.* **54**, 627-679.
Hartzell, S. H. (1978). *Geophys. Res. Lett.* **5**, 1-4.
Harvey, D. J. (1981). *Geophys. J. R. Astron. Soc.* **66**, 37-69.
Haskell, N. A. (1953). *Bull. Seismol. Soc. Am.* **43**, 17-34.
Heaton, T. H., and Helmberger, D. V. (1977). *Bull. Seismol. Soc. Am.* **67**, 315-330.
Heaton, T. H., and Helmberger, D. V. (1978). *Bull. Seismol. Soc. Am.* **68**, 31-48.
Incorporated Research Institutions for Seismology (1984). "Science Plan for a New Global Seismographic Network." California Inst. Technol., Pasadena.
Italian Explosion Seismology Group (1981). *Boll. Geof. Teor. Appl.* **23**, 297-330.
Jennings, P. C. (1983). *In* "Earthquakes: Observation, Theory and Interpretation" (H. Kanamori and E. Boschi, eds.), pp. 138-173. North-Holland Publ., Amsterdam.
Kanamori, H. (1979). *Bull. Seismol. Soc. Am.* **69**, 1645-1670.
Kanamori, H., and Given, J. W. (1982). *Phys. Earth Plan. Int.* **30**, 260-268.
Kerry, N. J. (1981). *Geophys. J. R. Astron. Soc.* **64**, 425-446.
Knopoff, L. (1964a). *Bull. Seismol. Soc. Am.* **54**, 431-438.
Knopoff, L. (1964b). *Rev. Geophys.* **2**, 625-660.
Knopoff, L., Aki, K., Archambeau, C. B., Ben-Menahem, A., and Hudson, J. A. (1964). *J. Geophys. Res.* **69**, 1655-1657.
Koyama, J. (1985). *Tectonophysics* **118**, 227-242.
Liao, A., Schwab, F., and Mantovani, E. (1978). *Bull. Seismol. Soc. Am.* **68**, 317-324.
Lyon-Caen, H. (1980). Ph.D. thesis, Univ. of Paris VII.
Maruyama, T. (1963). *Bull. Earthquake Res. Inst.* **41**, 467-486.
Mooney, H. M., and Bolt, B. A. (1966). *Bull. Seismol. Soc. Am.* **56**, 43-67.
Mueller, S. (1977). *In* "The Earth's Crust" (J. Heacock, ed.), pp. 289-317. Monogr. No. 20. Am. Geophys. Union, Washington, D.C.
O'Connell, R. J., and Budiansky, B. (1978). *Geophys. Res. Lett.* **5**, 5-8.
Panza, G. F. (1980). *In* "Mechanisms of Continental Drift and Plate Tectonics" (P. A. Davies and S. K. Runcorn, eds.), pp. 75-87. Academic Press, New York.
Panza, G. F. (1985). *J. Geophys.* **58**, 125-145.
Panza, G. F., and Calcagnile, G. (1974). *Geophys. J. R. Astron. Soc.* **40**, 475-487.
Panza, G. F., Schwab, F., and Knopoff, L. (1972). *Geophys. J. R. Astron. Soc.* **30**, 273-280.
Panza, G. F., Schwab, F., and Knopoff, L. (1973). *Geophys. J. R. Astron. Soc.* **34**, 265-278.
Papageorgiou, A. S., and Aki, K. (1983). *Bull. Seismol. Soc. Am.* **73**, 693-722.
Pestel, E. C., and Leckie, F. A. (1963). "Matrix Methods in Elastomechanics." McGraw-Hill, New York.
Romanowicz, B., Fels, J. F., and Karczewski, J. F. (1984). "Project Geoscope." I. P. G. and I. N. A. G. Paris, I. P. G., Strasbourg.
Schwab, F. (1970). *Bull. Seismol. Soc. Am.* **60**, 1491-1520.
Schwab, F., and Knopoff, L. (1971). *Bull. Seismol. Soc. Am.* **61**, 893-912.
Schwab, F., and Knopoff, L. (1972). *Methods Comput. Phys.* **11**, 86-180.
Schwab, F., and Knopoff, L. (1973). *Bull. Seismol. Soc. Am.* **63**, 1107-1117.
Schwab, F., Nakanishi, K., Cuscito, M., Panza, G. F., Liang, G., and Frez, J. (1984). *Bull. Seismol. Soc. Am.* **74**, 1555-1578.
Schwab, F., Cuscito, M., Panza, G. F., and Nakanishi, K. (1987). In preparation.
Slejko, D., and Renner, G. (1984). *In* "Finalità ed Esperienze della Rete Sismometrica del Friuli-Venezia Giulia," pp. 75-91. Regione Autonoma Friuli-Venezia Giulia, Trieste.

Spudich, P., and Frazer, L. N. (1984). *Bull. Seismol. Soc. Am.* **74**, 2061–2082.
Suhadolc, P., and Chiaruttini, C. (1986). *Proc. NATO Adv. Study Inst. Strong Ground Motion Seismol.*, Ankara (in press).
Suhadolc, P., and Panza, G. F. (1985). *J. Geophys.* **58**, 183–188.
Swanger, H. J., and Boore, D. M. (1978). *Bull. Seismol. Soc. Am.* **68**, 907–922.
Takeuchi, H., and Saito, M. (1972). *Methods Comput. Phys.* **11**, 217–295.
Thomson, W. T. (1950). *J. Appl. Phys.* **21**, 89–93.
Thrower, E. N. (1965). *J. Sound Vib.* **2**, 210–226.
Tolstoy, I. (1956). *J. Acoust. Soc. Am.* **28**, 1182–1192.
Watson, T. H. (1970). *Bull. Seismol. Soc. Am.* **60**, 161–166.

CHAPTER 5

# Techniques for Earthquake Ground-Motion Calculation with Applications to Source Parameterization of Finite Faults

Paul Spudich

*Office of Earthquakes, Volcanoes and Engineering*
*United States Geological Survey*
*Menlo Park, California 94025*

Ralph J. Archuleta

*Department of Geological Sciences*
*University of California, Santa Barbara*
*Santa Barbara, California 93106*

## I. Introduction

The term "forward modeling" of earthquake ground motions refers to the calculation of the ground motions that would result from a given earthquake occurring in a specified geologic environment. Such forward modeling can be used to predict ground motions for engineering design purposes and to determine earthquake source parameters by iterative modeling of observed ground motions. In this chapter we examine how ground motions can be calculated at locations near to large earthquakes, in which case the source cannot be approximated as a point in space or time. We will also show how these forward modeling techniques allow one to determine the evolution of slip on a fault in a manner consistent with observed ground-motion data. Our emphasis will be upon calculation techniques that have been used in geologic structures that vary in at least one dimension. Consequently, we will not dwell on examples of forward modeling of ground motions in uniform whole space or half-space velocity structures, except when they illustrate a principle applicable to more general velocity structures. We will concentrate on methods that are more rapid computationally than finite-element or finite-difference solutions to the problem. In addition we will emphasize techniques that model earthquakes as slip on fault surfaces, although volume sources will also be considered. We will present derivations of many of the relevant equations, although these derivations are intended only as tutorial sketches in which the most important concepts are emphasized and the details suppressed.

## A. Classification of Methods

For the purpose of later discussion, it will be helpful to classify some of the methods used to calculate ground motions caused by extended seismic sources. The main classification attribute will be whether point-source Green's functions are used and how they are used. Aside from illustrating the concepts behind the methods, this classification can be significant from a computational standpoint, depending upon the relative computational effort required to specify a source model and to propagate the resulting waves through the surrounding medium. Methods that use Green's functions that have been explicitly calculated will be called explicit Green's function integration methods. Methods called implicit Green's function integration methods will be those in which the Green's functions appear in the mathematics but are never actually calculated. Methods that do not use Green's functions at all will be called non-Green's function methods. The main difference between these methods is that for non-Green's function methods the action of the extended seismic source is inextricably linked to the propagation of the resulting waves to the observer. Every time the ground motions from a new extended source are desired, the entire calculation must be repeated. Examples of such work are the finite-element modeling of Archuleta and Frazier (1978) and the discrete wave number technique of Bouchon (1979). Explicit Green's function integration methods lie at the opposite end of the spectrum. In them, the wave propagation problem is solved once for a set of point sources, and these Green's functions are reused for every new extended source model. In implicit Green's function integration methods, the Green's functions appear in the mathematics, but there is no need ever to calculate them explicitly because of their simplicity.

## B. Seismic Representation Theorems

A representation theorem is a mathematical statement that relates an observable quantity, such as ground motion, to the parameters of an idealized model of a seismic source. The most commonly studied seismic sources are tectonic earthquakes, volcanic earthquakes, and explosions. Often earthquakes are idealized as dislocations on planar fault surfaces, and explosions are idealized as centers of compression, but these idealizations are simplifications of the actual situation. Explosions can result in the creation of permanent cavities and can be accompanied by nonlinear material behavior and shear deformation resulting from the relaxation of local tectonic stress. Earthquakes are similarly complex upon detailed examination of the source region. From mapping of surface rupture, geologic investigation of exhumed fault zones (Sibson, 1986), and exposures of source zones in mines (Gay and Ortlepp, 1974), we know that earthquakes can have regions in which the local rock deforms inelastically during the earthquake (White, 1973). They may have local volume changes (Gilbert and Dziewonski, 1975) possibly due to phase transitions or thermal expansion. The source zone generally has many internal surfaces across which slip occurs (Tchalenko and Ambraseys, 1970). These surfaces may or may not intersect each other. They may display roughness (i.e., deviation from a plane surface) on all spatial dimensions (Tchalenko, 1970; King, 1983). Dislocation on these surfaces may occur as cavitation and injection of fluids such as water vapor or magma (Julian, 1983). Frictional melting may take place on these surfaces during the earthquake (Sibson, 1980).

1. Volume Sources

Because actual earthquake and explosion sources show complexities that may not be properly represented by the simple idealizations mentioned earlier, we will start by discussing representation theorems valid for seismic sources occupying a volume. Such representations have already been used to calculate ground motions from an explosion (see, e.g., Bache *et al.*, 1982) but have not yet been applied to earthquakes.

Volume sources can be dealt with in two ways. The first of these uses the concepts of stress-free strain (Robinson, 1951) or stress glut (Backus and Mulcahy, 1976a). Backus and Mulcahy give a representation theorem valid for a source distributed in a volume $V$. The $k$th component of displacement at observation position $\mathbf{y}$ and time $t$ is

$$u_k(\mathbf{y}, t) = \int_{-\infty}^{\infty} dt' \iiint_V \Gamma_{pq}(\mathbf{x}, t') \, \partial_{x_q} G_{kp}(\mathbf{y}, t - t'; \mathbf{x}, 0) \, dV, \quad (1)$$

where $\mathbf{x}$ is a location within $V$, $\Gamma$ is the stress glut characterizing the seismic source, and $G_{kp}(\mathbf{y}, t - t'; \mathbf{x}, 0)$ is the $k$th component of displacement at position $\mathbf{y}$ and time $t - t'$ caused by an instantaneous force of unit impulse applied in the $p$ direction at position $\mathbf{x}$ and time $t = 0$ (i.e., the usual point-force Green's function), where the summation convention applies over repeated indices and where $\partial_{x_q} = \partial/\partial x_q$. The stress glut is related to the stress-free strain tensor $\mathbf{e}^F$ by

$$\Gamma = \mathbf{C} : \mathbf{e}^F, \qquad (2)$$

where $\mathbf{C}$ is the tensor of elastic constants.

To see how this representation could be used with a complicated source model, consider an initially quiescent "reference" earth model having known material properties and state of stress everywhere. For simplicity let us also assume that the stress–strain relation in our reference model is given by simple time-independent linear elasticity and that there is zero prestress. The Green's functions and elastic tensor in Eqs. (1) and (2) are those appropriate for the reference earth model. We characterize our source model by specifying the strain and stress fields $\mathbf{e}(\mathbf{x}, t)$ and $\sigma(\mathbf{x}, t)$ everywhere within the source volume $V$. Note that $\mathbf{e}$ cannot equal $\mathbf{C} : \sigma$; if it did, linear elasticity would prevail everywhere within the source volume, and no waves at all would be generated. Seismic waves are generated where the true stresses $\sigma$ in our source model differ from the stresses $\mathbf{C} : \mathbf{e}$ that would be expected in our reference earth model. This difference is the stress glut:

$$\Gamma(\mathbf{x}, t) = \sigma(\mathbf{x}, t) - \mathbf{C}(\mathbf{x}) : \mathbf{e}(\mathbf{x}, t). \qquad (3)$$

Because linear elasticity is usually assumed to prevail in the reference earth model, the region where stress glut is nonzero is identical to the region in which inelastic deformation occurs. Hence, regions where inelastic deformation occurs can be regarded as sources of seismic waves.

An arbitrarily complicated seismic source can be described by specifying its associated stress and strain fields $\sigma(\mathbf{x}, t)$ and $\mathbf{e}(\mathbf{x}, t)$. For realistic seismic sources specification of $\sigma$ and $\mathbf{e}$ can be quite difficult, since the action of a seismic source is governed by numerous laws of physics whose solution can be a formidable mathematical or computational problem. We shall use the term "dynamic" source models to refer to those models in which the appropriate laws of physics have been solved to obtain physically self-consistent $\sigma$ and $\mathbf{e}$. Models in which $\sigma$ and $\mathbf{e}$ are chosen that do not satisfy the appropriate physical laws will be called "kinematic" models, since the motion of the source is specified. From a historical standpoint, the distinction between kinematic and dynamic sources is not very important for volume earthquake sources, because almost no kinematic volume sources have been used for earthquakes thus far. This distinction is much more important for surface sources, such as slip on infinitesimally thin fault surfaces.

A second representation theorem has been used by Bache *et al.* (1982) for a volume source. In their method the source is replaced by the displacements and tractions it causes on an arbitrary surface $\Sigma$ enclosing the source (Fig. 1). These displacements and tractions can be calculated on

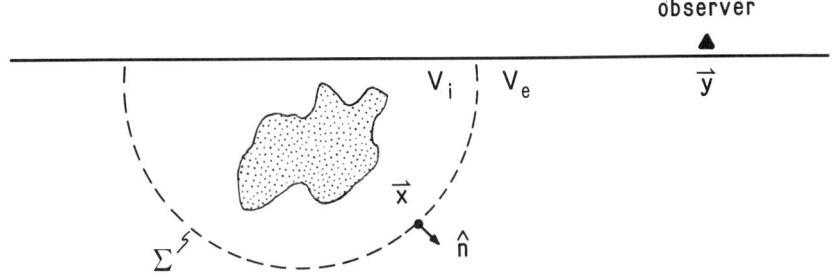

= zone of inelastic deformation

FIG. 1. Surface $\Sigma$ separates the regions of purely elastic behavior $V_e$ from a volume $V_i$ containing all inelastic behavior; $V_i$ can also contain elastic regions, and the exact position of the surface $\Sigma$ is chosen for computational convenience. The action of the source is represented by the displacements and tractions it causes on $\Sigma$.

$\Sigma$ for an arbitrarily complicated source by a finite-element or finite-difference method. Let us denote these displacements and tractions by $\mathbf{u}_s(\mathbf{x},t)$ and $\mathbf{T}_s(\mathbf{x}, t)$. We use the material properties in $V_e$ to define our reference earth model [see Bache *et al.* (1982) for details regarding the selection of a reference earth model], and we denote the Green's functions appropriate for this model by the vector $\mathbf{G}^j$, where the $i$th component of $\mathbf{G}^j$ is $G_{ij}$, as in Eq. (1), and $j$ is the point-force direction. $\mathbf{G}^j$ causes tractions $\mathbf{T}^j$ across a surface with normal $\hat{\mathbf{n}}$. We then have the representation

$$u_j(\mathbf{y}, t) = -\int_{-\infty}^{\infty} dt' \iint_\Sigma \mathbf{G}^j(\mathbf{x}, t'; \mathbf{y}, 0) \cdot \mathbf{T}_s(\mathbf{x}, t - t') - \mathbf{u}_s(\mathbf{x}, t')$$
$$\cdot \mathbf{T}^j(\mathbf{x}, t - t'; \mathbf{y}, 0) \, d\Sigma \tag{4}$$

for the geometry shown in Fig. 1. Both Eqs. (1) and (4) can be used for forward modeling of volume sources. If a finite-element or finite-difference method must be used to determine stress glut $\Gamma_{pq}$ in Eq. (1), then Eq. (4) is probably a more simple representation to use. This is because Eq. (4) involves a surface integral rather than a volume integral. However, Eq. (1) is preferable for doing the inverse problem because its kernel contains the source term we would like to resolve.

2. Surface Sources

Beginning with Aki's (1968) analysis of the 1966 Parkfield, California, earthquake, almost all modeling of earthquake sources and ground motions at local distances ($R < 200$ km) has made the rather restrictive assumptions that earthquakes occur as slip on a small number of infinitely thin, usually planar, fault surfaces, and that the slip direction is locally tangent to the fault surface. This model of earthquakes is mathematically and computationally much more convenient than a volume source model and has been adequate for obtaining a first-order understanding of the earthquake rupture process. It has been successfully used for waveform modeling of earthquake-generated ground motions at periods down to about $\frac{1}{5}$ of the faulting duration, which is approximately the length of the fault divided by the shear wave speed. In this chapter, we will concentrate on earthquake source models that make the surface source approximation, although we point out that the understanding of more realistic volume sources is a highly desirable and largely unpursued goal.

Neglecting the effects of self-gravitation and prestress, Backus and Mulcahy (1976b) show that the stress glut associated with a displacement discontinuity **s** on surface $\Sigma$ is

$$\Gamma_{ij} = C_{ijkl} n_k s_l, \tag{5}$$

where $\hat{\mathbf{n}}$ is a unit vector normal to $\Sigma$ pointing into the positive side of $\Sigma$, and the displacement discontinuity **s** is the difference in displacement between the positive and negative sides of the fault,

$$\mathbf{s}(\mathbf{x}, t) = \mathbf{u}(\mathbf{x}^+, t) - \mathbf{u}(\mathbf{x}^-, t), \tag{6}$$

where **x** is a point on surface $\Sigma$. Equation (5) is valid for displacement discontinuities both parallel and perpendicular to the fault. Inserting Eq. (5) into Eq. (1) gives the most commonly used representation

$$u_m(\mathbf{y}, t) = \int_{-\infty}^{\infty} dt' \iint_{\Sigma} C_{ijkl} n_k s_l \, \partial_{x_j} G_{mi}(\mathbf{y}, t - t'; \mathbf{x}, 0) \, d\Sigma, \tag{7}$$

which was presented by Maruyama (1963), Burridge and Knopoff (1964), and Haskell (1964). By using the reciprocity relation for Green's functions

$$G_{ij}(\mathbf{y}, t; \mathbf{x}, 0) = G_{ji}(\mathbf{x}, t; \mathbf{y}, 0), \tag{8}$$

Eq. (7) may be rewritten

$$u_m(\mathbf{y}, t) = \int_{-\infty}^{\infty} dt' \iint_{\Sigma} C_{ijkl} n_k s_l \partial_{x_j} G_{im}(\mathbf{x}, t - t'; \mathbf{y}, 0) \, d\Sigma. \tag{9}$$

The form of Eq. (7) is more intuitively satisfying than that of Eq. (9) because the Green's functions in Eq. (7) are those describing wave propagation from a point force applied on the fault surface to the observer location, whereas in Eq. (9) the point forces are applied at the observer location and the Green's functions are evaluated on the fault. However, Eq.

(9) is slightly more general than Eq. (7); Burridge and Knopoff (1964) derive Eq. (9) first and then obtain Eq. (7) by making an assumption of homogeneous boundary conditions. Moreover, Eq. (9) can be computationally more convenient than Eq. (7) depending upon the number of observation locations and the method for calculating Green's functions. For example, if a finite-element (FE) or finite-difference (FD) method is used to calculate Green's functions for a laterally heterogeneous seismic velocity structure, three Green's functions calculations must be performed for each observer location (one for each component $m$ of ground motion), whereas straightforward application of Eq. (7) would require three separate calculations for each point on $\Sigma$. For a laterally homogeneous velocity structure, Archuleta and Day (1980) show that use of Eq. (9) necessitates only two FE or FD calculations, regardless of the number of observers, due to the invariance of the Green's function under horizontal translation and to the cylindrical symmetry of the geometry. Spudich and Frazer (1984) point out that use of Eq. (9) is advantageous when using ray theory Green's functions, since the required ray tracing can be accomplished by simple shooting from the observer to the fault rather than by the two-point ray tracing needed to shoot from a point on the fault to each observation point. However, Eq. (9) offers no computational advantages over Eq. (7) if the desired number of observation locations is very large or if the so-called spectral methods (Chapman, 1978) are used to calculate Green's functions, because in these methods equal computation time is needed for the case of a single source depth and multiple receiver depths and vice versa. We note that Eq. (8) may also be applied to our volume source representations Eqs. (1) and (4).

Before proceeding, it is worthwhile to clarify some terminology. The terms "source" and "receiver" are ambiguous when applied to Eq. (9) because of the use of Green's function reciprocity. We will always use the term "observer location" or something similar to refer to point $\mathbf{y}$ in Eq. (9), which is the place where ground motions caused by the extended seismic source are to be evaluated. We will only use the term "receiver" to correspond to the first argument of either Green's function in Eq. (8). Hence, $\mathbf{y}$ is the receiver in Eq. (7) and $\mathbf{x}$ is the receiver in Eq. (9). The term "source" will refer to the location of the point force or point-moment tensor when discussing Green's functions and will otherwise denote the region of nonzero stress glut.

In order to simplify Eq. (9) further, we note that the stress tensor $\sigma_G$ associated with the Green's function in Eq. (9) is

$$\sigma_G^m(\mathbf{x}, t; \mathbf{y}, 0) = C_{ijkl}(\mathbf{x}) \, \partial_{x_j} G_{im}(\mathbf{x}, t; \mathbf{y}, 0), \tag{10}$$

where $m$ is the point-force direction. Hence the associated traction $\mathbf{T}^m$

exerted across the fault surface with normal $\hat{\mathbf{n}}$ is

$$\mathbf{T}^m(\mathbf{x}, t; \mathbf{y}, 0) = \hat{\mathbf{n}} \cdot \boldsymbol{\sigma}_G^m \tag{11}$$

and Eq. (9) may be written (Spudich, 1980) as

$$u_m(\mathbf{y}, t) = \int_{-\infty}^{\infty} dt' \iint_{\Sigma} \mathbf{s}(\mathbf{x}, t') \cdot \mathbf{T}^m(\mathbf{x}, t - t'; \mathbf{y}, 0) \, d\Sigma. \tag{12}$$

If we denote the Fourier transform of $f(t)$ by

$$f(\omega) = \int_{-\infty}^{\infty} f(t) e^{-i\omega t} \, dt, \tag{13}$$

then the frequency domain version of Eq. (12) is

$$u_m(\mathbf{y}, \omega) = \iint_{\Sigma} \mathbf{s}(\mathbf{x}, \omega) \cdot \mathbf{T}^m(\mathbf{x}, \omega; \mathbf{y}, 0) \, d\Sigma. \tag{14}$$

Although Eq. (14) is perfectly valid, the slip functions of greatest seismological interest have static offsets, and consequently their Fourier transforms are undefined. If we denote temporal derivations with an overdot, then slip velocity is $\dot{\mathbf{s}}$, and Eq. (14) may be written in a form more advantageous for numerical evaluation as

$$\dot{u}_m(\mathbf{y}, \omega) = \iint_{\Sigma} \dot{\mathbf{s}}(\mathbf{x}, \omega) \cdot \mathbf{T}^m(\mathbf{x}, \omega; \mathbf{y}, 0) \, d\Sigma, \tag{15}$$

where $\dot{u}_m$ is the $m$ component of ground velocity. The ground velocity time series may be obtained by inverse Fourier transformation of $\dot{u}(\omega)$. We have found it computationally most stable to obtain acceleration $\ddot{\mathbf{u}}(t)$ by inverse Fourier transformation of $i\omega\dot{\mathbf{u}}(\omega)$, and displacement $\mathbf{u}(t)$ by temporal integration of $\dot{\mathbf{u}}(t)$, thus avoiding difficulties with the $\omega = 0$ term. Although in later sections we will often refer to the slip function itself, it should be understood that, in practice, the slip rate function is used for computing ground motion when the method of calculation involves the frequency domain.

## II. The Fault Surface Integral

At present there is no clearly preferred method for calculating ground motions in the near-source region of an earthquake. A variety of methods have been used. Each has its own range of validity, and almost all are computationally expensive. In this section we will review a number of the methods, explain the basic physics behind them, and point out their strengths and weaknesses.

From Section I.B, the basic integral to be evaluated is

$$\dot{u}(\omega) = \iint_\Sigma \dot{\mathbf{s}} \cdot \mathbf{T} \, d\Sigma. \quad (16)$$

Calculation of near-source ground motions is generally difficult because for geologically realistic situations both terms in the integrand are computationally expensive to obtain. For example, the slip-rate vector $\dot{\mathbf{s}}$ may be obtained for a fault having heterogeneous strength and stress drop by three-dimensional FD, FE, or boundary-integral calculations, as discussed by Mikumo in Chapter 3 of this volume. The formidable problem of wave propagation in laterally heterogeneous media must be solved to obtain the traction Green's functions. If the surface integral in Eq. (16) is performed numerically and equal accuracy is desired for each frequency, the computational effort needed is proportional to the square of the frequency and to the cube of the frequency bandwidth desired in the calculated ground motions (we will elaborate on this statement later). Consequently, a number of approximations are often made to obtain $\dot{\mathbf{s}}$ and $\mathbf{T}$. In many cases, the particular approximations made to $\dot{\mathbf{s}}$ and $\mathbf{T}$ can interact to simplify the surface integral calculation. In the following sections, we will discuss some of these approximations.

A. SLIP RATE VECTOR $\dot{\mathbf{s}}(\mathbf{x}, t)$ FOR KINEMATIC MODELS

In many ways specification of the slip rate vector is the central element of a kinematic faulting model. The slip rate vector represents one's understanding of the behavior of the earthquake rupture. Based on waveform matching of synthetic ground motion and real ground motion (see, e.g., Archuleta and Day, 1980; Olson and Apsel, 1982; Hartzell and Helmberger, 1982; Hartzell and Heaton, 1983, 1986; Archuleta, 1984), the behavior of the earthquake source is relatively well modeled for frequencies up to 1–2 Hz. Above these frequencies the source is not well understood deterministically, and normally a statistical approach is taken (Haskell, 1966; Aki, 1967; Boore and Joyner, 1978; Andrews, 1981; Izutani, 1981; Boatwright, 1982; Koyama, 1985). This lack of understanding of the high-frequency behavior of an earthquake source is one of the major obstacles to accurate prediction of ground motions above 1.0 Hz. In the following we consider rupture models that have been successfully used at frequencies $\leq$ 1.0 Hz. We have chosen to discuss the slip rate vector rather than the slip vector because its Fourier transform is well behaved, i.e., we do not have to take into account the static offset inherent in the slip vector itself (see Section I.B). The slip rate vector has components only in the fault plane for shear faulting. Thus, if one chooses the convenient basis vectors

$\hat{\mathbf{x}}_1$ and $\hat{\mathbf{x}}_2$ in the directions along strike and downdip, respectively, the slip rate vector will have strike-slip and dip-slip components $\dot{s}_1$ and $\dot{s}_2$, respectively. Some of the parameters needed to describe the slip rate depend on the function chosen for the slip rate. Other parameters, primarily the rupture time, are directly associated with the slip rate but are not dependent on the function itself.

The most common functional form for the slip rate, a rectangle function $\Pi(t)$ (Bracewell, 1965, p. 52), is the Haskell (1964) model in which the slip at a given $\mathbf{x}$ is a linear ramp with a rise time $\tau(\mathbf{x})$ (Fig. 2a). In order to allow for a propagating rupture, different points on the fault will slip at different times. Thus, each point on the fault has a rupture time $t_s(\mathbf{x})$ at which the slip initiates. This parameter enters naturally into the slip rate function

$$\dot{\mathbf{s}}(\mathbf{x}, t) = [a_1(\mathbf{x})\hat{\mathbf{x}}_1 + a_2(\mathbf{x})\hat{\mathbf{x}}_2]\Pi\{[t - t_s(\mathbf{x})]/\tau(\mathbf{x})\}, \qquad (17)$$

where $a_1$ and $a_2$ are components of the slip-rate vector. This description of the slip rate depends on four basic parameters, $a_1(\mathbf{x}, t)$, $a_2(\mathbf{x}, t)$, $\tau(\mathbf{x})$, and $t_s(\mathbf{x})$.

There are a number of variations of the Haskell model. For example, Hartzell and Helmberger (1982) use a triangular slip rate function to compute synthetics for the 1979 Imperial Valley, California, earthquake. Because dynamic models of a propagating stress relaxation (Madariaga, 1976; Archuleta and Frazier, 1978; Day, 1982) show that the functional form of the slip rate closely resembles the form derived by Kostrov (1964) for a self-similar propagating stress drop, the Kostrov model (Fig. 2b) has become a prominent alternative to the Haskell model. Archuleta and Hartzell (1981), Boatwright (1981), and Campillo and Bouchon (1983) have

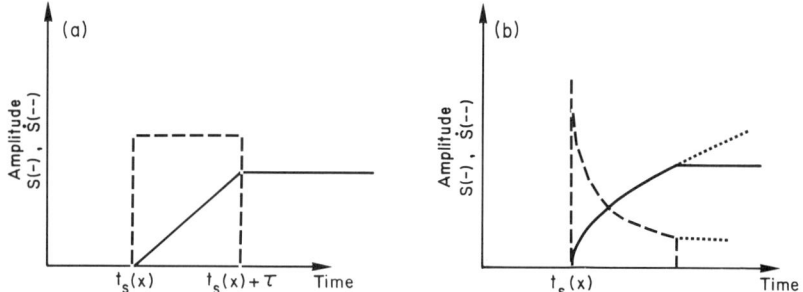

FIG. 2. Functional forms for slip and slip rate. (a) The Haskell slip function (solid line) and the slip rate function (dashed line) are shown for source point at position x. The rupture time function $t_s(\mathbf{x})$ and the rise time $\tau(\mathbf{x})$ are shown. (b) The modified Kostrov slip and slip rate functions are shown as the solid and dashed lines, respectively. The dotted line indicates the shape of the Kostrov functions before modification.

used a modified Kostrov function to generate high frequency synthetic seismograms. Although the functional forms of the Haskell model and the Kostrov model are quite different, the number of parameters needed to specify each is the same. Although the computation is not reduced, one could argue that the Kostrov dislocation better represents the physics of the earthquake mechanism. If only low frequencies are being considered, the two dislocation models are approximately the same because the singularity in the Kostrov slip rate at the arrival time of the rupture is eliminated by low passing the slip rate function (Archuleta and Hartzell, 1981). Both models are kinematic in that the dislocation is specified without regard to the forces acting in the process.

Given that the slip rate is specified with four parameters, a crucial observation is that two of the parameters $\dot{s}_1$ and $\dot{s}_2$ are linearly related to the ground motions, while $\tau$ and $t_s$ are not (Archuleta, 1984). This feature of the parameterization became the critical element in Archuleta's attempt to model the faulting of the 1979 Imperial Valley earthquake. He found that variations in the rupture time $t_s$ had a pronounced effect on the synthetic seismograms, whereas variations in the rise time $\tau$ and the slip rate amplitudes $s_1$ and $s_2$ cause either small or predictable changes. The effect on the synthetics of changing the slip rate amplitude is strongly governed by geometrical attenuation ($\approx R^{-1}$, $R^{-2}$, $R^{-4}$ for the far-, intermediate-, and near-field terms, respectively, where $R$ is the distance between the observer and the point on the fault). Thus, any change in the slip rate amplitude affects most the synthetics at stations closest to where the change in slip rate amplitude was made. This is not the case with the rupture time or the rise time. However, as Anderson and Richards (1975) showed, it takes a 300% change in rise time to compensate for a 17% change in rupture time. In section II.C.1 we show why rupture time variations have such a strong effect, which has also been discussed by Spudich and Oppenheimer (1986).

A primary effect is that when doing forward modeling of data, changing the slip rate amplitudes while holding the rupture time fixed may produce a better fit, i.e., one finds a local minimum of misfit in model space. Finding the global minimum, or even recognizing it, given that two of the parameters in the model are nonlinearly related to the data is a major unanswered question in forward modeling. This problem has been avoided to some extent in ground motion studies using linear inverse techniques by choosing a parameterization of $\dot{s}(\mathbf{x}, t)$ in which $t_s(\mathbf{x})$ does not appear explicitly. Spudich (1980) represented $\dot{s}$ as an arbitrary continuous function of position and time, in which case Eq. (15) is completely linear in $\dot{s}$. Using Spudich's parameterization, however, does not necessarily lead to solutions for $\dot{s}$ that show a conspicuous rupture front. To alleviate this problem,

Olson and Apsel (1982) and Hartzell and Heaton (1983, 1986) used a different, more restrictive parameterization of $\dot{s}$ that preserved the linearity of the problem while constraining all the slip to occur within a time window chosen *a priori*.

### B. Comments on Green's Functions Pertaining to the Fault Surface Integral

The calculation of Green's functions in various types of earth models has been a fundamental research area in seismology, and an enormous body of literature deals with the subject. In this section, we will limit ourselves to describing some of the methods that have been used for extended-source calculation, and we will comment primarily on those aspects of the methods that have a practical relevance in such calculations. These aspects are (1) the accuracy for realistic earth models, (2) the computational effort required by each, and (3) any structure in the mathematical expressions that facilitates the inclusion of an extended seismic source.

1. General Comment on Accuracy

For the modeling of observed earthquake seismograms, or for prediction of ground motions in a specific region, the first factor limiting the accuracy of any Green's function calculation is simple ignorance of the earth structure at the site. In order to calculate Green's functions accurately at 5 Hz for a particular region, it is necessary to know the 3D shear-velocity structure on a scale of a few hundred meters both horizontally and vertically. Since such a detailed knowledge of the earth's velocity structure may not be available in the near future, calculations of ground motions for real world situations are predestined to have errors. To our knowledge, no one has yet attempted to estimate these errors and carry them along as uncertainty estimates in subsequent calculations. Estimation of these errors would be relatively easy if microearthquake recordings were available in the study region. One could simply examine the spectrum of the difference between the observed microearthquake seismograms and synthetic seismograms calculated in the identical source–observer geometry for point dislocations. For frequencies below the microearthquake's corner frequency, the synthetic point-dislocation seismograms should match the observed seismograms. Probably it would be observed that the difference between the observed and synthetic seismogram spectra would increase as a function of frequency, corresponding to our progressively less accurate knowledge of earth structure on decreasing length scales. If quantitative measurements of these discrepancies were available, they could be used to estimate uncertainties in both forward modeling and inverse modeling of earthquake seismograms (Spudich, 1980).

A related approach was used by Archuleta (1984) in modeling the 1979 Imperial Valley earthquake seismograms. The shear wave velocity structure he used for calculating his Green's functions had been altered by trial-and-error modeling to improve the agreement between predicted ground motions and those observed for an aftershock of the 1979 event.

Another approach to the problem of inaccurate Green's functions is to use microearthquake seismograms as point dislocation Green's functions (Hartzell, 1978). Below the microearthquake corner frequency, these seismograms are exactly the Green's functions desired. A large amount of work has been done recently using this promising technique (Kanamori, 1979; Hartzell, 1982; Irikura, 1983; Imagawa et al., 1984; Munguia and Brune, 1984; Joyner and Boore, 1986), which will not be reviewed in this chapter. The main impediments to the use of this idea are that at present there are generally not enough microearthquake records available to simulate the rupture of a large fault, and in many cases it is desired to calculate ground motions at frequencies above the microearthquake corner frequencies, in which case the question of earthquake scaling relations enters the picture. We note that Hartzell (1978) used four aftershocks of the 1940 El Centro earthquake to simulate the mainshock record, and the simulation worked very well at periods of 5 sec and longer. This is quite a different style of simulation than the high-frequency syntheses of Irikura (1983). A second factor, related to the general question of accuracy of theoretical Green's functions, is an entirely subjective question, namely, what attributes of the ground motion are important to calculate accurately? If the ground motion calculation is being done to aid in the engineering design of a structure, it may be that it is most important to calculate peak ground velocity accurately, or perhaps the total duration of long-period ground motion, or the root-mean-square acceleration. Considerations like these definitely affect the choice of a Green's function calculation method. For example, close to a fault the peak ground velocities are generally caused by the direct S wave, so that use of simple ray theory Green's functions may be ideal for such calculations (Spudich and Frazer, 1984). On the other hand, at larger distances the duration of long-period motion is generally related to surface wave excitation, and use of ray theory Green's functions to predict the duration of long-period motions at large distances would be an utter disaster. In the remainder of this chapter, it is presumed that the limitations of the various Green's function methods are known.

2. Green's Functions in 2D and 3D Media with Arbitrary Velocity Structure

Currently, FE and FD methods are the only techniques available that can be used to calculate complete Green's functions, i.e., Green's functions containing all possible body and surface waves in general, laterally varying velocity structures. The chief limitation of such techniques is that the computational effort required for 3D models is proportional to $f_{max}^4$, where $f_{max}$ is the maximum frequency of interest in the calculation. The exponent of 4 results from the necessity of using a number of mesh points per linear distance that is proportional to frequency, and a time step that is inversely proportional to frequency. In addition, the needed storage goes up as $f_{max}^3$. Because of the large computational effort, FE and FD methods have not been used (to our knowledge) for ground motion calculations in 3D, laterally varying media. An additional characteristic of these methods is that the Green's functions they generate are specified as the motions of a given set of mesh points, meaning that integrals of quantities like $\mathbf{s} \cdot \mathbf{T}$ must be done purely numerically.

If only body waves are desired in laterally varying structures, considerable work that is applicable to the extended source problem has been done on asymptotic ray theory and related methods such as Gaussian beams [see Červený and Klimeš (1984), and other articles in the same issue]. The chief advantages of ray-based methods are their rapidity of calculation (which is independent of frequency), their accuracy at high frequency (except for the omitted waves, of course), and the simple form of the Green's functions, consisting primarily of propagating delta functions. As Bernard and Madariaga (1984) and Spudich and Frazer (1984) have shown, this simplicity in the Green's functions greatly simplifies the fault surface integral and facilitates their use with extended seismic sources.

Another Green's function method that may have great potential for handling topographic effects and scattering by basins is the use of ray theory in conjunction with the Kirchhoff–Helmholtz theory (Frazer and Sen, 1985; Sen and Frazer, 1985). Such a method is excellent for the reflection or transmission of body waves at irregular interfaces because it includes diffractions. In addition, it could be very easily used with extended sources by employing the isochrone integration approach of Bernard and Madariaga (1984) and Spudich and Frazer (1984). Such an approach would necessitate the use of isochrone integration on multiple surfaces of integration, as in Haddon and Buchen (1981), with one being the fault and the others being the base of the basin, etc.

3. Laterally Homogeneous Media

Most ground motion calculations for extended seismic sources have been performed in laterally homogeneous media because the theory for Green's functions has been most fully developed for such media, and because it is computationally feasible to calculate complete synthetic seismograms for such media on a routine basis. In this section, we will concentrate primarily on methods for complete synthetic seismograms. In most places the earth's material properties vary most strongly with depth, so that the assumption of lateral homogeneity seems reasonable. However, no one, to our knowledge, has ever examined quantitatively the degrees of misfit between real observed point-source seismograms and theoretical Green's functions calculated for a laterally homogeneous approximation to the local velocity structure.

The cylindrical symmetry and translational invariance of laterally homogeneous earth structures lead to simplifications of the mathematical expressions for the Green's functions. These simplifications have only partially been exploited to ease the integration of $\mathbf{s} \cdot \mathbf{T}$ over the fault surface. Before we can describe how the assumption of cylindrical symmetry interacts with integration of the Green's functions over the fault surface, it is necessary to present some of the mathematics describing waves in laterally homogeneous media. For simplicity and consistency of notation, we will refer primarily to Kennett (1983), who elaborates on the mathematics and gives proper credit to the original authors. The reader may also find Spudich and Ascher (1983) helpful.

*a. Arbitrary Variation of Velocity with Depth Using a 2D Finite-Element Solution.* While it is possible to use a 3D FE calculation to obtain Green's functions for a laterally homogeneous medium, Day (1977) has shown how Green's functions for the 3D geometry can be obtained from a less expensive 2D FE calculation in depth and epicentral distance. The advantages of his technique, aside from the obvious reduction of a 3D to a 2D calculation, are that Green's functions are obtained on a dense grid of points facilitating integration over a seismic source, and an arbitrary variation of velocity with depth is allowed in the medium. The basic idea of the method is that the displacements and stresses in an axisymmetric medium can be expanded in a Fourier series over azimuthal angle $\phi$. For example, displacement $\mathbf{u}$ can be expanded

$$\mathbf{u}(r, z, \phi, t) = \sum_n \mathbf{U}(r, z, t, n) e^{in\phi}, \qquad (18)$$

and the stresses can be expanded similarly. For each azimuthal order $n$, the expansion coefficient $\mathbf{U}$ can be obtained by a 2D FE calculation (see Day, 1977, for details). For a vertical point-force source, $\mathbf{U} \neq 0$ only for the $n = 0$ term, and for a horizontal point force $\mathbf{U} \neq 0$ only for $n = 1$. Hence, point-force Green's functions can be obtained for receivers at every point in a 3D axisymmetric medium by solution of two 2D FE calculations. Green's functions calculated in this manner were used by Archuleta and Day (1980) to do forward modeling of the 1966 Parkfield, California, earthquake.

*b. Arbitrary Variation of Velocity with Depth Using Other Methods.* By Fourier transformations of field quantities over azimuthal order, Day (1977) reduced a 3D problem to a series of 2D problems. By performing a Bessel transform over epicentral distance, the problem can be reduced further to a series of 1D problems. This is the most common approach to laterally homogeneous problems. Its advantage for extended source calculations is that the horizontal variation of Green's functions can be expressed analytically instead of purely numerically.

In a cylindrical coordinate system $(r, \phi, z)$, the Fourier transform of displacement and $z$ component of stress can be expanded as

$$\mathbf{u}(r, \phi, z, \omega) = \int_0^\infty k \sum_n [U\mathbf{R}_k^n + V\mathbf{S}_k^n + W\mathbf{T}_k^n]\, dk, \qquad (19)$$

and

$$\sigma_z(r, \phi, z, \omega) = \int_0^\infty k \sum_n [P\mathbf{R}_k^n + S\mathbf{S}_k^n + T\mathbf{T}_k^n]\, dk, \qquad (20)$$

where $k$ is a horizontal wave number and $n$ is an integer. The mutually orthogonal vector surface harmonics $\mathbf{R}_k^n$, $\mathbf{S}_k^n$, and $\mathbf{T}_k^n$ are given by Kennett [1983, Eq. (2.48)], and they are proportional to $J_n(kr)e^{in\phi}$ and its derivatives, where $J_n$ is the Bessel function of order $n$. The expansion coefficients $U, V, \ldots, T$ are functions of $z$, $k$, and $\omega$ that are to be determined. The continuity of $\mathbf{u}$ and $\sigma_z$ with depth and the equations of motion yield the coupled differential equations for the coefficients $U, V, \ldots, T$,

$$d_z \mathbf{b}_P = A_P \mathbf{b}_P + \mathbf{F}_P, \qquad (21)$$

and

$$d_z \mathbf{b}_H = A_H \mathbf{b}_H + \mathbf{F}_H, \qquad (22)$$

where

$$\mathbf{b}_P = [U, V, P, S]^T, \qquad (23)$$

and

$$\mathbf{b}_H = [W, T]^T. \qquad (24)$$

Here, $\mathbf{F}_P$ and $\mathbf{F}_H$ are inhomogeneous source terms, the matrices $A_P$ and $A_H$ are given by Kennett (1983, p. 30), the $T$ superscript indicates a transpose, and $d_z = d/d_z$. The system Eq. (21) corresponds to P–SV motion, and Eq. (22) corresponds to SH motion. We will refer to them as the P–SV and SH systems, respectively. Note that they are ordinary differential equations in one spatial dimension.

Because the elements of $\mathbf{b}$ are expansion coefficients for displacement and the $z$ component of stress, as in Eqs. (19) and (20), $\mathbf{b}$ is often called the displacement–stress vector; $\mathbf{b}(z)$ is a continuous function of depth, and the propagator matrix $P(z_2, z_1)$ can be used to relate $\mathbf{b}(z_1)$ to $\mathbf{b}(z_2)$:

$$\mathbf{b}(z_2) = P(z_2, z_1)\mathbf{b}(z_1) \qquad (25)$$

(Kennett, 1983, p. 40).

## 5. EARTHQUAKE GROUND-MOTION CALCULATION

Typically, for a choice of parameters $\omega$ and $k$, the P–SV and SH systems are solved subject to a free-surface boundary condition, a radiation condition at large depth, and a particular inhomogeneous source term $\mathbf{F}(z)$. Although it is possible to represent an extended source through the term $\mathbf{F}(z)$, this is not the approach usually taken, although it would be worth investigating. Bouchon (1979) and Chouet (1987) include an extended source in a related manner, which will be described in Section II.C.4. The usual approach is to assume existence of a point source at depth $z_s$, which gives rise to a forcing term of the form

$$\mathbf{F}(z) = \mathbf{F}_1 \delta(z - z_s) + \mathbf{F}_2 d_z \delta(z - z_s). \tag{26}$$

This source can be equivalently expressed as a discontinuity in $\mathbf{b}$ at depth $z_s$:

$$\mathbf{b}(z_s^+) - \mathbf{b}(z_s^-) = \mathbf{F}_1 + \omega A \mathbf{F}_2. \tag{27}$$

Kennett (1983, p. 95) gives expressions for the discontinuity in $\mathbf{b}$ caused by a point moment-tensor source. An alternate approach is to set $\mathbf{F}(z) = 0$ for all $z$ and solve the P–SV and SH systems with a given traction boundary condition at $z = 0$ (Spudich and Ascher, 1983; Olson et al., 1984). This procedure yields $\mathbf{b}$ at all depths for a surface source, and the spatial reciprocity relation for Green's functions can then be used to derive the free-surface displacements caused by buried sources.

Thus far we have been working in the frequency domain. Expansion of displacement and stress analogous to Eqs. (19) and (20) can be written in the time domain [see, e.g., Olson et al., 1984, Eq. (2.7)], and the time-domain versions of Eqs. (19) and (20) are given by Alekseev and Mikhailenko [1980, Eqs. (16–18)] and Olson et al. [1984, Eqs. (2.8a–2.8c)]. These are partial differential equations in the $(z, t)$ domain. Once Eqs. (21) and (22), or their time domain analogs, are solved for $U$, $V$, and $W$, the medium displacements are obtained by evaluation of Eq. (19).

When used to calculate Green's functions, the methods of Alekseev and Mikhailenko (1980), Spudich and Ascher (1983), and Olson et al. (1984) are in some ways well suited and in other ways poorly suited for extended seismic source calculations. Each of the methods solves the P–SV and SH systems on a dense grid of points in depth, which is very useful because the traction vector $\mathbf{T}$ must be known over the entire source. On the other hand, the computational effort they require rises very rapidly with frequency, limiting their utility to frequencies of a few hertz or less.

If complete seismograms are desired in the frequency band $(0, f_{\max})$, then the computational effort $C$ required by all these methods scales like

$$C \approx n_k n_f n_z \text{ (or } n_k n_t n_z) \approx f_{\max}^3,$$

where $n_k$, $n_f$, and $n_t$ are the number of wave numbers, frequencies, and time steps, respectively, for which the P–SV and SH systems are solved, and $n_z$ is the number of mesh points needed in depth to solve each system. Here, $C$ is proportional to $f_{max}^3$, for the several reasons. The requirement of complete seismograms implies the use of waves having all slownesses from zero to some maximum slowness $p_{max}$ usually greater than the slowness of the fundamental Rayleigh mode. Hence, the P–SV and SH equations must be solved for a range of wave numbers running from 0 to $k_{max} = 2\pi p_{max} f_{max}$. If the wave number sampling scheme is independent of frequency, as it is in Alekseev and Mikhailenko (1980), Spudich and Ascher (1983), and Olson et al. (1984), then the number of wave number sample points for which the P–SV and SH equations must be solved grows linearly with frequency. Similarly, the three papers mentioned above use a FD, collocation, and FE method, respectively, to solve the P–SV and SH equations. In order to solve these equations, each method requires a number of mesh points in depth (and hence computational effort) that grows linearly with frequency. Finally, the number of time steps needed or the number of frequencies for which the P–SV and SH solutions must be solved grows linearly with $f_{max}$.

While a cylindrical coordinate system is the natural one for point seismic sources in a laterally homogeneous medium, we shall see later that a Cartesian coordinate system offers certain advantages when extended seismic sources are considered. For that reason, let us consider how this same physical problem can be handled in Cartesian coordinates.

We start by expanding the displacement and $z$ component of stress in terms of an orthogonal set of basis functions

$$\mathbf{u}(x, y, z, \omega) = \int_{-\infty}^{\infty} \int_{-\infty}^{\infty} (U^c \mathbf{R}^c + V^c \mathbf{S}^c + W^c \mathbf{T}^c) \, dk_x \, dk_y, \qquad (28)$$

with an equation analogous to Eq. (20) for $\sigma_z$, where

$$\mathbf{R}^c = Y\hat{\mathbf{z}}, \qquad (29a)$$

$$\mathbf{S}^c = k_x^{-1} \hat{\mathbf{x}} \, \partial_x Y + k_y^{-1} \hat{\mathbf{y}} \, \partial_y Y, \qquad (29b)$$

$$\mathbf{T}^c = k_y^{-1} \hat{\mathbf{x}} \, \partial_y Y - k_x^{-1} \hat{\mathbf{y}} \, \partial_x Y, \qquad (29c)$$

$$Y(x, y) = \exp(-ik_x x - ik_y y), \qquad (30)$$

where $U^c$, $V^c$, and $W^c$ are functions of $z$, $k_x$, $k_y$, and $\omega$. One possible route to follow to obtain a solution for $U^c, \ldots, T^c$ would be to use the equations of motion to obtain an equation analogous to Eq. (21) involving the new expansion coefficients, i.e., let

$$\mathbf{b}^c = [U^c, V^c, W^c, P^c, S^c, T^c]^T, \qquad (31)$$

$$d_z \mathbf{b}^c = A^c \mathbf{b}^c + \mathbf{F}. \qquad (32)$$

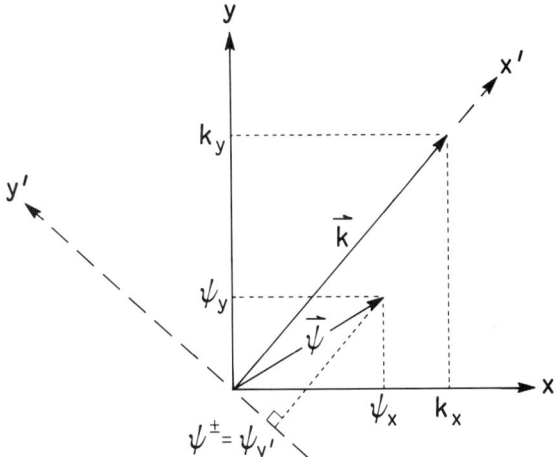

FIG. 3. A plane wave with wave number **k** propagates in the $x'$ direction; **k** lies in the $x-y$ plane. The vector potential $\Psi$ does not necessarily lie in the $x-y$ plane; $\Psi^{\pm}$ is the $y'$ component of $\Psi$.

However, because the P–SV and SH motions do not separate in a Cartesian coordinate system, $A^c$ is a $6 \times 6$ matrix that does not partition into a $4 \times 4$ P–SV and a $2 \times 2$ SH system like Eqs. (21) and (22). The $6 \times 6$ system [Eq. (32)] would have to be solved for every $(k_x, k_y)$ pair.

This problem may be reduced to a problem in which P–SV and SH separate by exploitation of the cylindrical symmetry. For every point $(k_x, k_y)$ in the wave number plane, we choose a new coordinate system $(x', y', z)$ with $x'$ oriented along the direction of the wave number vector $\mathbf{k} = k_x \hat{\mathbf{x}} + k_y \hat{\mathbf{y}}$ (Fig. 3). If we then let $k = |\mathbf{k}|$, and replace $U$, $V$, $W$, $P$, $S$, and $T$ in Eqs. (23) and (24) by $iu_z$, $u_{x'}$, $u_{y'}$, $i\sigma_{zz}$, $\sigma_{x'z}$, and $\sigma_{y'z}$, respectively, we find that Eqs. (21) and (22) hold in the rotated Cartesian system (Kennett, 1983, p. 36). They are then solved for every $(k_x, k_y)$ point yielding radial, transverse, and vertical displacements, and these results are rerotated back into the $(x, y, z)$ system for subsequent use. This technique has been used by Bouchon (1979) and Chouet (1982) for extended sources in layered media.

*c. Weakly Inhomogeneous Layers.* Thus far, we have restricted our comments to methods of solving the P–SV and SH equations for arbitrary variation of velocity with depth. To accommodate the generality of arbitrary variation, purely numerical methods (e.g., FE, FD, collocation) must be used to solve the equations. If the restrictions on the velocity profile are tightened to require piecewise-continuous functions with small velocity gradients $d_z\alpha$ and $d_z\beta$ between discontinuities ($\alpha$ being the P-wave velocity and $\beta$ being the S-wave velocity), then two benefits accrue from the standpoint of extended-source calculations. First, the variations of the Green's functions with depth [i.e., the terms $U, V, W$, etc., in Eq. (19)] can be written in terms of analytic functions such as Airy functions. Second, the computation times scale like $f_{max}^2$ rather than $f_{max}^3$, because the effort of solving the P–SV and SH equations, becomes independent of frequency. Because extended-source calculations have not been done with Green's function methods using the approximation of weakly inhomogeneous layers, we will not dwell on these methods. Cormier (1980), Kennett and Illingworth (1981), and Chapman and Orcutt (1985) describe useful techniques and their antecedents.

*d. Uniform Layers.* The theory for calculating synthetic seismograms is most fully understood for laterally homogeneous structures in which the material properties are assumed to be piecewise-constant functions of depth, i.e., uniform plane layers bounded by discontinuities. Historically, generalized ray theory has been widely used for calculating synthetic seismograms in layered media for both point dislocations (Helmberger and Malone, 1975; Helmberger and Harkrider, 1978; Liu and Helmberger, 1985) and for extended sources (Hartzell *et al.*, 1978; Heaton and Helmberger, 1979; Archuleta and Hartzell, 1981). Generalized ray theory does not yield complete synthetic seismograms for an inhomogeneous medium. Consequently, it can be safely used for velocity structures having only a few layers, but presents difficulties when used in complicated multilayered earth structures (Hartzell and Helmberger, 1982). For this reason, considerable effort has gone into developing theory for calculating complete synthetic seismograms in layered models. Chin *et al.* (1984a,b) review a number of the methods.

For the purpose of extended source calculations, the assumption of uniform layers has two relevant consequences. First, the expansion coefficients $U(z), V(z), \ldots, T(z)$ in Eqs. (19) and (20) can be expressed as linear combinations of simple exponential functions, rather than as more complicated Airy functions or functions obtained numerically on a grid of points in depth. Second, although not computationally inexpensive, it is at least feasible to calculate these Green's functions up to high frequencies, e.g., 25 Hz.

To elaborate on the first point above, we follow Kennett (1983, pp. 46–48). Within a uniform layer, the displacement-stress vector $\mathbf{b}_P$ from Eq. (21) can be related to the vector $\mathbf{v}$ of upgoing and downgoing P and S potentials through the $D_P$ matrix given by Kennett (1983, p. 48) as

$$\mathbf{b}_P = D_P \mathbf{v}_P, \tag{33}$$

where
$$v_P(z) = Q_P(z, z_0)v_P(z_0), \tag{34}$$
$$Q_P(z, z_0) = \text{diag}\{\exp[-i\omega q_\alpha(z - z_0)], \exp[-i\omega q_\beta(z - z_0)],$$
$$\exp[i\omega q_\alpha(z - z_0)], \exp[i\omega q_\beta(z - z_0)]\}. \tag{35}$$

Here, $q_\alpha$ and $q_\beta$ are vertical slownesses:
$$q_\alpha = (\alpha^{-2} - p^2)^{1/2}, \tag{36a}$$
$$q_\beta = (\beta^{-2} - p^2)^{1/2}, \tag{36b}$$
$$p = k/\omega. \tag{37}$$

For the SH problem, exactly analogous equations hold, with
$$Q_H(z, z_0) = \text{diag}\{\exp[-i\omega q_\beta(z - z_0)], \exp[i\omega q_\beta(z - z_0)]\}. \tag{38}$$

From the form of $Q_P$ and from Eq. (34), we can see that if we know $v$ at level $z_0$, which might be the top of a layer, we can determine $v$ anywhere within the layer by applying $Q_P$. Moreover, the phase advance given by the four terms in $Q_P$ corresponds to upgoing P and S waves and downgoing P and S waves, respectively. Hence, the elements of $v_P$ can be identified as upgoing and downgoing P and SV potentials, as
$$v_P = [\Phi^+, \Psi^+, \Phi^-, \Psi^-]^T. \tag{39}$$

The matrices $D_P$ and $D_H$ are independent of depth within a layer. Since we can obtain $v$ anywhere within a layer from knowledge of its value on a boundary, we can use $D$ to obtain $b$ anywhere within a layer. This is expressed by the construction of the propagator matrix for a uniform layer:
$$P(z, z_0) = D(z)Q(z, z_0)D^{-1}(z_0). \tag{40}$$

Because of the form of $Q$, Eqs. (35) and (38), $b$ will be linear combinations of terms containing $\exp[i\omega q(z - z_0)]$. The task of finding $b$ at each of the interfaces involves solving the P-SV and SH systems at each of the interfaces and is discussed in many of the papers already cited.

The simple exponential forms of the $v$ and $b$ vectors can be exploited in two ways for extended source calculations. The first way has been used by Bouchon and Aki (1977) and Bouchon (1979). When an extended seismic source is contained entirely within a layer, the upgoing and downgoing P and S potentials can be calculated directly for the source thanks to the simple form of $Q_P$, Eq. (35). These upgoing and downgoing source potentials are then used as a source term in the remainder of the synthetic seismogram calculation (this will be elaborated on in Section II.C.A.) The second way of exploiting the simple form of $v$ and $b$ has not yet been used.

In order to use the form of the representation theorem Eq. (15), the tractions at all depths must be derived for surface point sources. Equations (33)–(35) show how the tractions within a layer can be written analytically. For certain forms of the slip vector **s**, it may be possible to perform the vertical part of the fault surface integration analytically.

Computational effort usually goes like $f_{\max}^2$ or $f_{\max}$ (depending on the wave number integrations method) for methods using uniform layers. This compares to factors of $f_{\max}^3$ for methods that allow arbitrary variation of material properties with depth. The primary reason for the factor of $f_{\max}$ difference is that the number of layers remains fixed for all frequencies in uniform layer calculations, whereas the number of grid points in depth increases with frequency in methods such as the discrete wave number finite element (DWFE) method (Olson et al., 1984).

More specifically, the calculation time $C$ required by uniform layer methods generally scales like

$$C \approx n_f n_k n_s n_r n_l,$$

where $n_f$ and $n_k$ are the numbers of frequencies and wave numbers (or slownesses) at which Eqs. (21) and (22) are to be solved, $n_s$ and $n_r$ are the number of source and receiver depths desired, and $n_l$ is the number of layers. Here, $n_f$ is clearly a linear function of $f_{\max}$, and $n_s$, $n_r$, and $n_l$ are usually fixed for all frequencies. In some methods, such as Kennett (1980), Apsel and Luco (1983), and Yao and Harkrider (1983), it is very easy to place a source or receiver at an interface between layers, in which case

$$C \approx n_f n_k n_s n_l \quad \text{or} \quad n_f n_k n_r n_l,$$

where $n_k$ depends on the method used to do the inverse transform Eq. (19). Kind (1979) and Kennett (1980) use a trapezoidal rule in slowness, so for them $n_k$ is independent of frequency. Apsel and Luco (1983) and Frazer and Gettrust (1984) use a Filon's method in which $n_k$ probably grows roughly linearly with frequency. Bouchon (1981) and Yao and Harkrider (1983) use a discrete wave number summation in which $n_k$ is, in principle, independent of frequency but in practice is linear with frequency due to the convergence criterion used in the wave number integral. Because the functions $U, V, \ldots, T$ in Eqs. (19) and (20) become progressively more oscillatory functions of slowness as frequency increases (see, e.g., Figs. 6 and 7 of Spudich and Ascher, 1983), methods that use a fixed set of slowness points for all frequencies are likely to become inaccurate at sufficiently high frequencies.

A final note is appropriate regarding the "accuracy" of uniform layer Green's function methods when used to generate the seismic response of velocity structures having smooth gradients of velocity with depth. A uniform layer method will approximate a gradient with a stairstep velocity structure. At sufficiently high frequencies, the seismic wavelengths will shrink to a fraction of the layer thicknesses. The resulting Green's functions will then display considerable reverberative high frequency energy, corresponding to interlayer multiples, which looks gratifyingly like scattered waves in real seismograms. Methods like DWFE (Olson *et al.*, 1984) that are meant to handle gradient zones properly will yield high-frequency Green's functions having simple pulses and lacking all the interlayer multiples. In a certain sense, methods like DWFE can be thought of as using progressively thinner layers to achieve progressively better approximations of a velocity gradient as frequency increases. The DWFE result is the more accurate result for the desired gradient structure. However, it is unclear which Green's function method is better to use considering the scattering caused by lateral heterogeneities in the earth, which neither method can handle.

## C. Integrating over the Fault Surface

In this section we discuss how the integrals, Eqs. (12) and (15), have been performed by various investigators. Instead of presenting all of the relevant equations in complete detail, we choose to emphasize the concepts behind the methods. For explicit and implicit Green's function integration methods, the relevant concepts can largely be illustrated with a very simple line-source model of a rupturing fault. For non-Green's function methods, it will be necessary to abandon the line-source problem.

In our simplified problem, the analogs to Eqs. (12) and (15) are

$$u(t) = \int_{-\infty}^{\infty} dt' \int_0^L s(x, t') T(x, t - t') \, dx \tag{41}$$

$$u(\omega) = \int_0^L s(x, \omega) T(x, \omega) \, dx, \tag{42}$$

where $s$ and $T$ are the slip and traction functions on the line source, which extends from $x = 0$ to $x = L$. For some choices of slip and traction, their space and time dependence are separable. Thus, under this assumption the slip function may be written

$$s(x, t) = a_s(x) f_s[t - t_s(x)] = a_s(x) f_s(t) * \delta[t - t_s(x)], \tag{43}$$

where $x$ is a position along the line, $a_s$ is the slip amplitude, $t_s(x)$ is the initiation time of the slip at $x$, $f_s(t)$ is its time dependence, and the asterisk (*) indicates convolution. Similarly, the traction Green's function impinging upon the line may be given by

$$T(x, t) = a_T(x) f_T[t - t_T(x)] = a_T(x) f_T(t) * \delta[t - t_T(x)]. \tag{44}$$

Their temporal Fourier transforms are

$$s(x, \omega) = a_s(x) f_s(\omega) \exp[-i\omega t_s(x)], \tag{45}$$

$$T(x, \omega) = a_T(x) f_T(\omega) \exp[-i\omega t_T(x)]. \tag{46}$$

In the ensuing discussions, we will occasionally simplify further as necessary by assuming that $a_s(x) = a_T(x) = 1$ and $f_s(\omega) = f_T(\omega) = 1$. In particular, we may assume that the traction Green's function is a plane wave, in which case

$$t_T(x) = p_T x, \tag{47}$$

where $p_T$ is the component of the wave's slowness (1/velocity) along the $x$ axis, and we may assume that the source's propagation velocity $v_r$ (rupture velocity) is constant, so that

$$t_s(x) = x/v_r = p_r x, \tag{48}$$

where $p_r$ is rupture slowness (Fig. 5).

### 1. Ray Theory and Kinematic Sources

Both ray theory and Gaussian beam Green's functions can be calculated rapidly for laterally varying media. Bernard and Madariaga (1984) and Spudich and Frazer (1984) have combined ray theory Green's functions with extended source models, while Červený et al. (1987) and Cormier and Beroza (1987) have used Gaussian beams. For simplicity we will demonstrate how ray theory is used with extended sources. A slip function in the form of Eq. (43) is used in many kinematic source models (Section I.A). A Green's function given by Eq. (44) is appropriate for describing a propagating nondispersive pulse, such as the far-field term of the asymptotic ray theory Green's function. In this case $t_T(x)$ is simply the ray theory travel time from $x$ to the observer. Straightforward substitution of these two expressions into the representation theorem, Eq. (41), leads to an exact and rather simple integration formula,

$$u(t) = f_s(t) * f_T(t) * \int_0^L a_s(x) a_T(x) \delta[t - t_a(x)] \, dx, \tag{49}$$

where

$$t_a(x) = t_s(x) + t_T(x) \tag{50}$$

is the arrival time function, i.e., it is the time that a ray generated by the rupturing of $x$ arrives at the observer. Note that the integrand of Eq. (49) is nonzero only where the argument of the delta function is zero, i.e., only for the roots of the equation

$$t - t_a(x) = 0, \quad 0 \le x \le L. \tag{51}$$

## 5. EARTHQUAKE GROUND-MOTION CALCULATION

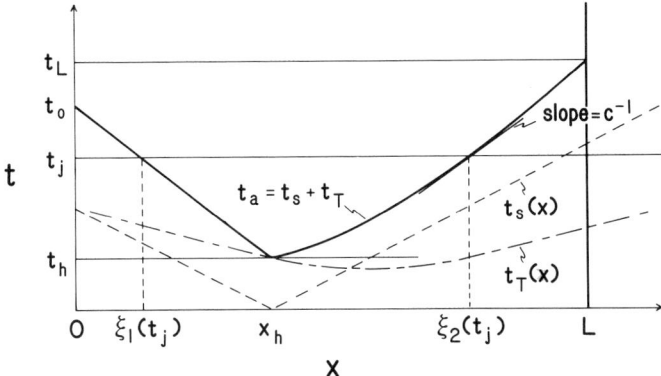

**FIG. 4.** The rupture time function $t_s(x)$ and the travel time function $t_T(x)$ are added to form the arrival time function $t_a$. For time $t = t_j$, Eq. (51) has two roots in $[0, L]$, $\xi_1(t_j)$ and $\xi_2(t_j)$.

Hence, the integral Eq. (49) may be converted to a sum over the roots of Eq. (51). The number $m(t)$ of roots in the interval $[0, L]$ will depend on $t$ and on the shapes of the $t_s$ and $t_T$ curves (Fig. 4). Let us denote the roots by $\xi_i(t)$, $i = 1, 2, \ldots, m(t)$. Generally, there will be only one root for unilateral rupture propagation or two roots for bilateral propagation, unless the rupture velocity exceeds the P- or S-wave speed somewhere in $[0, L]$, in which case there may be extra roots. Using the relation

$$\int_{-\infty}^{\infty} a(x)\delta[t - g(x)]\,dx = [a(x)|\partial_x g|^{-1}]|_{t-g(x)=0}, \qquad (52)$$

Eq. (49) becomes

$$u(t) = f_s(t) * f_T(t) * \sum_{j=1}^{m(t)} a_s[\xi_j(t)]a_T[\xi_j(t)]c[\xi_j(t)], \qquad (53)$$

where

$$c[\xi_j(t)] = |d_t[\xi_j(t)]|. \qquad (54)$$

The term $c$ is the velocity of the roots (Fig. 4), and $c^{-1}$ is exactly equal to the seismic directivity function for this line-source problem (Spudich and Frazer, 1984).

Bernard and Madariaga (1984) have introduced the idea of "critical points" to approximate Eq. (53) even further. Consider the behavior of $u(t)$ as $t$ increases from 0 to $\infty$. For $t < t_h$ (Fig. 4), $m(t) = 0$ and $u(t) = 0$. When $t$ exceeds $t_h$, $m(t)$ instantaneously jumps to 2 and $u(t)$ instantaneously takes on a finite value. As $t$ approaches $t_0$, the root $\xi_1(t)$ approaches 0. When $t$ exceeds $t_0$, $m(t)$ suddenly drops to 1 and $u(t)$ suffers another jump discontinuity. A similar set of phenomena occurs when $t$ exceeds $t_L$. Because the high-frequency part of $u(t)$ is dominated by its jump discontinuities, $u(t)$ can be approximated by pulses from the three places on the fault, $x = 0$, $x_h$, and $L$, associated with the discontinuities. These are the critical points. In our one-dimensional example, the critical points are the places where roots $\xi_1$ and $\xi_2$ either materialize or disappear. An additional critical point can occur at places where $c$ becomes singular, which will happen in the case of super-shear rupture velocity. We note in passing that seismogram amplitudes are linearly related to $c$, which is a nonlinear function of rupture time. This explains the strong influence of rupture time variation on seismogram amplitudes noted by Archuleta (1984).

For a two-dimensional fault, the result of Eq. (53) is easily generalized. If $\mathbf{x}$ is a position on the fault surface $\Sigma$, then the arrival time function is

$$t_a(\mathbf{x}) = t_s(\mathbf{x}) + t_T(\mathbf{x}). \tag{55}$$

The equation

$$t - t_a(\mathbf{x}) = 0 \tag{56}$$

is satisfied on the curve $\xi(t)$, which is an equal time contour [called an isochrone by Bernard and Madariaga (1984)] of the $t_a$ function. The term $c$ now becomes the velocity of an isochrone perpendicular to its length

$$c(\mathbf{x}) = |\nabla t_a(\mathbf{x})|^{-1}, \tag{57}$$

and the two-dimensional analog of Eq. (53) is

$$u(t) = f_s(t) * f_T(t) * I(t), \tag{58a}$$

$$I(t) = \int_{\xi(t)} a_s(\mathbf{x}) a_T(\mathbf{x}) c(\mathbf{x}) \, dl, \tag{58b}$$

where $l$ is the arc length along the curve $\xi(t)$. Each time point in the seismogram is obtained by doing a line integral along an isochrone. Evaluation of $I(t)$ in Eq. (58b) is quite simple for an arbitrarily complicated $t_a(\mathbf{x})$ using the algorithm of Spudich and Frazer (1984). A large fault $\Sigma$ is broken into a set of triangular subfaults. If the three vertices of a triangle are located at $\mathbf{x}_1$, $\mathbf{x}_2$, and $\mathbf{x}_3$, we define the quantities

$$g_i = a_s(\mathbf{x}_i) a_T(\mathbf{x}_i), \tag{59}$$

$$\tau_i = t_a(\mathbf{x}_i) \tag{60}$$

for $i = 1, 2, 3$, and we approximate $g$ and $\tau$ for $\mathbf{x}$ inside the triangle by linear interpolation between the values at the three vertices. Then $c(\mathbf{x})$ is constant within a triangle, isochrones are easily found straight lines cutting across the triangle, and the integrand $g(\mathbf{x})$ is a linear function along an isochrone and can be integrated exactly. Using this approach $I(t)$ is obtained separately for each triangular subfault; the subfaults may be dealt with in any sequence, and no complicated curve-following algorithm is needed to integrate along the isochrones. A typical calculation may involve 5000 subfaults.

Note that the result expressed in Eqs. (58a and b) is the exact result for the far-field Green's functions and is valid even when the observer is very close to the source. Spudich and Frazer (1984) note that the term "far-field" has been used both to describe certain terms in the Green's function and to describe a region, far from an extended seismic source, where the Fraunhofer approximation holds (Aki and Richards, 1980, pp. 804–805). They suggest the use of the terms far-source and near-source to describe the region where the Fraunhofer approximation is valid or invalid, respectively. Thus, Eqs. (58a and b) are the results valid in the near-source region for far-field Green's functions. What is missing from Eqs. (58a and b) is the near-field term of the Green's function. Because the near-field term is inherently lower in frequency than the far-field term, and because it decays more rapidly with distance, omission of this term leads to low-frequency errors at short epicentral distances. Farra *et al.* (1986) have investigated the importance of this omission. They have additionally extended the isochrone-integration technique to include multiple rays and maximum time pulses.

2. Point Source Summation

In performing the integrals Eqs. (12) and (15), an assumption frequently made in seismology is that a large earthquake can be simulated by a grid of point dislocations (Heaton and Helmberger, 1979; Archuleta and Hartzell, 1981; Bouchon, 1982; Heaton, 1982; Hartzell and Helmberger, 1982; Campillo and Bouchon, 1983; Liu and Helmberger, 1983). In this case the integrals over the fault surface in Eqs. (12) and (15) are replaced by a summation of all the point sources, i.e., Eq. (41) is approximated by

$$u(t) \approx \sum_{j=1}^{n} s(x_j, t) * T(x_j, t) \Delta x_j, \tag{61}$$

and Eq. (42) is approximated by

$$u(\omega) \approx \sum_{j=1}^{n} s(x_j, \omega) T(x_j, \omega) \Delta x_j, \tag{62}$$

where the interval $[0, L]$ has been broken into $n$ segments of width $\Delta x_j$ with sample point $x_j$ at the midpoint of segment $j$. Both Eqs. (61) and (62) are straightforward to compute and should yield identical results. For this

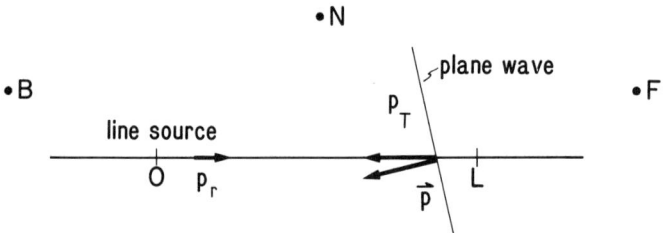

**FIG. 5.** A seismic source propagates from $x = 0$ to $x = L$ with slowness $p_r$. Plane waves emitted from distant observers at **B**, **N**, and **F** propagate with true slowness **p**, and the component of wave slowness along the $x$ axis is $p_T$.

reason we will refer to the point-source summation technique without always being specific about whether it is performed in the time or frequency domain.

The accuracy of the technique is more easily understood by analyzing the frequency-domain version. The point-source summation technique evaluates Eq. (62) by using the familiar midpoint rule for numerical quadrature. The accuracy of this approximation is highly dependent upon frequency and upon the observer's position with respect to the rupture propagation direction, as can be easily shown in our simple example. For simplicity, assume a plane wave, Eq. (47), and a constant rupture velocity, Eq. (48), and ignore amplitude variations and source time functions. Then

$$sT = \exp[-i\omega(p_r + p_T)x]. \tag{63}$$

If we are considering S waves from the rupture, and if the rupture velocity is close to the shear velocity, then from Fig. 5 it is clear that $p_r + p_T \approx 0$ for observers in the forward direction; $p_r + p_T \approx p_r$ for observers normal to the fault; $p_r + p_T \approx 2p_r$ for observers in the backward direction. In Fig. 6 it can be seen that the integrand becomes progressively more oscillatory as frequency increases or as the observer moves toward B. If the sample points $\{x_j\}$ in Eq. (62) are held fixed for all frequencies and observers, the computed value of $u(\omega)$ will be more accurate for low frequencies than high and for observers in the forward direction. In fact, this breakdown of the point-source approximation is well known (Hartzell *et al.*, 1978), and examples will be shown in Section III.A.1.

If evaluation of the integrand were easy, the most straightforward way to improve accuracy would be to decrease the sample spacing $\Delta x_j$ until the integral converged to a stable value. However, for dynamic source models, $s(x, t)$ can be very expensive to calculate, as can $T(x, t)$ for complete Green's functions. Hence, in general, we would like to perform Eq. (42) with as few expensive integrand evaluations as possible.

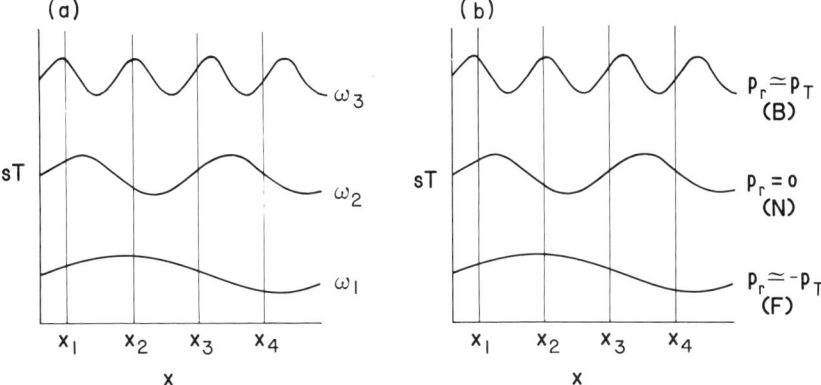

FIG. 6. Behavior of $sT$ as a function of position. (a) $p_r + p_T$ is held fixed and $sT$ is plotted for three differing frequencies (with vertical offsets). (b) The frequency $\omega$ is held fixed and $sT$ is plotted for observers in the direction of rupture (**F**), normal to the fault (**N**) and opposite the direction of rupture (**B**). See also Fig. 5. Here $x_1, \ldots, x_4$ are sample points for the numerical quadrature. Clearly the integrand becomes undersampled as $\omega$ becomes large and when $p_r + p_T$ becomes large.

## 3. Improvements on Point Source Summation Using Kinematic Slip Functions and Complete Green's Functions

Let us consider how the quadrature Eq. (62) can be improved under the assumptions that $s$ is easily calculated and $T$ is quite expensive to calculate. This case corresponds to the most commonly employed approximations used in actual ground motion modeling: $s$ is derived from a kinematic source model, while $T$ is a complete Green's function.

Under these assumptions, the basic approach has been to make the sample intervals $\{\Delta x_j\}$ as small as necessary to ensure convergence of the sum Eq. (62). Because the calculation of $s$ is easy, it is obtained exactly at each sample point. However, $T$ is calculated exactly only at a few widely spaced sample points, and it is derived at the intervening quadrature points $\{x_j\}$ by an interpolation scheme.

*a. Explicit Interpolation of Green's Functions in the Frequency Domain.* The method of Spudich (1981), which was used by Archuleta (1984) to model the 1979 Imperial Valley earthquake, interpolates $T$ explicitly in the frequency domain and works as follows. From Eqs. (45)–(48) we can see that $s$ and $T$ are basically oscillatory functions of position with spatial wavelengths $v_r/f$ for $s$ and $\beta/f$ for $T$, where $f$ is the frequency $\omega/2\pi$ and $\beta$ is the shear wave velocity. Because generally $v_r < \beta$, the spatial wavelength of $s$ is less than that of $T$ for equivalent $f$. Then, $T$ may be specified by exact calculation of its values on a grid of points $\{x_{T_j}\}$ with spacing

$$\Delta x_T(x, f) = \beta(x)/(mf), \tag{64}$$

where $m$ is the desired number of sample points per shear wavelength. Similarly, $s$ is specified by exact calculation on a grid of points $\{x_{s_j}\}$ with spacing

$$\Delta x_s(x, f) = v_r(x)/(nf), \tag{65}$$

where $n$ is a desired number of sample points. Note that the sample spacing decreases as frequency increases (Fig. 6), and that it can also vary depending on local S-wave and rupture velocities. In addition $n$ can be chosen larger than $m$ to improve accuracy. Consequently, the method of Spudich (1981) can be regarded as a crudely adaptive quadrature with the adaptive rules built in *a priori*. Equations (64) and (65) cannot be used for very low frequencies because $s$ and $T$ are not constant over the fault at $\omega = 0$. In practice some maximum allowable $\Delta x_s$ and $\Delta x_T$ are specified and used when Eqs. (64) and (65) specify greater $\Delta x_s$ and $\Delta x_T$.

Trapezoidal-rule quadrature of the product $sT$ is performed separately for each frequency, with the quadrature points being the sample points $\{x_{s_j}\}$ of $s$, and with the values of $T$ at the points $\{x_{s_j}\}$ derived from the exact values at $\{x_{T_j}\}$ by linear interpolation (Fig. 7). Two strengths of this procedure are that the error of the quadrature is independent of frequency (unlike direct point-source summation) and that the method for interpolating $T$ is valid regardless of the types of seismic waves that comprise $T$. The

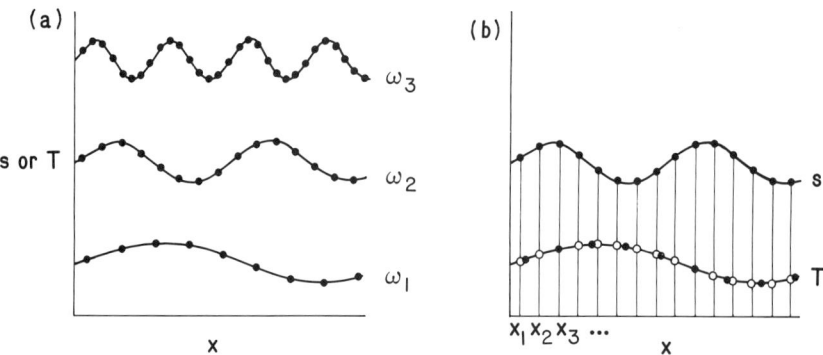

FIG. 7. (a) Both $s$ and $T$ oscillate as a function of position $x$ on the fault. They are both evaluated exactly at sample points whose spacing decreases with increasing frequency. (b) Here, $s$ and $T$ do not have the same wavelength on the fault; $s$ is evaluated exactly at quadrature points $x_1, x_2, \ldots$, while values of $T$ at the quadrature points are determined by linear interpolation between nearby exact values. The closed circle (●) indicates values calculated exactly and the open circle (○) indicates values derived by linear interpolation.

method's chief disadvantages are that computational effort and storage are proportional to the cube of the frequency bandwidth for two-dimensional faults. For any single frequency $f$, the number of quadrature points on the fault is proportional to $(\Delta x_T)^{-2}$ or $f^2$. Hence, to obtain seismograms in the frequency band $(0, f_{\max})$, computing effort and Green's function storage scale like $f_{\max}^3$. This scaling limits the frequency band in which the adaptive method may be applied. We note in passing that we have also tried to integrate $sT$ by Gaussian quadrature and by analytic integration of a piecewise cubic spline passed through the samples of $sT$, but have found the trapezoidal rule to be more dependable. While it is true that other integration methods converge more quickly than trapezoidal rule as sample spacing decreases, the obverse of that statement is also true; other methods diverge more rapidly as sample spacing increases. Our experience is that trapezoidal rule gives the most reliable results in massive computations that operate on the verge of undersampling. However, we have not investigated other quadrature methods exhaustively.

   *b. Explicit Interpolation of Green's Functions in the Time Domain.* At present there are time-domain Green's function interpolation methods that avoid the $f_{\max}^3$ storage requirement for Green's functions at the expense of a less accurate interpolation scheme. Some investigators (see, e.g., Hartzell *et al.*, 1978; Heaton and Helmberger, 1979; Heaton, 1982; Hartzell and Helmberger, 1982; Liu and Helmberger, 1983) perform an explicit time-domain point-source summation of the form

$$u(t) = \int_0^L a_s(x) f_s(t) * \delta[t - t_s(x)] * T(x, t) \, dx, \tag{66}$$

$$u(t) = f_s(t) * \int_0^L a_s(x) T[x, t - t_s(x)] \, dx$$

$$\approx f_s(t) * \sum_{j=1}^n a_s(x_j) T[x_j, t - t_s(x_j)] \Delta x_j, \tag{67}$$

where Eq. (43) has been used to define $s(x, t)$ but $T(x, t)$ has been left general. Because the slip time function $f_s$ is independent of position, it has been factored out of the sum. Because $T$ is expensive to calculate, it is obtained on a sparse grid of points $\{x_{T_i}\}$ and its value at quadrature points $x_j$ is obtained by linear interpolation that incorporates a particular physical assumption about $T$. Hartzell *et al.* (1978) assume that $T$ is composed of a set of dispersive pulses that can be separated in time and have characteristic phase velocities. In their method each pulse in a seismogram is interpolated individually. For a half-space there are only four pulses, P, S, SP, and the Rayleigh wave. Suppose that exact Green's functions are known at $x_{T_i}$ and

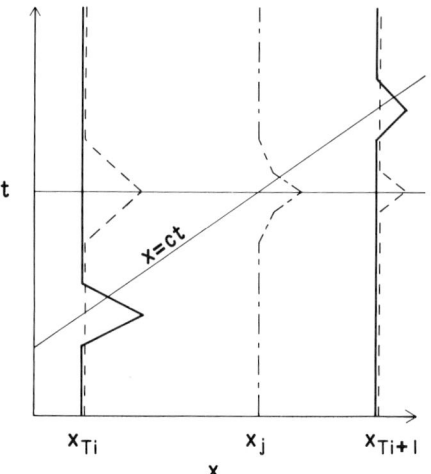

FIG. 8. Exact Green's functions calculated at $x_{T_i}$ and $x_{T_{i+1}}$. Each contain a pulse traveling at velocity $c$. Exact seismograms are appropriately stretched or compressed so that the pulses align horizontally. Linear interpolation can then be done to obtain a Green's function at $x_j$ (———, exact; ---, stretched; ----, interpolated).

$x_{T_{i+1}}$, and we desire to interpolate between these two points to obtain an approximate Green's function at $x_j$. Suppose also that we interpolate a single pulse that arrives at $x_{T_i}$ at time $t_i$ and at $x_j$ at time $t_j$. Then the interpolation rule of Hartzell et al. (1978) is

$$T(x_j, t) = \frac{t_i}{t_j}\left|\frac{t_j - t_{i+1}}{t_i - t_{i+1}}\right| T\left[x_{T_i}, \frac{t_i}{t_j}t\right] + \frac{t_{i+1}}{t_j}\left|\frac{t_i - t_j}{t_i - t_{i+1}}\right| T\left[x_{T_{i+1}}, \left(\frac{t_{i+1}}{t_j}\right)t\right]. \tag{68}$$

The exact Green's functions are first stretched or compressed in the time domain. They are weighted by a factor that introduces a $R^{-1}$ amplitude behavior in addition to the usual linear interpolation weighting factor (Fig. 8). This procedure works very well for a half-space (Fig. 9) but introduces pulse shape distortions for nondispersive waves (Fig. 8).

In many cases the largest pulse in a Green's function is a body wave similar to the direct S wave, in which case the time stretching used by Hartzell et al. (1978) can be replaced by a simple time shift that aligns the important pulse (Heaton and Helmberger, 1979; Heaton, 1982; and Hartzell and Helmberger, 1982). In this technique the $R^{-1}$ scaling is also

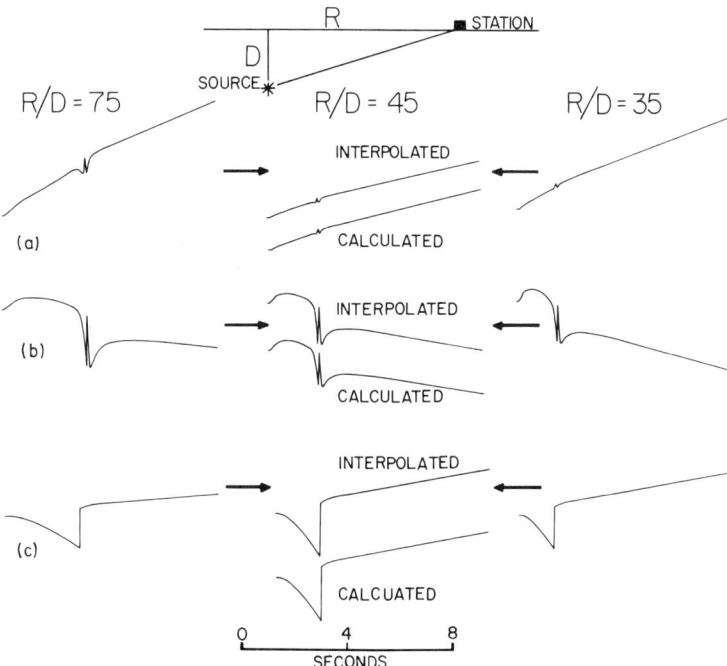

FIG. 9. Examples of linearly interpolated Green's functions for (a) radial ($R$), (b) vertical ($z$), and (c) azimuthal ($\phi$) components of displacement for a strike-slip double-couple source, from Hartzell et al. (1978). The interpolated Green's functions are compared with those calculated directly by a Cagniard–deHoop method. (Reprinted with permission from the Bulletin of the Seismological Society of America.)

omitted, in which case Eq. (68) becomes

$$T(x_j, t) = \left|\frac{t_j - t_{i+1}}{t_i - t_{i+1}}\right| T(x_{T_i}, t + t_i - t_j)$$
$$+ \left|\frac{t_{i+1} - t_j}{t_i - t_{i+1}}\right| T(x_{T_{i+1}}, t + t_{i+1} - t_j), \quad (69)$$

which corresponds to shifting the time series the same amount as indicated in Fig. 8 without stretching the seismograms. The implementation of both Eqs. (68) and (69) becomes quite complicated if pulses having differing phase velocities cross, and if the exact Green's functions do not consist of distinct arrivals, the methods break down. In practice, this interpolation is generally used assuming that the S wave is the only significant arrival, which is usually true near buried seismic sources. Interpolation using S-wave phase velocities will cause errors for waves traveling at other velocities, such as P waves and surface waves. For interpolation when a mixture of waves is present, slant-stacking techniques may prove useful (Cabrera and Levy, 1984). The advantage of the procedures presented in Eqs. (68) and (69) is that for calculations in the frequency band $(0, f_{\max})$, the storage space required for the Green's functions is proportional to $f_{\max}$, which is much more manageable than $f_{\max}^3$. In addition a minimal number of expensive theoretical Green's function evaluations are needed. However, the actual performance of the point-source summation Eq. (61) is not facilitated by this interpolation.

*c. Implicit Green's Function Extrapolation Using Temporal Convolution.* Using the assumption that the Green's function is dominated by the S wave, Apsel *et al.* (1981) have developed a simple procedure, using isochrone integration, which simultaneously extrapolates the Green's functions and includes the effects of rupture propagation. Suppose we have calculated $T(x, t)$ exactly on a set of sample points $\{x_{T_j}\}$, and we wish to approximate $T(x, t)$ at nearby points simply by shifting $T(x_{T_j}, t)$ forward or backward in time so that the S wave of the interpolated seismogram always arrives at $x$ at time $t_T(x)$, as in Eq. (44). We may also want to modify the amplitude of the interpolated seismogram by some factor $A(x, x_{T_j})$ to account for geometric spreading or other factors. Then our extrapolated Green's function may be written

$$T(x, t) = A(x, x_{T_j})T(x_{T_j}, t) * \delta[t + t_T(x_{T_j}) - t_T(x)], \quad (70)$$

with $A(x_{T_j}, x_{T_j}) = 1$. Note that Eq. (70) involves only $x_{T_j}$. For this reason Eq. (70) is an extrapolation. Let our integration interval $[0, L]$ be broken into $n$ domains of width $\Delta x_{T_j}$, $j = 1, 2, \ldots, n$, with each domain covering the interval $[x_{L_j}, x_{G_j}]$. Let the slip be given by

$$s(x, t) = a_s(x)f_s(t) * \delta[t - t_s(x)], \quad (71)$$

where $t_s(x)$ is the time that rupture initiates at point $x$, and $f_s(t)$ is the slip time function, which in this simple example we fix to be independent of position. From Eq. (41) we then have

$$u(t) \approx f_s(t) * \sum_{j=1}^{n} T(x_{T_j}, t) * I_j(t), \quad (72)$$

$$I_j(t) = \int_{x_{L_j}}^{x_{G_j}} a_s(x) A(x, x_{T_j}) \delta[t - t_{a_j}(x)] \, dx, \quad (73)$$

where

$$t_{a_j}(x) = t_T(x) - t_T(x_{T_j}) + t_s(x) \quad (74)$$

is an arrival time function analogous to Eq. (50). Thus, applying the same trick of Eq. (52) to Eq. (73) gives

$$I_j(t) = \sum_{k=1}^{m(t)} a_s[\xi_k(t)] A[\xi_k(t), x_{T_j}] c[\xi_k(t)], \tag{75}$$

where $\{\xi_k(t)\}$ are the $m(t)$ roots of $t - t_{a_j}(x) = 0$ in the interval $[x_{L_j}, x_{G_j}]$, and where

$$c[\xi_k(t)] = |d_t \xi_k(t)|. \tag{76}$$

Thus, $I_j(t)$ is a convolution filter that is applied to the point-source Green's function $T(x_{T_j}, t)$ in Eq. (72) to correct it for the effect of rupture propagation across the domain $[x_{L_j}, x_{G_j}]$, using the extrapolation rule of Eq. (70) for Green's functions. As long as the extrapolation rule is valid (i.e., as long as most of the energy in the Green's functions propagates at the S-wave velocity), use of Eqs. (72) and (75) will lead to a synthetic seismogram that is independent of grid size $\Delta x_T$. Of course, phases traveling at other velocities, such as surface waves, will not be interpolated properly. For the amplitude correction factor $A$, Apsel et al. (1981) use

$$A(x, x_{T_j}) = \left[r(x_{T_j})/r(x)\right]^P, \tag{77}$$

where $r(x)$ is the distance from the observer to the point $x$ on the fault, and $P$ is approximately 2.0. However, they point out that the amplitude correction is of considerably less importance than the time correction, particularly at distances greater than 5 or 10 km. Equation (72) assumes that the slip function $f_s(t)$ is the same over the entire fault. If the slip time function is independent of position within each domain $[x_{L_j}, x_{G_j}]$ but varies from domain to domain, then Eq. (72) may be rewritten

$$u(t) \approx \sum_{j=1}^{n} f_s(x_{T_j}, t) * T(x_{T_j}, t) * I_j(t). \tag{78}$$

If further spatial variability of $f_s$ is desired, then the domain $[x_{L_j}, x_{G_j}]$ can be subdivided into many smaller subdomains in which $f_s$ is position independent, and $I_j(t)$ can be written as a sum of terms, each similar to Eq. (75) and each convolved with its own $f_s(t)$. This would be a computational inexpensive way of allowing $f_s(t)$ to vary within $[x_{L_j}, x_{G_j}]$. Similar to the explicit time-domain Green's function interpolation methods, this implicit extrapolation method requires a relatively small number of exact Green's function evaluations, and the necessary Green's function storage grows linearly with frequency bandwidth. The computational effort of calculating $I_j(t)$ in Eq. (73) and its two-dimensional analog Eq. (58b) probably scales like $f_{max}^2$ for a two-dimensional fault because the necessary time sampling interval goes like $f_{max}^{-1}$, as does the length of the line integral segment $dl$ in Eq. (58b).

### 4. A Non-Green's Function Method Using Simple Kinematic Source Models in a Layered Medium

Thus far all the methods that we have discussed that use complete Green's functions have been explicit Green's function integration methods. In all these methods, the slip models and the Green's function $T$ are calculated independently of each other. Bouchon and Aki (1977) and Bouchon (1979) introduced a non-Green's function method in which an extended seismic source is explicitly used for the source term for wave propagation in a layered medium. This is made easy for them by their use of Cartesian coordinates in a homogeneous layer. In the following discussion we follow the derivation by Chouet (1987). We saw earlier in a laterally homogeneous medium how the effect of a point seismic source at depth can be included as a discontinuity in the displacement-stress vector **b** at depth $z$. When an extended seismic source is enclosed within a uniform layer, the effect of the extended source may be included in a different way by determining the stresses and displacements it causes on the overlying and underlying layer interfaces. Propagator matrices are then used to relate these source-induced stresses and displacements to those at the free surface and in the underlying half-space. Once this is done, the wave propagation through the layers is accomplished using the method of Dunkin (1965). While the use of Cartesian coordinates is not essential to the method, it simplifies the integrals over the sources that are used to determine the stresses and displacements it causes.

We start with a layered velocity structure, as in Fig. 10 and the Cartesian coordinate system of Fig. 3. The ground displacement **u** may be written

$$\mathbf{u}(x, y, z, \omega) = \int_{-\infty}^{\infty} \int_{-\infty}^{\infty} \mathbf{u}(k_x, k_y, z, \omega) \exp(-ik_x x - ik_y y) \, dk_x \, dk_y. \tag{79}$$

As discussed in Section II.B.2.b, to obtain $\mathbf{u}(k_x, k_y, z, \omega)$ at a particular point $(k_x, k_y)$ involves the solution of the P-SV and SH problems for a plane wave with wave number $\mathbf{k} = k_x \hat{\mathbf{x}} + k_y \hat{\mathbf{y}}$. This yields radial, transverse, and vertical displacements $u_{x'}$, $u_{y'}$, and $u_z$, which are all functions of $k_x$, $k_y$, $z$, and $\omega$. Displacements in the $x$ and $y$ directions are obtained by

$$u_x = (k_x/k)u_{x'} - (k_y/k)u_{y'}, \tag{80a}$$

and

$$u_y = (k_y/k)u_{x'} - (k_x/k)u_{y'}, \tag{80b}$$

where $k = |\mathbf{k}|$.

We follow Chouet (1987) in obtaining $u_{x'}$, $u_{y'}$, and $u_z$ for an extended source. Consider first the P-SV problem and define the displacement-stress

4. EARTHQUAKE GROUND-MOTION CALCULATION        241

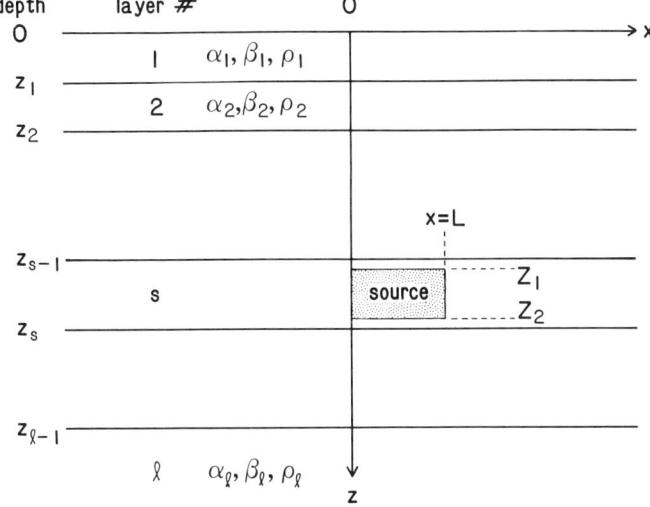

FIG. 10. Cross section of a homogeneously layered earth model. Material properties are constant within each layer. The source is a strike-slip fault in the $x$–$z$ plane extending from $x = 0$ to $x = L$ and bounded by depths $Z_1$ and $Z_2$. Slip is uniform over the fault, and rupture propagates in the $+x$ direction as a vertical line with velocity $v_r$.

vector **b** to be $[u_{x'}, u_z, \sigma_{zz}, \sigma_{zx'}]^T$. We note that this definition is slightly different from Kennett's. Following the layer numbering scheme of Fig. 10, let $\mathbf{b}_n(z)$ be the displacement-stress vector at depth $z$ in layer $n$. Assume for the moment that we know $\mathbf{b}_s^0(z_{s-1})$ and $\mathbf{b}_s^0(z_s)$, which are the displacement-stress vectors above and below the source caused by upgoing and downgoing waves, respectively, generated by the source. The total displacement-stress vector at depth $z_{s-1}, \mathbf{b}_s(z_{s-1})$, is a sum of the upgoing waves from the source, $\mathbf{b}_s^0(z_{s-1})$, and reverberated waves coming from other depths. We relate **b** at the top of the source layer to **b** at the free surface:

$$\mathbf{b}_s(z_{s-1}) = P_{s-1}P_{s-2} \cdots P_1 \mathbf{b}_1(0), \tag{81}$$

where we define $P_n$ to be the propagator across the $n$th layer,

$$P_n = P_n(z_{n-1}, z_n). \tag{82}$$

The **b** at the bottom of the source layer is related to **b** in layer $l$, the underlying half-space, by

$$\mathbf{b}_l(z_{l-1}) = P_{l-1}P_{l-2} \cdots P_{s+1} \mathbf{b}_{s+1}(z_s) \tag{83}$$

and from Eq. (33)

$$\mathbf{v}_l(z_{l-1}) + D_l^{-1} \mathbf{b}_l(z_{l-1}). \tag{84}$$

Our boundary conditions are a stress-free surface at $z = 0$,

$$\mathbf{b}_1(0) = [u_{x'}, u_z, 0, 0]^T, \tag{85}$$

no upgoing waves in layer $l$,

$$\mathbf{v}_l(z_{l-1}) = [0, 0, \Phi^-, \Psi^-]^T, \tag{86}$$

and continuity of $\mathbf{b}$ across all interfaces,

$$\mathbf{b}_{i+1}(z_i) = \mathbf{b}_i(z_i), \quad i = 1, \ldots, s-1, s, \ldots, l-1. \tag{87}$$

We note that Eq. (87) shows explicitly that in Chouet's (1987) formulation the source contribution does not cause a discontinuity in $\mathbf{b}$ at an interface. Then

$$\mathbf{b}_s(z_s) - \mathbf{b}_s^0(z_s) = P_s[\mathbf{b}_s(z_{s-1}) - \mathbf{b}_s^0(z_{s-1})], \tag{88}$$

i.e., the propagator connects only the reverberated components of the waves at levels $z_{s-1}$ and $z_s$. Skipping the details of the subsequent derivation, which are found in Chouet (1987), we finally obtain

$$\begin{bmatrix} u_{x'}(k_x, k_y) \\ u_z(k_x, k_y) \end{bmatrix} = R_{11}^{-1}[Q_{11}Q_{12}\mathbf{b}_s^0(z_{s-1}) - P_{11}P_{12}\mathbf{b}_s^0(z_s)] \tag{89}$$

for the radial displacement $u_{x'}$ and vertical displacement $u_z$ caused by source terms $\mathbf{b}_s^0(z_{s-1})$ and $\mathbf{b}_s^0(z_s)$. Here,

$$P = D_l^{-1}P_{l-1} \cdots P_{s+1}, \tag{90a}$$

$$Q = P_{l-1} \cdots P_s, \tag{90b}$$

and

$$R = D_{l-1}^{-1}P_{l-1} \cdots P_1, \tag{90c}$$

and each of these $4 \times 4$ matrices is partitioned into four $2 \times 2$ matrices, e.g.,

$$R = \begin{bmatrix} R_{11} & R_{12} \\ R_{21} & R_{22} \end{bmatrix}. \tag{91}$$

The matrix $R_{11}^{-1}$ may be written

$$R_{11}^{-1} = \tilde{R}_{11}/|R_{11}|, \tag{92}$$

where $|R_{11}|$ is the determinant of $R_{11}$ and is the secular function for the layered medium and $\tilde{R}_{11}$ is the transpose of the cofactor of $R_{11}$. Of course, direct evaluation of $|R_{11}|$ and products $\tilde{R}_{11}Q_{11}$ and $\tilde{R}_{11}P_{11}$ lead to well-known instabilities (Chin et al., 1984b). Bouchon (1979) and Chouet (1981, 1987) use the method of Dunkin (1965) to evaluate these terms. Equation

## 4. EARTHQUAKE GROUND-MOTION CALCULATION

(89) is used to obtain $u_{x'}$ and $u_z$, and $u_{y'}$ is found using a similar equation for the SH problem. Displacements in the $x$–$y$–$z$ coordinate system for a plane wave with wave number **k** are derived from Eqs. (80a and b), and the total ground motion is a sum of all the plane wave components of Eq. (79). We comment later on how this integral is performed.

We now turn to the derivation of $\mathbf{b}_s^0(z_{s-1})$ and $\mathbf{b}_s(z_s)$, the displacement-stress vectors resulting above and below the source from the upgoing and downgoing source radiation. We follow the approach of Bouchon (1979) and integrate the upgoing and downgoing P and S potentials and then use Eq. (33).

The displacements **u** in the medium are related to the potentials $\Phi$ and $\Psi$ by

$$\mathbf{u} = \nabla \Phi + \nabla \times \Psi, \quad \nabla \cdot \Psi = 0, \quad (93)$$

where $\Psi = (\Psi_x, \Psi_y, \Psi_z)^T$ (Fig. 3), and $\Phi$ and $\Psi$ are solutions to the wave equations

$$\nabla^2 \Phi = 1/\alpha^2 \, \partial_t^2 \Phi, \quad (94a)$$

$$\nabla^2 \Psi = 1/\beta^2 \, \partial_t^2 \Psi. \quad (94b)$$

Within a uniform layer, they may be written as plane waves of the form

$$\Phi(\omega) = A \exp(-ik_x x - ik_y y \pm i\nu z), \quad (95a)$$

$$\Psi(\omega) = \mathbf{B} \exp(-ik_x x - ik_y y \pm i\gamma z), \quad (95b)$$

where

$$\nu = \left(k_\alpha^2 - k_x^2 - k_y^2\right)^{1/2}, \quad \operatorname{Im} \nu \leq 0, \quad (96a)$$

and

$$\gamma = \left(k_\beta^2 - k_x^2 - k_y^2\right)^{1/2}, \quad \operatorname{Im} \gamma \leq 0, \quad (96b)$$

are vertical wave numbers with $k_\alpha = \omega/\alpha$ and $k_\beta = \omega/\beta$ being wave numbers along the direction of propagation of P and S waves, respectively. The plus and minus signs correspond to upward and downward propagation, respectively.

A point dislocation can be represented by an equivalent double couple of point forces. If $\Phi^x, \Psi^x; \Phi^y, \Psi^y;$ and $\Phi^z, \Psi^z$ are the potentials radiated by point forces applied at $(x_0, y_0, z_0)$ in the $x$, $y$, and $z$ directions, respectively, then the potentials radiated by a point dislocation at $(x_0, y_0, z_0)$ having the same mechanism as our extended source (Fig. 10) are

$$\Phi_p(x, y, z) = m(x_0, z_0)\left[\partial_{y_0}\Phi^x(x, y, z) + \partial_{x_0}\Phi^y(x, y, z)\right] \quad (97a)$$

and

$$\Psi_p(x, y, z) = m(x_0, z_0)[\partial_{y_0}\Psi^x(x, y, z) + \partial_{x_0}\Psi^y(x, y, z)], \quad (97b)$$

where $m(x_0, z_0)$ is the moment density per unit area of the source. Equations (97a and b) are valid for a left-lateral, strike-slip dislocation. Similar terms can be found for other dislocation mechanisms. Hence, to find the total potential radiated by an extended source, we integrate Eqs. (97a and b) over the fault surface. If slip on the fault is given by $s(x_0, z_0, \omega)$, then the radiated compressional potential is

$$\Phi(x, y, z, \omega) = \mu_s \int_{Z_1}^{Z_2} \int_0^L s(x_0, z_0, \omega)[\partial_{y_0}\Phi^x + \partial_{x_0}\Phi^y] \, dx_0 \, dz_0, \quad (98)$$

with a similar expression for $\Psi$. The expressions for the potentials radiated by a point force in the $x$ direction are of the form

$$\Phi^x \approx (k_x/\nu)\exp[-ik_x(x-x_0) - ik_y(y-y_0) - i\nu|z-z_0|], \quad (99a)$$

and

$$\Psi_z^x \approx (k_y/\gamma)\exp[-ik_x(x-x_0) - ik_y(y-y_0) - i\gamma|z-z_0|] \quad (99b)$$

(Bouchon, 1979, Eq. (13)), with similar expressions for other components of $\Psi$ and point-force directions. The absolute value signs in Eqs. (99a and b) ensure that we have downgoing waves below the source depth $z_0$ and upgoing waves above. For simplicity we define

$$e_p = \exp[-ik_x(x-x_0) - ik_y(y-y_0) - i\nu|z-z_0|] \quad (100a)$$

and

$$e_s = \exp[-ik_x(x-x_0) - ik_y(y-y_0) - i\gamma|z-z_0|]. \quad (100b)$$

Because of the form of the potentials, Eqs. (99a and b), spatial differentiation to obtain the potentials caused by force couples is easy, e.g.,

$$\partial_{x_0}\Phi^x = ik_x\Phi^x. \quad (101)$$

Hence, the compressional potential radiated from the extended source of Eq. (98) will be a sum of terms of the form

$$\Phi(k_x, k_y, z, \omega) \approx F(\omega, k_x, k_y)I_F, \quad (102)$$

where

$$I_F = \int_{Z_1}^{Z_2} \int_0^L \mu s e_p \, dx_0 \, dz_0, \quad (103)$$

and where $F(\omega, k_x, k_y)$ is a product of terms like the $k_x/\nu$ in Eq. (99a), and the $ik_x$ in Eq. (101). Similar expressions can be written for $\Psi$. We separate $F$ from the integrated source term $I_F$ to emphasize that $I_F$ is the two-dimensional Fourier transform of the slip distribution. If the slip function $s$ is sufficiently simple, $I_F$ can be obtained analytically. For example, Bouchon (1979) accomplishes this by using

$$s(x_0, z_0, \omega) = (D/i\omega)\exp(-ix_0/v_r), \tag{104}$$

where $D$ is a constant dislocation. The exponential phase factor corresponds to a unilateral rupture in the $x$ direction with speed $v_r$. Chouet (1982) treats the problem of an expanding circular tensile crack, and in $I_F$ he uses

$$s(r, \theta, \omega) = (D/i\omega)\exp(-i\omega r/v_r), \tag{105}$$

where $r$ and $\theta$ are a cylindrical coordinate system with the origin at the center of the crack. For this choice of $s$, $I_F$ can also be integrated analytically. For some choices of $s$, the integration over one dimension of the fault can be done analytically, whereas the other dimension must be integrated numerically. Chouet (1981), studying a fluid driven tensile fracture, uses an opening function that varies with $x_0$ but is independent of $z_0$:

$$s(x_0, z_0, t) = D(x_0, t), \tag{106}$$

where $D(x_0, t)$ is the output from a 2D finite-difference simulation of the fracture process. In this case the analytic integral over $z_0$ is straightforward, but the integral over $x_0$ must be performed numerically. Chouet performs that integral by using the rectangle rule quadrature. Other applications of this technique can be found in Chouet (1983, 1985). An alternative method for such integrals might be a generalized Filon method (Frazer and Gettrust, 1984).

Equation (102) and analogous equations give us the source-induced $\Phi$ and $\Psi$ at depths $z_{s-1}$ and $z_s$. We can use Eqs. (33) and (39) to write the source terms needed in Eq. (89),

$$\mathbf{b}_s^0(z_{s-1}) = D_P \mathbf{v}(z_{s-1}) \tag{107}$$

and

$$\mathbf{b}_s^0(z_s) = D_P \mathbf{v}(z_s), \tag{108}$$

where

$$\mathbf{v}(z_{s-1}) = [\Phi^+(z_{s-1}), \Psi^+(z_{s-1}), 0, 0]^T, \tag{109}$$

and

$$\mathbf{v}(z_s) = [0, 0, \Phi^-(z_s), \Psi^-(z_s)]^T. \tag{110}$$

The component $\Psi^\pm$ of $\Psi$ associated with the P-SV propagation having wave number $\mathbf{k}$ is the component of $\Psi$ lying in the $x$–$y$ plane perpendicular to **k** (Fig. 3):

$$\Psi_{y'} = (k_x \Psi_y - k_y \Psi_x)/k, \tag{111a}$$

$$\Psi^+ = \Psi_{y'}(z_{s-1}), \tag{111b}$$

$$\Psi^- = \Psi_{y'}(z_s). \tag{111c}$$

Here, $\Phi^+$ is given by evaluating Eq. (102) for $z = z_{s-1}$, and $\Phi^-$ is obtained from Eq. (102) with $z = z_s$.

The derivation of $\mathbf{b}_s^0$ for the SH problem is slightly different from that for the P–SV problem, because it is not necessary to use the SH form of Eq. (33) to transform from potentials to displacement. If we denote the components of SH displacement above and below the source as $u_{y'}^+$ and $u_{y'}^-$, respectively, then

$$u_{y'}^\pm = i[k\Psi_z \pm (\gamma/k)(k_x \Psi_x + k_y \Psi_y)] \tag{112}$$

(Bouchon, 1979, Eq. (23)), and

$$\mathbf{b}_s^0(z_{s-1}) = u_{y'}^+ [1, i\mu_s \gamma_s]^T, \tag{113}$$

$$\mathbf{b}_s^0(z_s) = u_{y'}^- [1, -i\mu_s \gamma_s]^T \tag{114}$$

(Chouet, 1987).

A final comment is warranted on the evaluation of the Fourier transform over the wave number (Eq. 79). Any of a variety of quadrature rules could be used (Frazer and Gettrust, 1984). Bouchon and Aki (1977) and Bouchon (1979) use the rectangle rule quadrature to convert Eq. (79) to the form

$$\mathbf{u}_d(x, y, z, \omega) \approx \left[4\pi^2/L_x L_y\right] \sum_{n_x = -\infty}^{\infty} \sum_{n_y = -\infty}^{\infty} \mathbf{K}_F(n_x, n_y), \tag{115}$$

where

$$\mathbf{K}_F(n_x, n_y) = \mathbf{u}(k_x, k_y, z, \omega) \exp(-ik_x x - ik_y y), \tag{116}$$

$$k_x = 2\pi n_x/L_x, \qquad k_y = 2\pi n_y/L_y, \tag{117}$$

and where the $d$ subscript indicates the result of a discrete sum. Their conversion of the continuous integral to a discrete sum has interesting properties. In this particular problem, use of the rectangle rule happens to give the exact answer to a different but related physical problem, one in which there is a grid of identical sources separated by distances $L_x$ and $L_y$ in the $x$ and $y$ directions. Consequently $\mathbf{u}_d(x, y, z, t)$ will be identically equal to the result for a single source $\mathbf{u}(x, y, z, t)$ for times $t$ up until the wave from the nearest multiple source arrives at the observer. Hence, this time can be made arbitrarily large by choosing $L_x$ and $L_y$ sufficiently large.

Another way to look at the properties of the discrete sum is to recognize that Eq. (79) is a spatial Fourier transform because of the exponential term. By using the sum, Eq. (115), we are implicitly saying that $\mathbf{K}_F(k_x, k_y)$ is a sum of Dirac delta functions of the form $\sum\sum \delta(k_x - 2\pi n_x/L_x)\delta(k_y - 2\pi n_y/L_y)$, where $k_x$ and $k_y$ are continuous variables. Hence its Fourier transform must be periodic in space. The use of the discrete transform is in no way necessary for including the effects of an extended source.

The conversion from integrals over $k_x$ and $k_y$ to the discrete sum can be performed at any stage in the derivation. We have delayed its introduction until after derivation of the potentials from the extended source; Bouchon and Aki (1977) and Bouchon (1979) introduced it earlier and expressed the point force potentials $\Phi^x, \Psi^x; \Phi^y, \Psi^y;$ and $\Phi^z, \Psi^z$ as discrete sums. The actual evaluation of the sum is the last step in obtaining the extended-source seismograms.

From the standpoint of extended-source modeling, the most useful consequence of this method of inverse transformation is that it yields the ground motions at a dense grid in $x$ and $y$ of observer locations. By contrast, all of the implicit and explicit Green's function integration methods we have previously discussed yield ground motions at a small set of observer locations. Hence, this method would be quite useful for studies of ground motions over large regions (Bouchon, 1980a,b; Chouet, 1981, 1983, 1985).

## III. Examples of Finite Fault Calculations

### A. HASKELL MODEL IN A WHOLE SPACE

Although a Haskell model of rupture occurring in a uniform whole space was never considered, even by Haskell himself, to be a realistic model of an earthquake, it conveniently illustrates a few of the principles presented earlier. Figure 11 shows our test geometry. The fault is a 1 km × 1 km square lying in the $x_1$–$x_2$ plane, centered on $x_1 = 0$, $x_2 = 0$. The rupture front is a line parallel to the $x_2$ axis, and it advances at a uniform velocity in the $+x_1$ direction. The slip vector is uniform everywhere and is parallel to the $x_1$ axis. Five observer locations are considered: two in the forward direction ($\mathbf{F}_1$ and $\mathbf{F}_2$), two normal to the fault ($\mathbf{N}_1$ and $\mathbf{N}_2$), and one in the backward direction (**B**). Except when stated otherwise, rupture velocity in all these examples is 3.0 km/sec, slip velocity is 1.0 cm/sec, rise time is 0.12 sec, and the medium parameters are $\alpha = 6.0$ km/sec, $\beta = 3.4$ km/sec, and $\rho = 2.5$ gm/cm³.

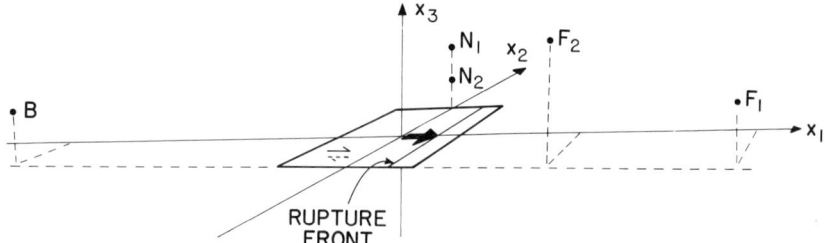

**FIG. 11.** Geometry of fault and observer locations $F_1$ (3.0, −0.5, 0.5), $F_2$ (1.5, −0.5, 1.0), $N_1$ (0, 0.5, 0.43), $N_2$ (0, 0.5, 0.2), and **B** (−3.0, −0.5, 0.5) for a Haskell model in a wholespace. The fault is 1 km × 1 km and lies in the $x_1$–$x_2$ plane. The rupture front is a line parallel to the $x_2$ axis that advances with constant velocity in the $+x_1$ direction. Slip is parallel to the $x_1$ axis and has constant amplitude on the fault.

### 1. Point-Source Summation

The characteristic behavior of the point-source summation method becomes immediately obvious if we examine the form of **s** and **T** on the fault surface. Suppose we wish to simulate our Haskell model by using a grid of 30 × 30 point sources on the fault and using the form of the representation theorem, Eq. (15). For an observer at $F_1$ and for $\omega = 20\pi$ sec$^{-1}$, Re$[T_1^1(\mathbf{x}, \omega; F_1, 0)]$ is shown in Fig. 12, sampled on the 30 × 30 grid. Because the Green's function is dominated by the far-field S wave at this

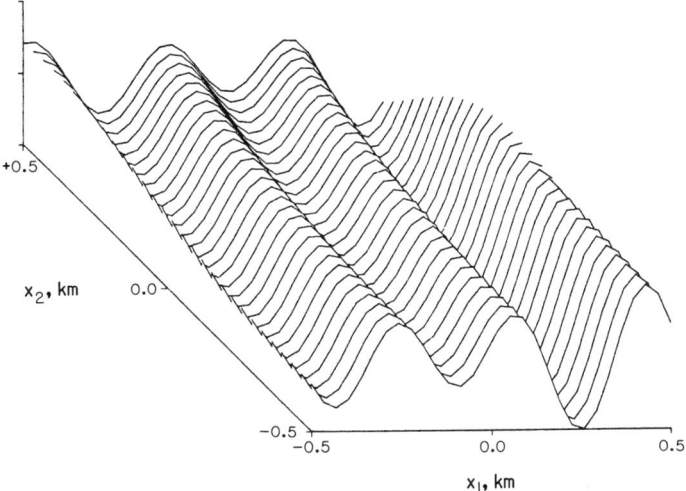

**FIG. 12.** Real part of the $x_1$ component of traction resulting on the fault in Fig. 11 caused by the application of a point force in the $x_1$ direction at the observer at $F_1$. The result shown is the 10-Hz spectral component, evaluated on a square grid of 900 points on the fault.

distance, the traction should be of the form of Eq. (46), and lines of equal travel time should be lines of equal phase. Consequently, we expect the lines of equal phase to form circles concentric about $\mathbf{F}_1$, which is what Fig. 12 shows. Slip $\mathbf{s}$ is of the form of Eq. (45), with lines of equal phase being lines of equal rupture time (Fig. 13). The dot product of these two functions, i.e., the integrand of Eq. (14) is shown in Fig. 13. As promised in Section II.C.1 for an observer in the forward direction, the integrand is smoother than either $\mathbf{s}$ or $\mathbf{T}$. It is well sampled by the $30 \times 30$ grid, and

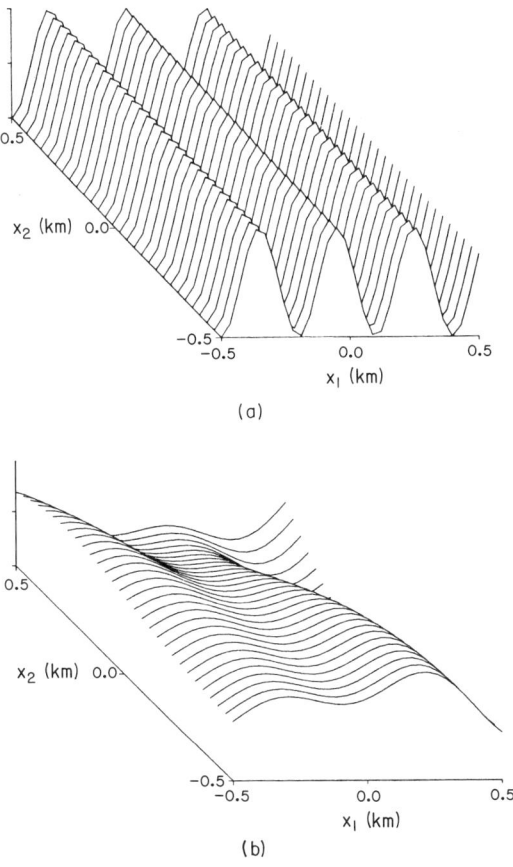

**FIG. 13.** (a) Real part of the 10-Hz component of the slip spectrum, shown as a function of position on the fault. The rupture front is a line that advances at 3.0 km/sec left to right. (b) Real part of the product of the function shown in (a) with the traction function of Fig. 12. This scalar function is integrated over the fault surface to obtain the 10-Hz component of the ground-motion spectrum at $\mathbf{F}_1$. The functions in (a) and (b) are both evaluated on a square grid of 900 points, which is sufficiently dense to represent $\mathbf{sT}$ accurately.

consequently it can be integrated accurately by the midpoint rule. If we drop the rupture velocity to 1.0 km/sec, both **s** and **s** · **T** become considerably more oscillatory, and the 30 × 30 grid no longer provides an adequate sampling of **s** · **T** (Fig. 14). The sparse sampling leads to an inaccurate quadrature result. In the time domain seismograms, these numerical inaccuracies are manifested as individual pulses from each of the point sources. These inaccuracies are greater for observers in the backward direction than for those in the forward direction (Fig. 15). As mentioned earlier, the numerical errors associated with point-source summations first become

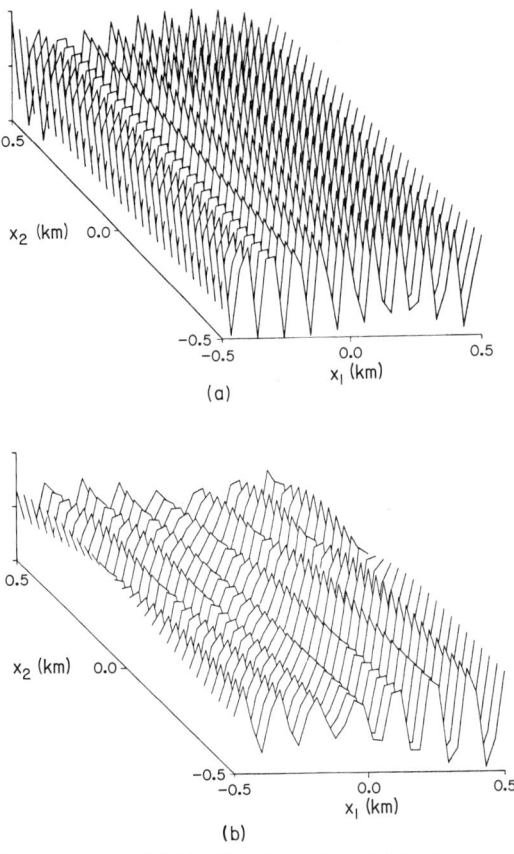

**FIG. 14.** (a) The same as panel (a) in Fig. 13 but for a 1.0 km/sec rupture velocity. Note that the spatial wavelength of the function is much shorter and that it is undersampled by the 900 point grid. (b) The real part of the product of the function shown in (a) with the traction function of Fig. 12. This product is undersampled solely because of the poor sampling of the slip function.

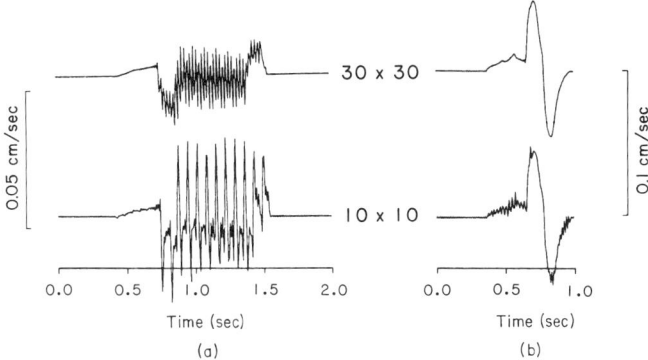

FIG. 15. Particle velocity in the $x_3$ direction at observer locations **B** (a) and $F_2$ (b), calculated by point-source summation in the frequency domain. These seismograms are unfiltered and are calculated in the 0–63 Hz band. Seismograms in the upper row are calculated for a square grid of 900 point sources on the fault; those in the lower row are for a grid of 100 point sources. The effect of discrete point sources is much more apparent in the backward direction (observer at **B**, Fig. 11) than in the forward direction ($F_2$).

evident at high frequencies. At low frequencies, for which the integrands have long spatial wavelengths, the point source summation can be quite accurate, as can be seen for the flatness of the zero levels in the seismograms in Fig. 15.

2. Adaptive Integration

The inaccuracies of a point-source summation generally result from the undersampling of **s**. In Fig. 12 it is clear that the grid of 900 points is quite adequate for sampling **T** at 10 Hz regardless of the rupture velocity. Although it may be necessary to calculate **s** and **s** · **T** on a much denser grid to assure an accurate result, values of **T** on an arbitrarily dense grid can be obtained by bilinear interpolation between the sample points shown in Fig. 12. Consider momentarily an example calculation concocted to exacerbate the problems of point-source summation. We again use a Haskell model in a whole space, but we boost the medium velocities to $\alpha = 7.0$ km/sec, $\beta = 4.0$ km/sec, and $\rho = 2.8$ gm/cm³. We drop the rupture velocity to 0.6 km/sec and shorten the rise time to 0.01 sec. Seismograms calculated in the 0–10 Hz band for observers at $F_2$, $N_1$, and **B** (Fig. 11) are shown in Fig. 16. The discreteness of the 16 × 16 grid of point sources is apparent at all azimuths but especially in the backward direction. If the adaptive sampling rules, Eqs. (64) and (65), are used with $m = n = 6$, **T** is calculated on a 5 × 5 grid and **s** on a 6 × 6 grid at the lowest frequencies. At 10 Hz **T** is calculated on a 16 × 16 grid, and **s** is calculated on a 104 × 104 grid. By

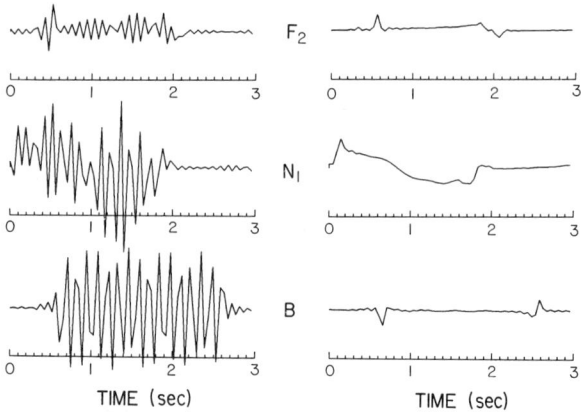

**FIG. 16.** The $x_3$ component of velocity in the 0–10-Hz band for observers at $\mathbf{F}_2$, $\mathbf{N}_1$, and $\mathbf{B}$ (Fig. 11) for a 0.6 km/sec rupture velocity. The left column consists of seismograms calculated by point-source summation of a grid of $16 \times 16$ sources. The seismograms in the right column result from summations in which the number of grid points was adjusted for each frequency to ensure 6 grid points per wavelength. The traction function was sampled on a $5 \times 5$ grid at the lowest frequency and on a $16 \times 16$ grid at the highest frequency. The slip function was sampled on a $6 \times 6$ grid at the lowest frequency and a $104 \times 104$ grid at the highest frequency. Such dense sampling of **s** was necessitated by the very low rupture velocity.

using an adaptive grid, a much more accurate answer is obtained at high frequencies, compared to a point-source summation, and **T** is evaluated ~ 50% fewer times. Because the adaptive technique attempts to distribute errors equally over the entire frequency band, at low frequencies the adaptive technique is less accurate than the point-source summation in this case. This feature can be seen by examining the seismograms for the observer at $\mathbf{N}_1$ after 2 sec in Fig. 16.

### 3. A Haskell Model and Far-Field Green's Functions

In Section II.C.1 we used far-field ray theory Green's functions with kinematic rupture models to obtain high-frequency seismograms in the near-source region of a fault. We now show an example of such a calculation for an observer at $\mathbf{N}_2$ (Fig. 11). The observer at $\mathbf{N}_2$ is well within one source dimension of the fault. In this example we return to the original test problem except that the rise time is 0.1 sec. Figure 17 compares the isochrone integration with a frequency-domain point-source summation using only the $R^{-1}$ terms (discarding the $R^{-2}$ and $R^{-4}$ terms) for a dislocation in a whole space [Aki and Richards, 1980, Eq. (4.30)]. The isochrone-integration seismogram is unfiltered, and the point-source summation seismogram is bandpass filtered from 0 to 30 Hz using a zero

# 4. EARTHQUAKE GROUND-MOTION CALCULATION

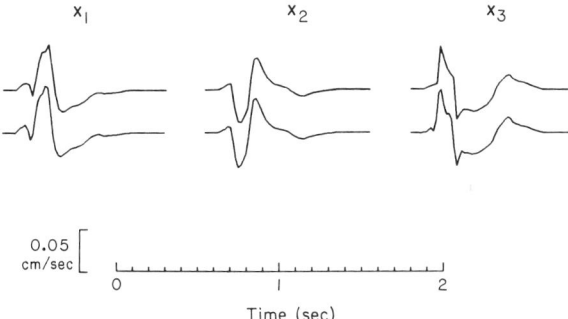

FIG. 17. Demonstration that the isochrone integration technique yields the exact result for far-field Green's functions in the near-source region of a fault. The three columns contain seismograms for the $x_1$, $x_2$, and $x_3$ components of velocity at observer $N_2$ (Fig. 11). The upper trace of each pair is the isochrone integration result obtained directly in the time domain. The lower trace of each pair is the result obtained by frequency domain summation of a 30 × 30 grid of point sources (far-field terms only). The lower traces have been filtered by a filter that is flat in the 0–20-Hz band and has a cosine-squared rolloff in the 20–30-Hz band. Only the $R^{-1}$ radiation terms for a point dislocation have been used in the lower traces.

phase-shift filter having a cosine-squared ($\cos^2$) taper from 20 to 30 Hz. The two results are essentially identical except for minor differences caused by the filtering of the point-source summation result. In this example the magnitude of the neglected $R^{-2}$ and $R^{-4}$ terms is comparable to that of the $R^{-1}$ term.

## B. More Complicated Sources in a Vertically Inhomogeneous Medium

### 1. Comparisons Between Ray Theory and Exact Seismograms

Because the isochrone integration method using ray theory can be so much faster computationally than methods using complete Green's functions, it is worthwhile to investigate its performance in situations more realistic than a Haskell model in a whole space. Bernard and Madariaga (1984) have compared ray theory seismograms to synthetic seismograms calculated by Bouchon (1982) to simulate the Gilroy 6 recording of the 1979 Coyote Lake earthquake (Fig. 18). Bouchon's synthetics were calculated by a point-source summation of complete Green's functions for the layered medium. While the ray theory result lacks the near-field wave that causes the long-period ramp before the arrival of the S wave from the origin, the far-field term reproduces the 2.5-sec-long main pulse surprisingly

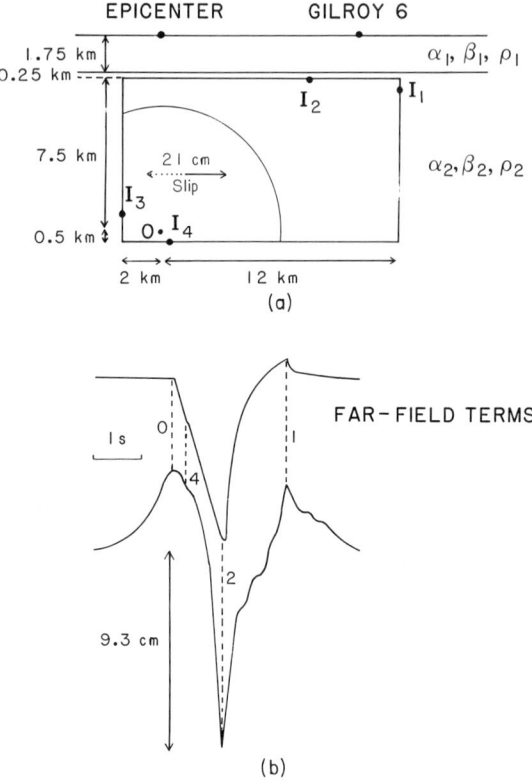

**FIG. 18.** Comparison of isochrone integration seismograms and complete seismograms for a simulation of the 1979 Coyote Lake earthquake. (a) Test geometry. A vertical section of the earth containing the fault is shown. The earth model consists of a uniform layer overlying a uniform half-space containing the fault, where O is the hypocenter. The rupture front is an expanding circle limited by the edges of the rectangular fault. $I_1$, $I_2$, $I_3$, and $I_4$ are critical points where isochrones touch the edges of the fault, thus radiating high-frequency pulses. Displacement, perpendicular to the fault, is calculated at station Gilroy 6. (b) The upper trace (asymptotic method) is the result of isochrone integration (far-field terms). The lower trace [Bouchon's method (1982)] is the result of point-source summation of complete Green's functions, including all near-field terms. Here, 0, 1, 2, and 4 show the locations of $S$ pulses associated with the hypocenter and critical points.

well. Its higher frequency features correlate with those in the complete seismogram, as they should.

In the next example, we show what may be one of the most significant limitations on the use of ray theory for extended-source calculations. Ray theory does not correctly predict the phase shifts, diffracted P wave, and whispering gallery mode that in some cases accompany an SV wave

4. EARTHQUAKE GROUND-MOTION CALCULATION 255

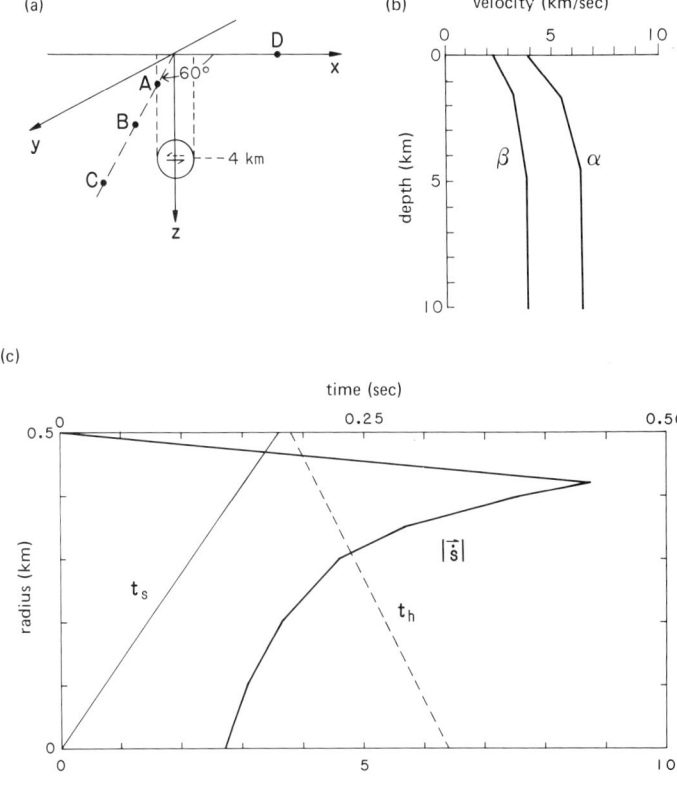

FIG. 19. (a) Test geometry. A circular fault 1 km in diameter with its center at a depth of 4 km lies in the $x$–$z$ plane. Observers at **A**, **B**, and **C** are located in the $x$–$y$ plane (free surface) at an azimuth 60° off strike, at epicentral distances of 2.0, 5.0, and 10.0 km, respectively. Observer **D** is at an epicentral distance of 5.0 km along strike. (b) The earth velocity structure consists of two gradient zones overlying a uniform half-space (Table I). (c) The slip model is pure strike-slip and is circularly symmetric about the hypocenter. Rupture time is shown by (———), healing time (- - -), and slip rate (———). Note that the fault slips for 0.32 sec at its center and 0.01 sec at its periphery.

incident at the free surface (Spudich and Frazer, 1984). The test situation, Fig. 19, largely mimics that of Campillo and Bouchon (1983). Rupture initiates at a 4-km depth and expands uniformly to fill a circle of 0.5-km radius. Healing then initiates at the periphery and propagates inward to the origin. In our case the slip-rate time function is a rectangle (see Section II.A) rather than the Kostrov function used by Campillo and Bouchon. The slip-rate, rupture time, and healing time on the crack as a function of radius

TABLE I. EARTH MODEL PARAMETERS[a]

| Depth (km) | α(km/sec) | β(km/sec) | ρ(g/cm$^3$) |
|---|---|---|---|
| 0.0 | 4.0 | 2.3 | 2.6 |
| 1.5 | 5.5 | 3.2 | 2.8 |
| 4.5 | 6.3 | 3.65 | 2.9 |
| ∞ | 6.3 | 3.65 | 2.9 |

[a] Linear interpolation is used to derive model parameters for depths not given in this table.

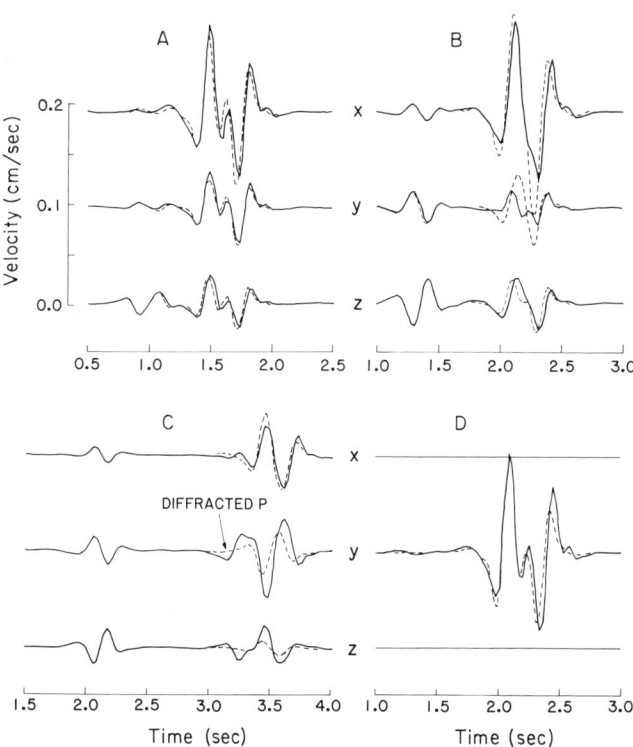

FIG. 20. Comparison of isochrone integration (ray theory, ----) and complete (——) seismograms for the test in Fig. 19, for observers at **A** ($r = 2$ km), **B** ($r = 5$ km), **C** ($r = 10$ km), and **D** ($r = 5$ km). All seismograms have been band pass filtered from 1–10 Hz using a noncausal filter. The three components of ground velocity are shown for each observer. The P waves agree well at all observers, and S waves agree well at short distance (**A**) and where only SH energy is present (**D**); S-wave discrepancies are apparent at **B** and **C** due to phase shifts, the diffracted SP wave, and a whispering gallery mode associated with SV waves incident at the free surface. These effects are not included in the ray theory result.

are shown in Fig. 19. Slip everywhere on the fault is parallel to the $x$ axis. Our earth model is given in Table I. Figure 20 shows the $x$, $y$, and $z$ components of ground velocity observed at **A**, **B**, **C**, and **D** in Fig. 19. The dashed line is the ray theory result, and the solid line is the result obtained using the adaptive integration method of Spudich (1981) and complete Green's functions calculated with the DWFE method of Olson *et al.* (1984). All seismograms have been bandpass filtered from 1 to 10 Hz using a zero phase-shift filter having a cosine ($\cos^1$) taper from 1 to 3 Hz and from 7 to 10 Hz. The ray theory result was obtained using about 1% of the computer time required for the complete result. The two methods agree quite well at **A** (2-km epicentral distance), showing that the near-field term is not important in this case. The P waves match well at all epicentral distances, but the S-wave match deteriorates at **B** and **C** due to the complexities associated with SV at the free surface. A diffracted P wave, preceding the S wave, can be seen at 3.2 sec in the $y$ and $z$ components at observer **C**. These discrepancies are clearly associated with the SV wave because the ray theory S wave at observer **D**, which is on an SV node, agrees well with the complete synthetic.

These examples show that ray theory works very well when used within its range of validity. Because the largest motions in earthquakes are generally associated with S waves, future work in ray theory should concentrate upon finding a solution for the SV problems, which does not affect the simplicity of the isochrone formalism.

2. Modeling Strong Motion Records from the 1979 Imperial Valley Earthquake

In this section we discuss some of the numerical considerations involved in a realistic extended-source calculation. The example we will use is from the work of Archuleta (1984), who determined a faulting model consistent with the data for the 1979 Imperial Valley earthquake using DWFE (Olson *et al.*, 1984) to calculate the Green's functions and the adaptive method of Spudich (1981) to evaluate the fault surface integral. We will not dwell on the source parameters or the interpretation of the model, all of which are found in Archuleta (1984).

The objective was to find a rupture model that predicted synthetic seismograms consistent with the data. The data were the three-component particle velocity records at 16 stations, all within 23 km of the Imperial Valley fault. The medium had a well-determined P-wave velocity structure and a reasonably well-determined S-wave velocity structure (see Table II). The fault had a length of at least 35 km and a downdip width of 13 km. Faulting at the surface was primarily right-lateral strike-slip with a minor amount of dip-slip motion.

TABLE II. ELASTIC PARAMETERS FOR THE IMPERIAL VALLEY[a]

| Depth (km) | $\alpha$(km/sec) | $\beta$(km/sec) | $\rho$(g/cm$^3$) |
|---|---|---|---|
| 0.0 | 1.70 | 0.40 | 1.8 |
| 0.4 | 1.80 | 0.70 | 1.8 |
| 5.0 | 5.65 | 3.20 | 2.5 |
| 11.0 | 5.85 | 3.30 | 2.8 |
| 11.0 | 6.60 | 3.70 | 2.8 |
| 12.0 | 7.20 | 4.15 | 2.8 |

[a] Between any two successive depths linear interpolation is used to determine intermediate values. A half-space exists for depths greater than 12 km. Different elastic parameters at the same depth indicate a discontinuity.

The basic numerical constraints on generation of the Green's functions using the DWFE method were determined by five parameters: the maximum length of time of the synthetics ($t_{max}$), the maximum frequency of interest ($f_{max}$), the maximum distance between any observer and the most distant point on the fault ($r_{max}$), the maximum P-wave velocity in the medium ($\alpha_{max}$), and the minimum shear-wave velocity in the medium ($\beta_{min}$). For the Imperial Valley data, Archuleta (1984) selected the following values: $t_{max}$ = 30 sec, $f_{max}$ = 1.0 Hz, $r_{max}$ = 35 km, $\alpha_{max}$ = 7.2 km/sec, and $\beta_{min}$ = 0.4 km/sec.

The first step of the problem is to obtain traction Green's functions **T** on a grid of points on the fault. This grid must be sufficiently dense to allow accurate integration of $\dot{\mathbf{s}} \cdot \mathbf{T}$ over the fault surface. As mentioned earlier, the DWFE method obtains expansion coefficients $U(z_i, t_j, k)$, $V(z_i, t_j, k), \ldots, T(z_i, t_j, k)$, $i = 1, \ldots, N_z$, $j = 1, \ldots, N_t$, where $N_z$ is the number of mesh points and $N_t$ is the number of time steps, by using a finite-element technique to solve the time-domain form of Eqs. (21) and (22) for each wave number $k$. The DWFE method adjusts its mesh spacing $\Delta z$ in depth to give 6 mesh points per shortest shear wavelength:

$$\Delta z(z) = \beta(z)/(6f_{max}), \tag{118a}$$

$$z_i = \sum_{j=1}^{i} \Delta z(z_{j-1}), \quad z_0 = 0, \quad i = 1, \ldots, N_z. \tag{118b}$$

Because traction Green's functions are ultimately calculated at the depths $\{z_i\}$, this variable spacing is completely consistent with the sampling criterion of Eq. (64). For the Imperial Valley case, with a fault extending to a 13-km depth, Green's functions for depth $z_i$ with $1 \le i \le 22$ were calculated and saved. A time step of about 0.015 sec was chosen to satisfy

4. EARTHQUAKE GROUND-MOTION CALCULATION 259

the Courant stability criterion for the finite-element calculation and to give $N_t = 2048$ for a 30-sec-long time series.

The maximum number of wave numbers is simply

$$N_k = k_{max}/\Delta k, \qquad (119)$$

where

$$k_{max} = 2\pi f_{max}/(0.9\beta_{min}), \qquad (120)$$

and

$$\Delta k \approx 2\pi/R_{max}, \qquad (121)$$

where

$$R_{max} = (\alpha_{max} t_{max} + r_{max})/2. \qquad (122)$$

The quantity $(0.9\beta_{min})$ represents the minimum phase velocity of interest, the velocity of the fundamental mode Rayleigh wave in the shallowest part of the medium. If the time window of interest does not include such slowly traveling waves, then the number of wave numbers can be reduced appropriately. Here, $R_{max}$ represents the radius at which a reflecting boundary must be placed such that reflected energy from this boundary does not arrive within the time window $[0, t_{max}]$ being synthesized. Substituting the values used for the Imperial Valley, we found $N_k = 348$. In practice, DWFE optimizes itself in the course of the computation so that this number is the upper limit with the actual number of wave numbers being less, especially if the fault surface does not penetrate into the lowest velocity ($\beta_{min}$) material.

The computation of $U$, $V$, $W$, $P$, $S$, and $T$ and their Fourier transformation into the frequency domain required 3.5 hr of computation time (all computation times will be those measured on a VAX11/780 with a VMS operating system). The components $U$, $V$, $W$, $P$, $S$, and $T$ are saved as functions of $(z, k, \omega)$. The number of frequencies $N_f$ saved after transforming to the frequency domain is simply $f_{max}/\Delta f$ where $\Delta f = 1.0/t_{max}$. Thus, only 31 frequencies, including zero frequency, are saved. The advantage of saving $U, \ldots, T$ in the $z, k, \omega$ domain is that they embody the exact medium response. Green's functions for any epicentral distance and point-force orientation can be obtained from them by use of Eq. (19). The disadvantage is the amount of storage, amounting to approximately $6N_f N_k N_z$ complex words (about 1.5 million in this case).

The next step in the procedure is to calculate for all observers the tractions $\mathbf{T}^m(\mathbf{x}, \omega; \mathbf{y}_i, 0)$, as in Eq. (15), where $m$ is the point-force direction, $\mathbf{x}$ is the position on the fault, and $\mathbf{y}_i$ is the location of observer $i$. This is accomplished by using Eqs. (19) and (20) and by further differentiating the displacements to get strains, from which are derived stresses and tractions.

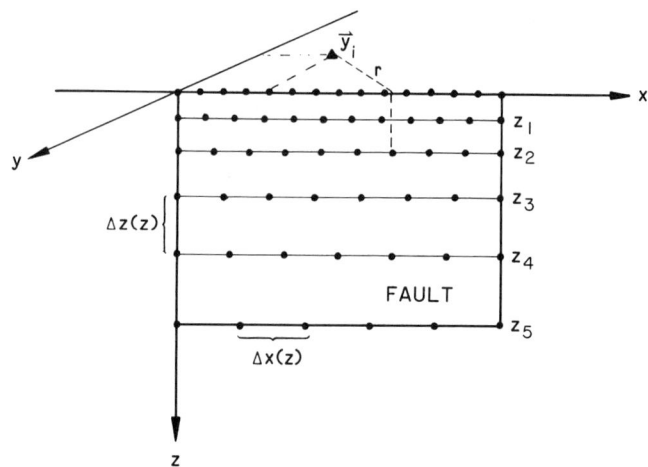

**FIG. 21.** Spacing of sample points $x_T$ where $T$ is sampled on the fault, and $z_0, z_1, \ldots,$ are the depths at which $U, V, \ldots, T$ are obtained using DWFE. Horizontal spacing of sample points varies with local S-wave velocity, which is a function of depth (Table II). Here, $y_i$ is the location of observer $i$, and $r$ is the cylindrical radius for which Eq. (19) must be evaluated to obtain the traction at the indicated point.

Here, **T** is calculated on a variably spaced mesh of points on the fault, in accordance with Eq. (64). For simplicity we assume the fault lies in the $x-z$ plane (Fig. 21), and we let $\mathbf{x}_{ij} = x_i\hat{\mathbf{x}} + z_j\hat{\mathbf{z}}$ be the $ij$th traction sample point on the fault. The grid depths used by DWFE in Eq. (118b) are $\{z_j\}$ and

$$\Delta x(z, f) = \beta(z)/6f, \tag{123a}$$

$$x_i = \sum_{j=1}^{i} \Delta x(z_{j-1}, f), \qquad z_0 = 0. \tag{123b}$$

This leads to a horizontal spacing of points on the fault that varies as a function of frequency and depth. For each of the observers used by Archuleta (1984), the evaluation of **T** (e.g., Fig. 22) for all $\mathbf{x}_{ij}$ and all frequencies took about 2 hr of CPU time, and storage of **T** required about 0.4 million complex words, which were saved for use with subsequent slip models.

The slip rate function was specified using four parameters: strike-slip rate amplitude, dip-slip rate amplitude, slip duration, and rupture time, on a 182-point grid that was spaced equally every 2.5 km along-strike and 1.0-km downdip. Thus, each faulting model consisted of 728 independent parameters. This grid of 182 points was not the same as the grid of quadrature points $\{x_{s_j}\}$ (Section II.C.3.a). The sample spacing in both $x$

## 4. EARTHQUAKE GROUND-MOTION CALCULATION

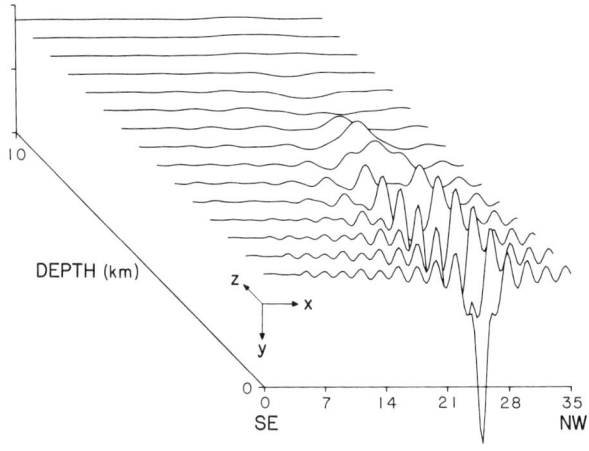

FIG. 22. Real part of the $x$ component of traction resulting on the Imperial fault from the application of a point force at $x = 24.9$ km, $y = -1.0$ km, corresponding to Station E07 of the El Centro array (Archuleta, 1984). The 0.51-Hz component of the traction spectrum is shown. Because the point force is quite close to the fault, the traction peaks strongly nearest the point force. The wavelength of the traction is smallest at zero depth, where the shear velocity is minimum, and the wavelength increases with depth.

and $z$ for the quadrature is given by Eq. (65). Using $f = 0.5$ Hz and $m = 8$, on the shallowest part of the fault the rupture velocity might be 0.5 km/sec, implying a 0.125-km spacing for quadrature points. Deeper on the fault, where the rupture velocity might be 3.0 km/sec, the quadrature point spacing is about 0.75 km. Because quadrature points almost never coincide with the 182 points upon which the rupture model parameters are specified, the four rupture parameters are obtained at quadrature points by bilinear interpolation. The spectrum of the slip rate, $\dot{s}(\mathbf{x}_s, \omega)$ is easily calculated from the four interpolated parameters. The integration of the product of the Green's function and slip model ($\dot{\mathbf{s}} \cdot \mathbf{T}$) over the fault for 31 frequencies required 15 min of computation time per observer. Because it is generally insufficient to determine the validity of a particular rupture model by looking at the resulting synthetic seismograms at only one observer, synthetic seismograms were calculated at a minimum of six observers for each rupture model. Thus, each model took about 1.5 hr of CPU time.

While the initial computation time needed to obtain $U$, $V$, $W$, $P$, $S$, and $T$ can be large, in fact, the inverse Bessel transform for each observer required nearly the same amount of time. The actual computation time for doing the integration and inverse Fourier transform to compute a particle velocity at a given observer is rather small by comparison. However, in the trial-and-error procedure of trying to find agreement between the synthetics and the data, it may take a large number (~ 300 in the case of the Imperial Valley earthquake) of trial rupture models before a satisfactory fit is found. In this case, the total computation time for the integration over the fault plane becomes the costly part of the problem. Although the overall computation time is nontrivial, synthetics, which incorporate the vertical hetero-

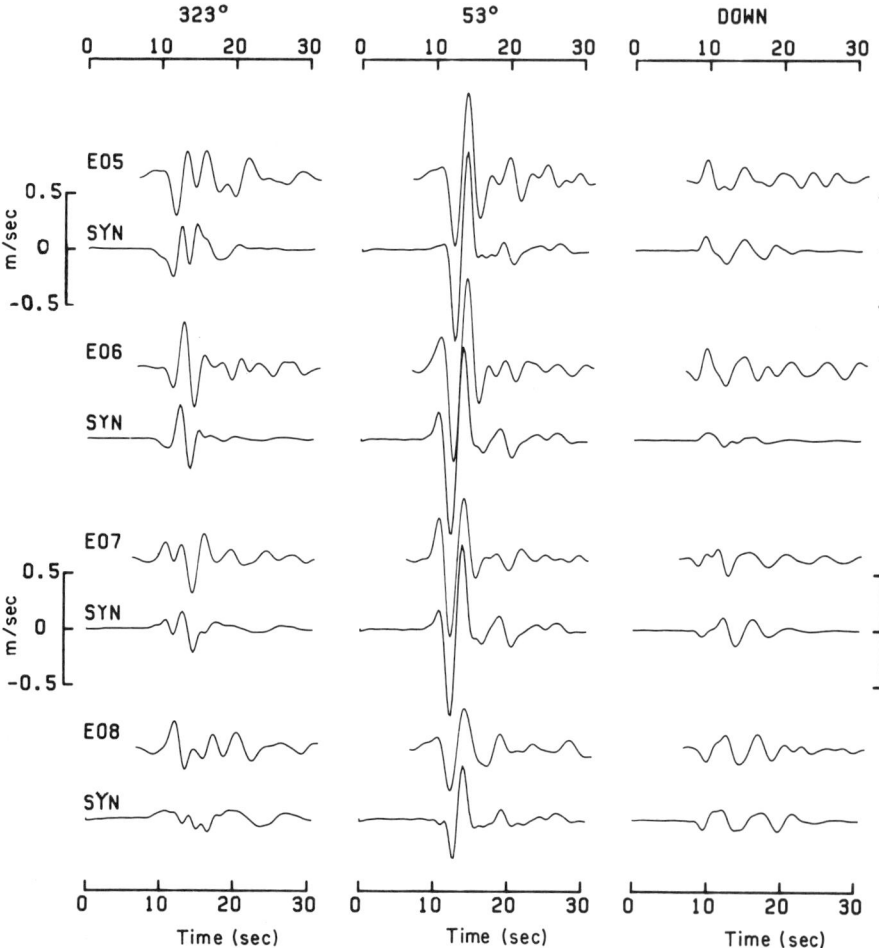

**FIG. 23.** Comparison of synthetic particle velocities (lower trace of each pair) with the data for Stations E05, E06, E07, and E08. Each column represents a different component of motion. The 323° and the 53° are the horizontal components with positive motion directed along the azimuth, measured clockwise from north. The data and synthetics have been filtered to 0.5 Hz. (From Archuleta, 1984; published by the American Geophysical Union.)

geneity of the velocity structure and the spatial and temporal finiteness of the faulting, can be computed that are in agreement with the data (Fig. 23).

ACKNOWLEDGMENTS

The authors wish to thank Dr. Bernard Chouet for his explanations on the inner workings of his and Dr. Bouchon's methods. Drs. Chouet, S. H. Hartzell, and T. H. Heaton provided helpful reviews of this paper. We thank Dr. Allen Olson for allowing us to use and modify his DWFE code.

This research was supported, in part, by the Office of Nuclear Regulatory Research, U.S. Nuclear Regulatory Commission.

REFERENCES

Aki, K. (1967). *J. Geophys. Res.* **72**, 1217–1231.
Aki, K. (1968). *J. Geophys. Res.* **73**, 5359–5376.
Aki, K., and Richards, P. G. (1980). "Quantitative Seismology: Theory and Methods." Freeman, San Francisco, California.
Alekseev, A. S., and Mikhailenko, B. G. (1980). *J. Geophys.* **48**, 161–172.
Anderson, J. G., and Richards, P. G. (1975). *Geophys. J. R. Astron. Soc.* **42**, 347–373.
Andrews, D. J. (1981). *J. Geophys. Res.* **86**, 10821–10834.
Apsel, R. J., and Luco, J. E. (1983). *Bull. Seismol. Soc. Am.* **73**, 931–952.
Apsel, R. J., Frazier, G. A., Jurkevics, A., and Fried, J. C. (1981). *Geol. Surv. Open-File Rep. (U.S.)* No. 81-276.
Archuleta, R. J. (1984). *J. Geophys. Res.* **89**, 4559–4585.
Archuleta, R. J., and Day, S. M. (1980). *Bull. Seismol. Soc. Am.* **70**, 671–689.
Archuleta, R. J., and Frazier, G. A. (1978). *Bull. Seismol. Soc. Am.* **68**, 541–572.
Archuleta, R. J., and Hartzell, S. H. (1981). *Bull. Seismol. Soc. Am.* **71**, 939–957.
Bache, T. C., Day, S. M., and Swanger, H. J. (1982). *Bull. Seismol. Soc. Am.* **72**, 15–28.
Backus, G., and Mulcahy, M. (1976a). *Geophys. J. R. Astron. Soc.* **46**, 341–361.
Backus, G., and Mulcahy, M. (1976b). *Geophys. J. R. Astron. Soc.* **47**, 301–329.
Bernard, P., and Madariaga, R. (1984). *Bull. Seismol. Soc. Am.* **74**, 539–558.
Boatwright, J. (1981). *Bull. Seismol. Soc. Am.* **71**, 69–94.
Boatwright, J. (1982). *Bull. Seismol. Soc. Am.* **72**, 1049–1068.
Boore, D. M., and Joyner, W. B. (1978). *Bull. Seismol. Soc. Am.* **68**, 283–300.
Bouchon, M. (1979). *J. Geophys.* **84**, 3609–3614.
Bouchon, M. (1980a). *J. Geophys. Res.* **85**, 356–366.
Bouchon, M. (1980b). *J. Geophys. Res.* **85**, 367–375.
Bouchon, M. (1981). *Bull. Seismol. Soc. Am.* **71**, 959–971.
Bouchon, M. (1982). *Bull. Seismol. Soc. Am.* **72**, 745–759.
Bouchon, M., and Aki, K. (1977). *Bull. Seismol. Soc. Am.* **67**, 259–277.
Bracewell, R. (1965). "The Fourier Transform and Its Applications." McGraw-Hill, New York.
Burridge, R., and Knopoff, L. (1964). *Bull. Seismol. Soc. Am.* **54**, 1875–1888.
Cabrera, J. J., and Levy, S. (1984). *Geophysics* **49**, 1915–1932.
Campillo, M., and Bouchon, M. (1983). *Bull. Seismol. Soc. Am.* **73**, 83–96.

Červený, V., and Klimeš, L. (1984). *Geophys. J. R. Astron. Soc.* **79**, 119–134.
Červený, V., Klimeš, L., Pleinerová, V., and Pšenčík, I. (1987). *Geophys. J. R. Astron. Soc.* Submitted.
Chapman, C. H. (1978). *Geophys. J. R. Astron. Soc.* **54**, 481–518.
Chapman, C. H., and Orcutt, J. A. (1985). *Rev. Geophys.* **23**, 105–164.
Chin, R. C. Y., Hedstrom, G., and Thigpen, L. (1984a). *J. Comput. Phys.* **54**, 18–56.
Chin, R. C. Y., Hedstrom, G., and Thigpen, L. (1984b). *Geophys. J. R. Astron. Soc.* **77**, 483–502.
Chouet, B. (1981). *J. Geophys. Res.* **86**, 5985–6016.
Chouet, B. (1982). *J. Geophys. Res.* **87**, 3868–3872.
Chouet, B. (1983). *J. Volcanol. Geotherm. Res.* **19**, 367–379.
Chouet, B. (1985). *J. Geophys. Res.* **90**, 1881–1893.
Chouet, B. (1987). *Bull. Seismol. Soc. Am.* Submitted.
Cormier, V. F. (1980). *Bull. Seismol. Soc. Am.* **70**, 691–716.
Cormier, V. F., and Beroza, G. C. (1987). *Bull. Seismol. Soc. Am.* Submitted.
Day, S. M. (1977). Ph.D. Thesis, Univ. of California, San Diego.
Day, S. M. (1982). *Bull. Seismol. Soc. Am.* **72**, 1881–1902.
Dunkin, J. W. (1965). *Bull. Seismol. Soc. Am.* **55**, 335–358.
Farra, V., Bernard, P., and Madariaga, R. (1986). *In* "Earthquake Source Mechanics," (S. Das, J. Boatwright, and C. Scholz, eds.), pp. 121–130. Geophys. Monogr. 37 of the Am. Geophys. Union, Washington, D.C.
Frazer, L. N., and Gettrust, J. F. (1984). *Geophys. J. R. Astron. Soc.* **76**, 461–481.
Frazer, L. N., and Sen, M. K. (1985). *Geophys. J. R. Astron. Soc.* **80**, 121–147.
Gay, N. C., and Ortlepp, W. D. (1974). *Bull. Geol. Soc. Am.* **90**, Part I, 47–58.
Gilbert, F., and Dziewonski, A. (1975). *Philos. Trans. R. Soc. London, Ser. A* **278**, 187–269.
Haddon, R. A., and Buchen, P. W. (1981). *Geophys. J. R. Astron. Soc.* **67**, 587–598.
Hartzell, S. (1978). *Geophys. Res. Lett.* **5**, 1–4.
Hartzell, S. (1982). *Bull. Seismol. Soc. Am.* **72**, 2381–2388.
Hartzell, S., and Heaton, T. (1983). *Bull. Seismol. Soc. Am.* **73**, 1553–1583.
Hartzell, S., and Heaton, T. (1986). *Bull. Seismol. Soc. Am.* **76**, 649–674.
Hartzell, S., and Helmberger, D. (1982). *Bull. Seismol. Soc. Am.* **72**, 571–596.
Hartzell, S., Frazier, G., and Brune, J. (1978). *Bull. Seismol. Soc. Am.* **68**, 301–316.
Haskell, N. A. (1964). *Bull. Seismol. Soc. Am.* **54**, 1811–1841.
Haskell, N. A. (1966). *Bull. Seismol. Soc. Am.* **56**, 125–140.
Heaton, T. H. (1982). *Bull. Seismol. Soc. Am.* **72**, 2037–2062.
Heaton, T. H., and Helmberger, D. V. (1979). *Bull. Seismol. Soc. Am.* **69**, 1311–1341.
Helmberger, D. V., and Harkrider, D. G. (1978). *In* "Modern Problems in Elastic Wave Propagation" (J. Miklowitz and J. D. Achenbach, eds.), pp. 499–518. Wiley, New York.
Helmberger, D. V., and Malone, S. (1975). *J. Geophys. Res.* **80**, 4881–4888.
Imagawa, K., Mikami, N., and Mikumo, T. (1984). *J. Phys. Earth* **32**, 317–338.
Irikura, K. (1983). *Bull. Disaster Prev. Inst., Kyoto Univ.* **33**, 63–104.
Izutani, Y. (1981). *J. Phys. Earth* **29**, 537–558.
Joyner, W. B., and Boore, D. M. (1986). *In* "Earthquake Source Mechanics," (S. Das, J. Boatwright, and C. Scholz, eds.), pp. 269–274. Geophys. Monogr. 37 of the Am. Geophys. Union, Washington, D.C.
Julian, B. R. (1983). *Nature (London)* **303**, 323–325.
Kanamori, H. (1979). *Bull. Seismol. Soc. Am.* **69**, 1645–1670.
Kennett, B. L. N. (1980). *Geophys. J. R. Astron. Soc.* **61**, 1–10.
Kennett, B. L. N. (1983). "Seismic Wave Propagation in Stratified Media." Cambridge Univ. Press, London and New York.

Kennett, B. L. N., and Illingworth, M. R. (1981). *Geophys. J. R. Astron. Soc.* **66**, 633–675.
Kind, R. (1979). *J. Geophys.* **45**, 373–380.
King, G. C. P. (1983). *Pure Appl. Geophys.* **121**, 761–816.
Kostrov, B. V. (1964). *J. Appl. Math. Mech.* (*Engl. Transl.*) **28**, 1077–1087.
Koyama, J. (1985). *Tectonophysics* **118**, 227–242.
Liu, H.-L., and Helmberger, D. V. (1983). *Bull. Seismol. Soc. Am.* **73**, 201–218.
Liu, H.-L., and Helmberger, D. V. (1985). *Bull. Seismol. Soc. Am.* **75**, 689–708.
Madariaga, R. (1976). *Bull. Seismol. Soc. Am.* **66**, 639–666.
Maruyama, T. (1963). *Tokyo Daigaku Jishin Kenkyusho Iho* **41**, 467–486.
Munguia, L., and Brune, J. N. (1984). *Geophys. J. R. Astron. Soc.* **79**, 747–771.
Olson, A. H., and Apsel, R. J. (1982). *Bull. Seismol. Soc. Am.* **72**, 1969–2002.
Olson, A. H., Orcutt, J. A., and Frazier, G. A. (1984). *Geophys. J. R. Astron. Soc.* **77**, 421–460.
Robinson, K. (1951). *J. Appl. Phys.* **22**, 1045.
Sen, M. K., and Frazer, L. N. (1985). *Geophys. J. R. Astron. Soc.* **82**, 415–438.
Sibson, R. H. (1980). *J. Geophys. Res.* **85**, 6239–6247.
Sibson, R. H. (1986). *Annu. Rev. Earth Planet. Sci.* **14**, 149–175.
Spudich, P. (1980). *Geophys. Res. Lett.* **7**, 717–720.
Spudich, P. (1981). *EOS Trans. Am. Geophys. Union* **62**, 960.
Spudich, P., and Ascher, U. (1983). *Geophys. J. R. Astron. Soc.* **75**, 101–124.
Spudich, P., and Frazer, L. N. (1984). *Bull. Seismol. Soc. Am.* **74**, 2061–2082.
Spudich, P., and Oppenheimer, D. (1986). In "Fifth Maurice Ewing Symposium on Earthquake Source Mechanics," Series 6 (S. Das, J. Boatwright, and C. Scholz, eds.), pp. 285–296. Am. Geophys. Union, Washington, D.C.
Tchalenko, J. S. (1970). *Bull. Geol. Soc. Am.* **81**, 1625–1640.
Tchalenko, J. S., and Ambraseys, N. N. (1970). *Bull. Geol. Soc. Am.* **81**, 41–60.
White, S. H. (1973). *Nature (London)* **244**, 276–278.
Yao, Z. X., and Harkrider, D. G. (1983). *Bull. Seismol. Soc. Am.* **73**, 1685–1699.

## CHAPTER 6

# Path Effects in Strong Motion Seismology

JOHN E. VIDALE
AND
DONALD V. HELMBERGER

*Seismological Laboratory*
*California Institute of Technology*
*Pasadena, California 91125*

## I. Introduction

Records of earthquake- and explosion-induced motions are studied by seismologists for several reasons: (1) to learn the details of earthquake and explosion sources, (2) to determine earth structure, and (3) from the knowledge of sources and structures, to predict the shaking at the surface of the earth from earthquakes and explosions. This chapter explores methods of computing the motions that result from elastic waves propagating through complex structures. These methods aid in the investigation of all three topics. In particular, we will apply these methods to understand the effect of laterally varying near-surface structures on ground motions.

Strong ground-motion waveforms recorded by seismic stations on soft rock sites are generally more complicated than those recorded on hard rock sites. For example, a typical aftershock sequence recorded on a hard rock site displayed in Fig. 1 may be contrasted with a typical recording from a soft rock or basin site displayed in Fig. 2. Note that in this case the timescales, instrument response, size of the earthquake, and component of motion plotted are different between the two figures. Unfortunately, from a modeling point of view most earthquakes occur in complicated geologic settings. Also, cities and strong motion instruments tend to be in basins. Therefore, interpretation of most strong motion records requires consideration of propagation through complex structures.

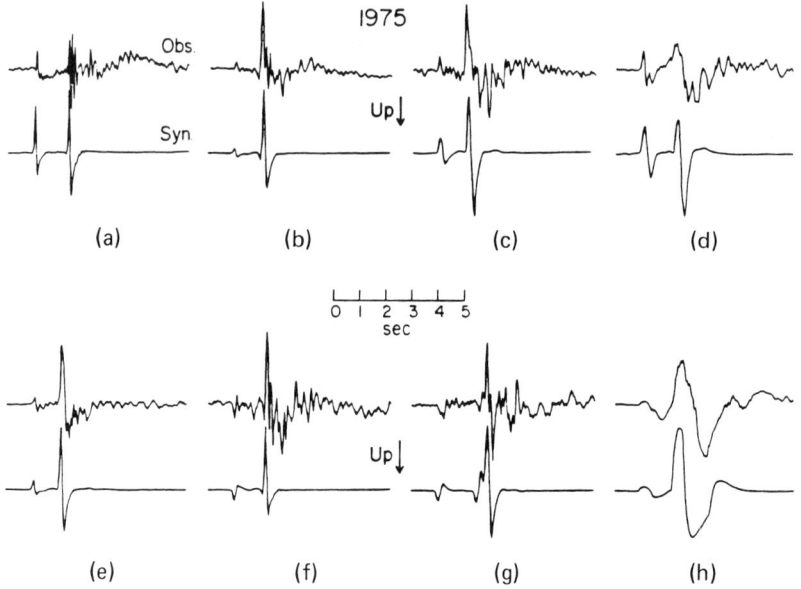

**FIG. 1.** Seismograms and synthetics for eight aftershocks of the 1975 Oroville, California, earthquake, displaying P and S waves with duration versus magnitude dependence recorded on relatively hard rock sites: (a) $M_L = 3.2, \tau = 0.1$; (b) $M_L = 3.1, \tau = 0.2$; (c) $M_L = 3.3, \tau = 0.3$; (d) $M_L = 4.0, \tau = 0.4$; (e) $M_L = 3.2, \tau = 0.2$; (f) $M_L = 3.5, \tau = 0.3$; (g) $M_L = 3.6, \tau = 0.3$; (h) $M_L = 4.9, \tau = 1.4$. (After Cohn et al., 1982.)

The seismograms displayed in Fig. 1 were modeled by assuming a half-space velocity structure with a point shear dislocation as the source. The source time history was described by a trapezoidal shape with duration $\tau$. Most of the events have depths greater than the range to the station. The compressional and shear direct body waves constitute most of the record,

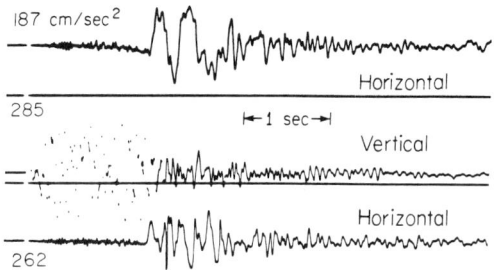

**FIG. 2.** Typical accelerograms for a shallow Imperial Valley event, recorded at a soft rock site at Brawley airport, October 16, 23:16, 1979, $M_l = 4.9$.

and the increase in source duration as magnitude increases is apparent (Cohn *et al.*, 1982).

The motions displayed in Fig. 2 are from an aftershock of the October 15, 1979, Imperial Valley earthquake. Note that the P waves are observed almost exclusively on the vertical component in contrast to the S waves that dominate the horizontal motions. The lack of a strong SV phase on the vertical component shows that the ray is almost vertical. This can be explained by the strong near-surface velocity gradients combined with very low surface velocity (Liu, 1984; Liu and Helmberger, 1985). Modeling these waveforms has proven difficult because the details of structure are not known.

Some progress has been made in modeling longer period waveforms, particularly from nuclear blasts. Fewer parameters suffice to describe explosions than earthquakes and they are generally better known (depth, origin time, and source time function). Section II reviews some aspects of explosion modeling. Section III develops a source description for use in two-dimensional finite-difference or finite-element methods that can produce approximate point-source seismograms that include the effects of complex structures. Section IV presents applications of the methods developed in this chapter for which the need for allowing lateral variation in the structures becomes apparent.

## II. Explosion Sources

### A. EXPERIENCE FROM MODELING EXPLOSION DATA

Over the last few years there has been considerable progress in the interpretation of strong motions observed from nuclear explosions. We know the origin time and source position, and there are often several events fired at a given test site (see Fig. 3). Thus, propagational effects are more easily separated from source properties. For example, in the profile of the Milrow data presented in Fig. 4, we note that the radial component of the P wave is larger than the vertical component at station M01, but the reverse holds for the other stations. This same feature is observed for the Cannikin event and is probably caused by some complexity in velocity structure near the station. Burdick (1983) models the near-field ground motions of the type displayed in Fig. 4 and shows that at ranges greater than 5 km, the earth may be treated as an elastic medium.

There are numerous ground motion records from closer ranges with large accelerations that indicate that the earth behaves in a nonlinear and nonelastic fashion. An example of behavior at these ranges is displayed in

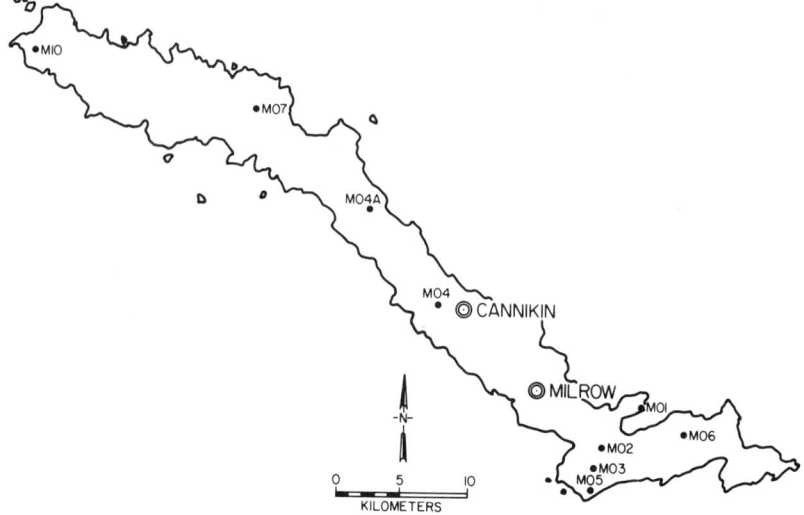

FIG. 3. Location of two Amchitka nuclear tests and of the near-field strong motion instruments deployed to record them.

Fig. 5. Ground motions predicted by linear elastic models compare poorly with the observations. Note the uniform negative slope in velocity between the peak positive value and peak negative value referred to as the N wave (Rodean, 1971). This feature is caused by free-fall of the top layer of the earth and the instruments that therefore experience an acceleration of $-1.0$ g. This zone of nonlinear behavior extends to a range of several kilometers, depending on the size of the explosion, and is called the spall zone. Better comparisons between observations and current theory occur outside this region.

In Fig. 6, some of the records modeled by Burdick (1983) are compared with synthetic seismograms. These synthetics are generated by a hybrid technique in which generalized ray theory (GRT) is used to compute body waves; the Rayleigh waves are computed by summing normal modes. Synthetic ground motions produced by the two techniques are spliced together. The motions computed using GRT are tapered after $\sim 2$ sec and are combined with the motion computed from the modes that have been tapered at the early, body-wave part of the record. Summing the two tapered responses yields the synthetic seismograms displayed in Fig. 6. This response contains only the fundamental mode and the direct rays plus first multiples. No converted phases are included in the calculation. This procedure for modeling these near-field records proved both efficient and enlightening. However, as with any approximate method, it is necessary to check it

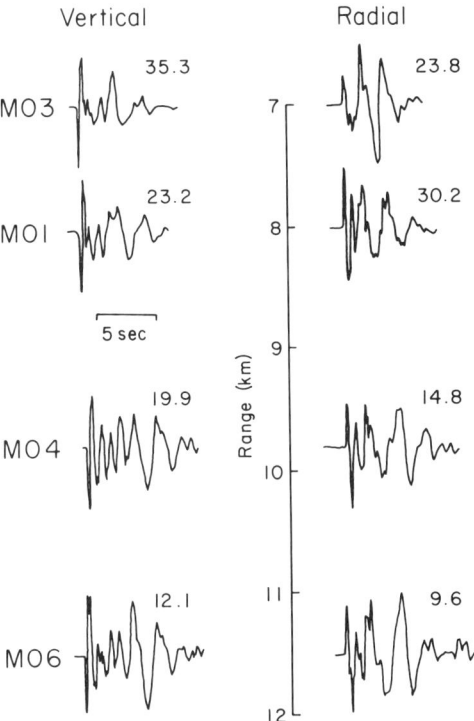

**FIG. 4.** A profile of the velocity records for MILROW. The peak amplitude is given in centimeters/second.

against a more exact method. We compare the hybrid results with those of the wave-number–frequency numerical integration (WI) approach (Apsel, 1979). Comparisons of the results of this technique against GRT plus modes are given in Fig. 7. The P-wave crustal model (Table I) consists of 9 layers derived by Burdick (1983) by fine tuning the model proposed by Engdahl (1972). This model predicts the P-wave travel times well. The S-wave velocity structure, which is also given in Table I, was added to model the Rayleigh wave arrivals by Burdick (1983). The reduced displacement potential (RDP), which is a source time function, of Helmberger and Hadley (1981) is used; namely,

$$\Psi(t) = \Psi_\infty \{1 - e^{-Kt}[1 + Kt + 0.5(Kt)^2 - B(Kt)^3]\}. \tag{1}$$

In this section, the corner frequency parameter $K$ is set at 9 sec$^{-1}$, and the overshoot parameter $B$ is set at a value of 1.

The seismograms shown in Fig. 7 are the calculated synthetic velocity

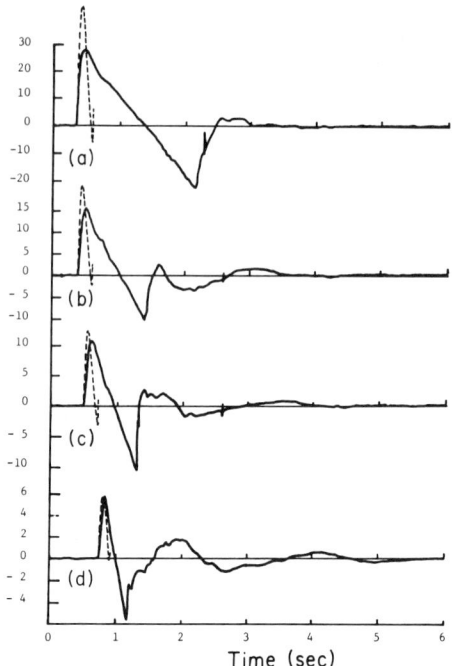

FIG. 5. A comparison of the observed velocity pulse rise times (solid curves) of MILROW from the inside of the spall zone with the predictions of the model (dashed curves) developed from the data from the near-field elastic region: (a) S–0, $\Delta = .08$ km; (b) S–2, $\Delta = .57$ km; (c) S–4, $\Delta = 1.22$ km; (d) S–8, $\Delta = 2.4$ km. (After Burdick, 1983.)

records for ranges of 6, 8, 10, and 12 km. This spans the range of distances for which there are records from Amchitka. The waveforms of the WI and the GRT plus modes seismograms match for both the body- and surface-wave pulses, and the relative amplitude and timing of the two methods agree. The results from a two-dimensional finite-difference (FD) code used with the line-to-point source mapping procedure described in Section III are also included in Fig. 2.5. The FD method is also used in Section II.C to investigate the effect of near-surface structure on synthetic seismograms for the Amchitka blast. The waveforms of the body waves and the surface waves match well between the FD and the WI methods. The amplitudes of the body waves are too large relative to the surface waves for the FD method. This is an artifact of using a 2D rather than a 3D wave equation, as discussed in Section II.B. The GRT method can be easily generalized to a more complicated laterally varying structure as discussed in Section II.B.

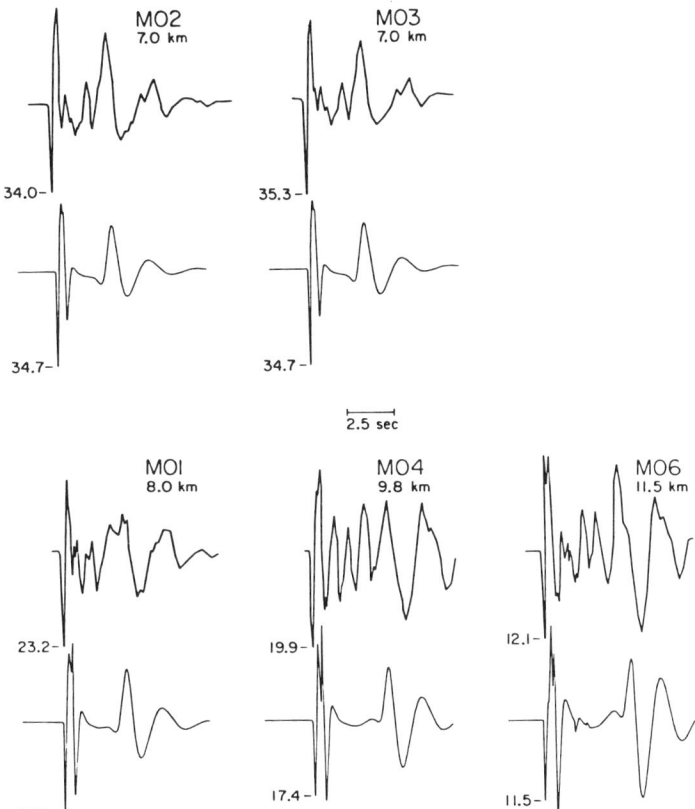

FIG. 6. Synthetic vertical velocity records computed from the model of Burdick (1983) (bottom) are compared to observations from MILROW (top). The peak velocity in centimeters/second is indicated. (After Burdick, 1983.)

B. REVIEW OF NUMERICAL TECHNIQUES

Boundary-value problems involving complicated geometries have generated considerable interest in recent years. Two approaches have gained popularity: one is based on approximations near the wave front or generalizations of ray theory, the other is based on the full wave equation using finite-difference, finite-element, and boundary value integral methods. Ray theory has been developed by Chapman (1978) with the development of the Wentzel, Kramers, Brillouin, and Jeffreys (WKBJ) method and the Gaussian beam approach [Cerveny et al. (1982), and more recently Madariaga and Papadimitriou (1985), among others]. Finite-difference techniques using

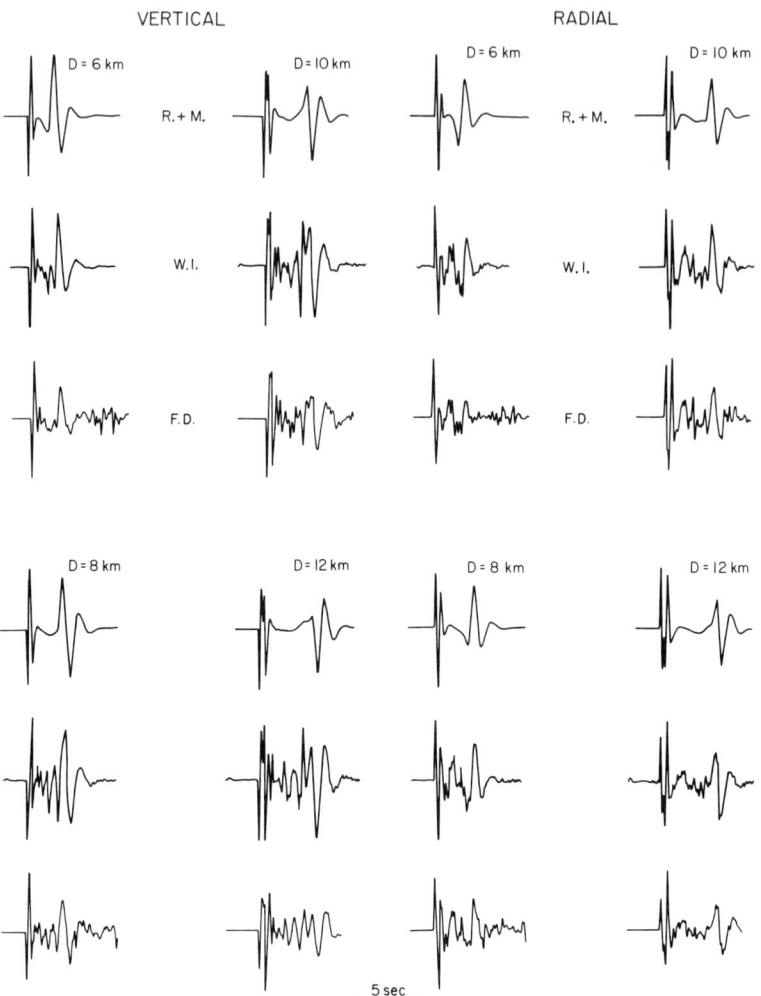

FIG. 7. Comparison for layered velocity model listed in Table I of the results from the generalized rays plus modes methods, the wave number integration method, and the finite-difference method. The source time function is from Helmberger and Hadley (1981) with corner frequency $K = 9.0$ Hz and overshoot parameter $B = 1.0$.

transparent boundary conditions in elastic models are only now becoming economically viable at frequencies of interest [see Clayton and Engquist (1980) and Vidale *et al.* (1985)]. The source descriptions to allow approximate explosion and double-couple point source seismograms to be made from 2D FD grids are developed in Vidale *et al.* (1985) for SH waves and in

## 6. PATH EFFECTS IN STRONG MOTION SEISMOLOGY

TABLE I. VELOCITIES IN THE LAYERED AMCHITKA MODEL

| P-wave velocity (km/sec) | S-wave velocity (km/sec) | Density (g/cm$^3$) | Layer thickness (km) |
|---|---|---|---|
| 3.4 | 1.7 | 2.3 | 0.2 |
| 3.7 | 1.9 | 2.4 | 0.6 |
| 4.2 | 2.1 | 2.4 | 0.5 |
| 4.6 | 2.3 | 2.5 | 0.5 |
| 4.9 | 2.8 | 2.6 | 0.7 |
| 5.1 | 2.9 | 2.7 | 0.5 |
| 5.9 | 3.3 | 2.7 | 6.0 |
| 6.9 | 4.0 | 2.8 | 28.0 |
| 8.2 | 4.7 | 3.2 | $\infty$ |

this work for P-SV waves. Regan and Glover (1985) use a finite-element code and Bard and Bouchon (1985) and Bard and Tucker (1985) use the Aki-Larner technique on similar problems. The ray and more full wave approaches should be used together to model data because rays are more accurate at high frequencies, while the FD method computes the full solution but becomes very expensive at high frequencies.

In this section (II.B) we discuss the GRT method and then the FD method for explosive sources. In Section III.A, shear dislocation sources are described.

We begin by finding the displacement due to an explosive point source in a whole space. At some distance outside the elastic radius from an explosion we require the displacement potential to satisfy the behavior of a spherical wave, that is,

$$\phi_p = (1/R)f(t - R/\alpha), \qquad (2)$$

where $f$ defines a source description such as Eq. (1), $t$ is time, $\alpha$ is the compressional wave velocity, and $R$ is the distance between the idealized shot point and the receiver (see Fig. 8a). The Laplace transformed solution

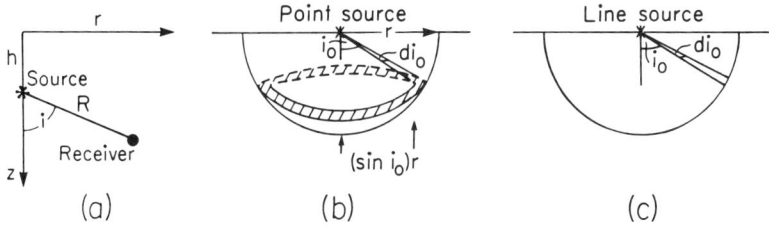

FIG. 8. (a) Conventions for cylindrically symmetric geometry. (b) Diagram showing energy with takeoff angle $i$ in the range $i_0 < i < i_0 + di$ for point source. (c) Diagram showing energy with takeoff angle $i$ in the range $i_0 < i < i_0 + di$ for line source.

in cylindrical coordinates (Strick, 1959) is

$$\overline{\phi}_p(r, z, s) = f(s)\frac{2}{\pi}(s)\,\text{Im}\int_\delta^{i\infty+\delta} K_0(spr)e^{-s\eta|z-h|}\frac{p\,dp}{\eta}, \tag{3}$$

where $s$ is the transform variable over time, $K_0$ the modified Bessel function, $\eta = \sqrt{1/\alpha^2 - p^2}$, $h$ is the source depth, $z$ the receiver depth, and $\delta$ a small number that serves to keep the integration off the complex axis. Using an asymptotic expansion of $K_0$ and keeping the first term only, we obtain

$$\overline{\phi}_p(r, z, s) = \sqrt{\frac{2}{\pi r s}}\,\text{Im}\int_0^{i\infty+\delta}\frac{\sqrt{p}}{\eta}e^{-s(pr+\eta|z-h|)}\,dp, \tag{4}$$

where we have assumed that $f(s) = 1/s$ or that $f(t)$ is a step function. This first-term approximation is valid for large $spr$ and is not valid directly above the source (small $p$), at very small ranges (small $r$), or at very long period (small $s$). This approximation is also used in Section III for earthquake sources (see Eq. 24). This integral occurs often in GRT and has been solved by the Cagniard–de Hoop technique [see de Hoop (1960) or Helmberger (1983)]. One defines $t = pr + \eta|z - h|$ and requires $t$ to have real values along a contour in the complex $p$ plane. Thus,

$$\phi_p(r, z, t) = \sqrt{\frac{2}{r}}\,\frac{1}{\pi}\left[\frac{1}{\sqrt{t}} * J(t)\right], \tag{5}$$

where

$$J(t) = \text{Im}\left(\frac{\sqrt{p}}{\eta}\frac{dp}{dt}\right) \tag{6}$$

or more simply

$$J(t) = \text{Re}(\sqrt{p})\frac{H(t - R/\alpha)}{(t^2 - (R^2/\alpha^2))^{1/2}}$$

and

$$p = \frac{r}{R^2}t + i\left(t^2 - \frac{R^2}{\alpha^2}\right)^{1/2}\frac{|z-h|}{R^2}. \tag{7}$$

Note that at times near $t = R/\alpha$

$$p = (\sin i)/\alpha \equiv p_0 \tag{8}$$

or $p_0$ becomes the geometric ray parameter (see Fig. 8a). Substituting Eq. (6) into Eq. (5), assuming $p = p_0$, we obtain, as expected,

$$\phi_p = (1/R)H(t - R/\alpha).$$

From the work of Gilbert and Knopoff (1961) we also have the solution for the line source excitation as

$$\phi_L(r, z, t) = \operatorname{Im}\left(\frac{1}{\eta}\frac{dp}{dt}\right), \tag{9}$$

or more explicitly

$$\phi_L = H(t - R/\alpha)/(t^2 - R^2/\alpha^2)^{1/2}, \tag{10}$$

where the line source is perpendicular to the $r$ and $z$ axes (see Fig. 8a). Thus, we can find the point source solution $\phi_p$ from the line source solution $J$ as shown in Eq. (5) if $J = \sqrt{p}\,\phi_L$ is used. We use this trick to study the effect of laterally heterogeneous structure with the FD method. The displacements derived from the potential in Eq. (9) are used to drive the FD grid, as described in Section III.D. Line source vertical and horizontal seismograms $\tilde{W}(t)$ and $\tilde{Q}(t)$ extracted from the FD grid are transformed to point source vertical and radial seismograms $\mathbf{W}(t)$ and $\mathbf{Q}(t)$ as

$$\mathbf{W}(t) = \frac{\sqrt{2}}{r}\frac{1}{\pi}\left[\frac{H(t)}{\sqrt{t}} * \tilde{W}(t)\right], \tag{11a}$$

$$\mathbf{Q}(t) = \frac{\sqrt{2}}{r}\frac{1}{\pi}\left[\frac{H(t)}{\sqrt{t}} * \tilde{Q}(t)\right]. \tag{11b}$$

The absence of the $\sqrt{p}$ in Eq. (9) means the solution is approximate. The mapping between the line-to-point source method discussed here breaks down near $p_0 = 0$, that is, directly above and below the source (Vidale et al., 1985). In the case of a laterally varying structure, each arrival in a record may have a different ray parameter $p$, but we can only correct for a constant $\sqrt{p_0}$. The result is that the importance of vertically as opposed to horizontally traveling energy is overemphasized in the line source case when compared to the point source case. Thus, the body waves are too large relative to the surface waves in the FD seismograms in Fig. 7.

The additional $\sqrt{p_0}$ in the point source or three-dimensional solution can be explained in terms of geometrical spreading (see Figs. 8b and c). The energy with takeoff angle $i$ between $i_0$ and $i_0 + di_0$ for the point source becomes

$$E_P \propto [(2\pi \sin i_0\, r) r\, di_0]/2\pi r^2 \quad \text{or} \quad E_P \propto \sin i_0\, di_0,$$

while for the line-source

$$E_L \propto (r\, di_0)/2\pi r \quad \text{or} \quad E_L \propto di_0.$$

Since energy is proportional to the square of the amplitude, we obtain the $\sqrt{p_0}$ dependence.

### (a) Line Source

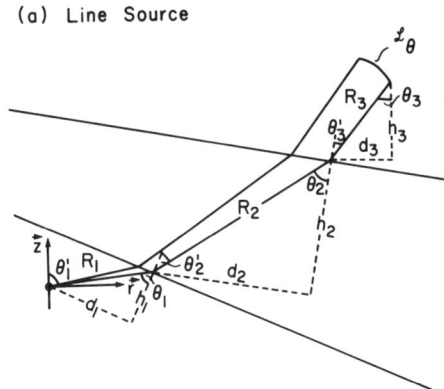

### (b) Point Source

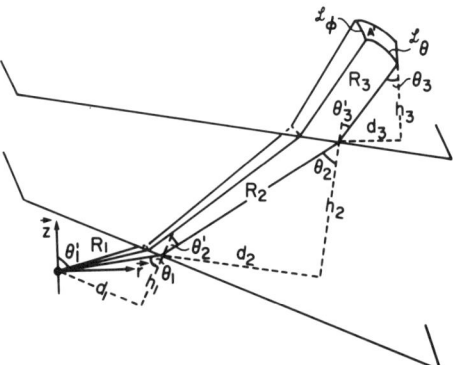

FIG. 9. Diagram displaying the geometric spreading of rays encountering noncoplanar interfaces.

The GRT formalism easily handles multilayers where the interfaces can be flat or locally dipping (Helmberger *et al.*, 1986). For instance, for the problem setup displayed in Fig. 9, we would replace Eq. (6) by

$$J(t) = \text{Im}\big((\sqrt{p}/\eta_1)T_{12}(p_1)T_{23}(p_2)\,dp/dt\big), \qquad (12)$$

where

$$p_1 = (\sin\theta_1)/\alpha_1 \qquad \text{and} \qquad p_2 = (\sin\theta_2)/\alpha_2$$

are the local ray parameters, and $T_{12}$ and $T_{23}$ are the transmission coefficients. The travel time is defined by

$$t = \sum_{i=1}^{3}(p_i d_i + \eta_i h_i), \qquad (13)$$

where

$$\eta_i = ((1/\alpha_i^2) - p_i^2)^{1/2}.$$

In general, $dp/dt$ must be determined numerically by first inverting for $p(t)$ or finding the Cagniard contour. Performing the first-motion approximation about $p_0$, which is equivalent to the saddle point approximation (Helmberger, 1983), we obtain

$$\phi_p = H(t - R/\alpha) \operatorname{Re}(T_{12} \cdot T_{23}) \left(\frac{A_0}{A}\right)^{1/2} \tag{14}$$

as a high-frequency solution; $A_0$ is defined as a small cross-sectional area of the beam at unit distance, and $A$ is the projected area at the receiver (Hong and Helmberger, 1978). The area $A$ is determined numerically by simply shooting rays out around the small cone $A_0$ and approximating the area of the receiver, see Fig. 9. The method, called "glorified optics," can be used in 2D and 3D problems and has proven useful in explaining receiver effects (Cohn et al., 1982).

We return to the effects of the locally dipping structure in Section IV.A. Before leaving this section on explosions, we examine the effect of interference between up- and downgoing energy on strong motion attenuation and use the FD method to find the effect of lateral variations in structure.

## C. Strong Motion Attenuation for Explosions

One customary approach for the assessment of strong motion hazards is to plot peak velocity or acceleration as a function of range. A plot of such data for nuclear explosions is quite spectacular as displayed in Fig. 10 (Burdick, 1983). The observed vertical velocities are shown as circles and the solid line represents the predictions from the elastic model discussed earlier. Note the peak velocity at S2 is 480 cm/sec.

The observed rise times of the vertical velocity records are shown in Fig. 5. The theoretical seismograms are plotted on the same scale, both in amplitude and with the same origin time. Although the agreement is good, these results need to be interpreted cautiously. Many complex processes are occurring in the spall zone, and the agreement between the model predictions and the data could be fortuitous. The results suggest that the vertical component of motion can be adequately modeled by nonlinear absorption along raypaths extending into the elastic near-field region as long as the stresses are compressive, see Burdick (1983) for more details on this subject. The behavior in which we are particularly interested is the shape of the amplitude fall-off curve beyond 2.5 km, which may be shown to be

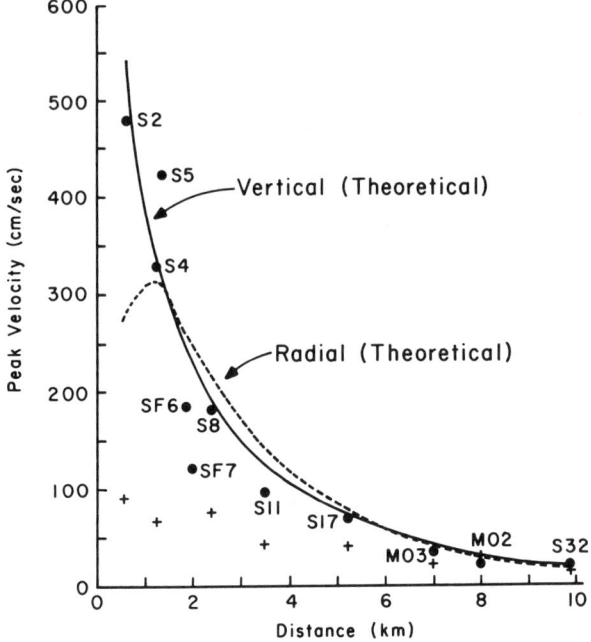

FIG. 10. A comparison of the observed peak velocity from inside the spall zone of MILROW with the predictions of the model developed from the data from the near-field elastic region; vertical (observed, ●), radial (observed, +), $\psi_\infty = 1.4 \times 10^{11}$, $K = 9$, and $B = 1$.

controlled by the crustal structure. The amplitude may be influenced strongly by interaction between the upgoing rays and the downgoing rays.

Before simulating this effect, it is perhaps easier to gain some insight from the systematics predicted for a simple gradient problem (see Fig. 11, in Helmberger et al., 1986). We suppose that a velocity model such as a linear gradient can be chosen such that a step function source will look like a step at receiver positions 1 through 4. Secondly, we break the gradient into a stack of homogeneous layers for computational reasons.

The simulation of the step response can be achieved by summing the contributions from three energy paths, namely, the direct energy, the energy reflected from just below the source or reference plane, and the diving ray energy. A diving energy path can be computed by summing generalized rays or by applying the WKBJ method. It is important to include the transmission coefficient across the reflecting plane in computing this response. The diving ray contributes little energy until large ranges, since the transmission coefficient is 0 for large ray parameters. At the nearest

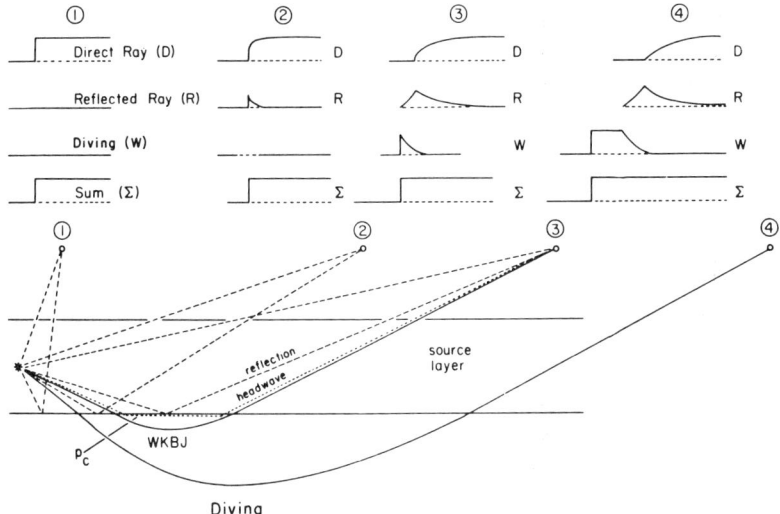

FIG. 11. Schematic picture of ray paths and summation response for a layered stack that approximates a smooth velocity gradient. Reflected and headwave paths for case 4 are omitted.

distance, position 1, the direct ray dominates. The reflected path contributes more as the critical angle is approached. At still larger distances such as position 3, a head wave along the bottom of the reference interface develops and is followed by the critically reflected pulse. The head wave contribution is included in the reflected response since it is associated with the reflected generalized ray. At larger ranges, the WKBJ contribution becomes increasingly dominant. Note that the WKBJ response stops contributing at the same time as the head wave starts, because the transmission coefficient drops to zero. Combining the generalized rays and WKBJ response eliminates the truncation phase and avoids the turning point breakdown of the WKBJ theory.

Figure 12 shows synthetic seismograms for the MILROW body waves computed using generalized ray theory. The ray summations for different groupings of rays at a given distance are shown in each row. The first sum in each row shows only the upgoing P. This is a single ray leaving the source upward and proceeding directly to the receiver. The next sum contains rays that depart downward from the source and turn back up in the crust before arriving at the receiver. The third seismogram is the sum of the first two, and the final sum contains rays of the pP type. At the closest distance, the upgoing ray dominates and pP has virtually no effect. At larger ranges the amplitude of the diving P rapidly becomes larger than for the upgoing ray. Note how the sum of upgoing and downgoing P appears to

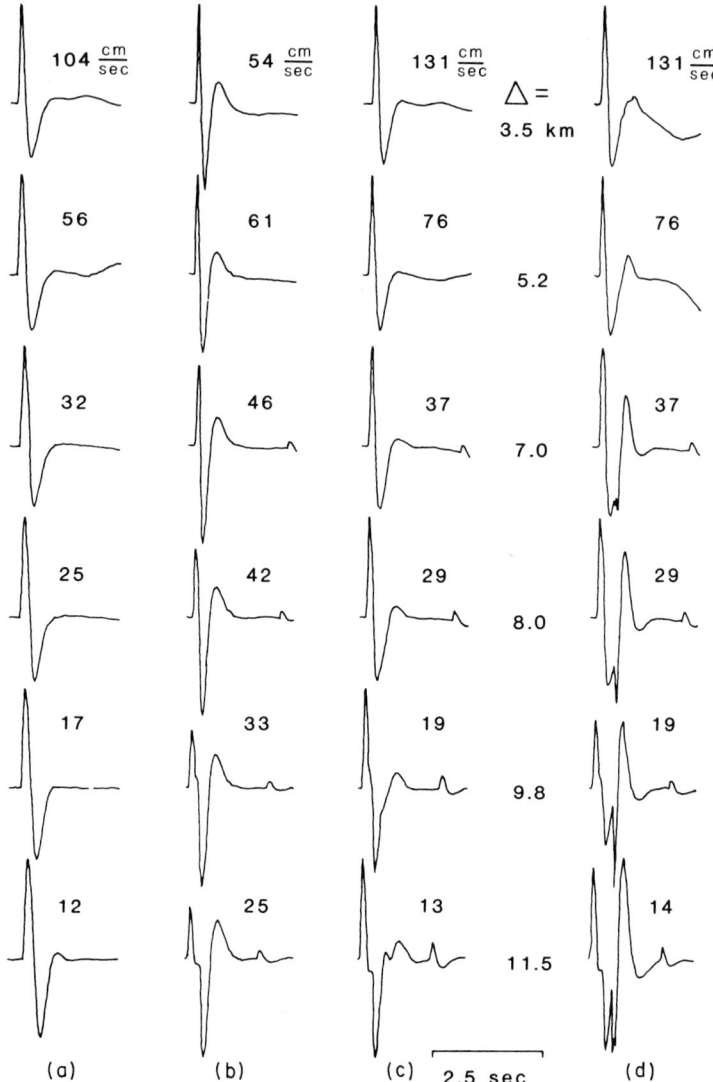

**FIG. 12.** The interaction of the generalized rays which make up the near-field P wave as it develops with distance. Column (a) is just the upgoing generalized ray. Column (b) is the sum of the downgoing direct P rays. Column (c) is the sum of the first two. Column (d) includes rays of the pP type.

have only a single arrival, while its amplitude is determined mainly by the downgoing P. The pP phase begins to emerge at 7 km and becomes more pronounced with distance. At 7 km pP reflects at nearly the halfway distance of 3 km. Thus, it may reflect outside the spall zone. This may explain why near-field pP is unaffected by spall.

The synthetic seismogram predictions shown in Fig. 12 are compared to the observed MILROW body waves in Fig. 13. Burdick (1983) constructed his model to fit the lower 4 seismograms, so the close match is to be

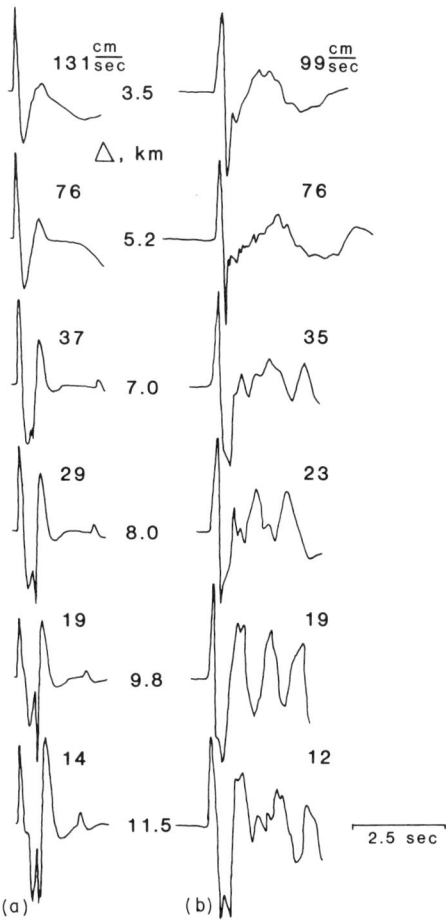

FIG. 13. A comparison of the synthetic (a) and observed (b) P waves for MILROW from the near-field elastic region. Burdick (1983) based his model on the lower 4 records but did not consider the upper two. Note the close fit in the observed and predicted amplitude.

expected. However, he was unaware of the upper two records, and his model fits them as well. A similar comparison is shown for CANNIKIN in Fig. 14.

The delicate interaction of the upgoing and downgoing wave fields to produce the composite P wave indicates that amplitudes at these ranges might be particularly sensitive to lateral variations in structure. If the timing of either the upgoing or downgoing ray is perturbed slightly, the amplitude of the composite arrival changes rapidly with range. Changes in source symmetry so that upgoing and downgoing radiation have unequal strength could also drastically change the signal amplitude. Since the Burdick (1983) models predict amplitude well, we can infer that MILROW

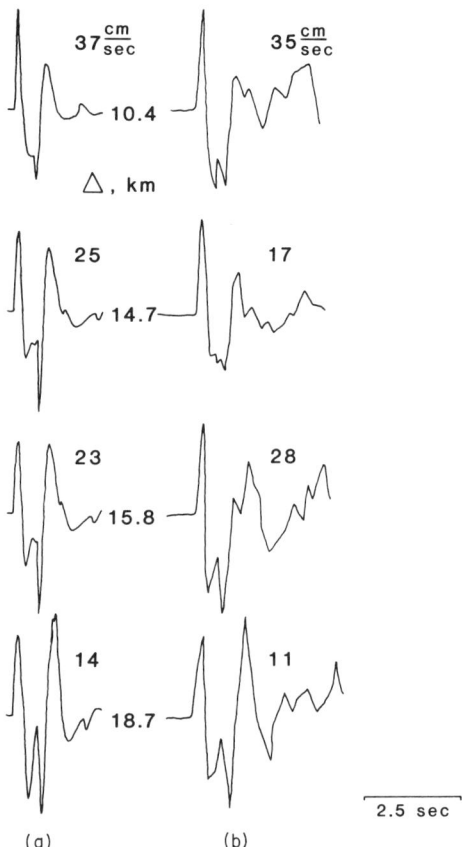

FIG. 14. A comparison of the synthetic (a) and observed (b) P waves from the near-field elastic region of CANNIKIN.

and CANNIKIN have relatively isotropic sources and that lateral variations in the Amchitka crust did not have a strong effect.

D. FD MODELING OF STRUCTURAL EFFECTS

Burdick (1983) by necessity modeled the records of the Amchitka blast with a layered structure. It is instructive to investigate the effect of several perturbations to the layered model by the FD method. Orphal *et al.* (1970) display geologic cross sections from the blast to the various stations and report density and shear and compressional wave velocities for rocks from Amchitka. The three models tested are reasonable given the available information about Amchitka structure.

The effect of a fault with a 1-km offset of the velocity seismograms is compared with the result from the flat-layered model described in Table I in Fig. 15. Compared to the flat-layered case, the Rayleigh waves in the faulted model are larger and earlier. The waves are earlier because the fault has lifted material to the surface faster and larger because the layers that dip down away from the source tend to convert body waves to surface waves (see Section IV.A). The body-wave arrival changes in a less predictable fashion, since it consists of a fairly unstable combination of interfering arrivals, as described in Section II.C. The peak amplitudes of the body waves fluctuate by ~ 50% and the radial velocities appear to fluctuate more than the vertical velocities.

The effect of a soft rock site next to a hard rock site is compared with the flat-layered model in Fig. 16. The soft material has compressional wave velocity $\alpha = 2.0$ km/sec, shear wave velocity $\beta = 1.0$ km/sec, and density $\rho = 1.8$ g/cm$^3$; the hard material has $\alpha = 4.6$ km/sec, $\beta = 2.3$ km/sec, and $\rho = 2.3$ g/cm$^3$; and the rest of the layer, which is 200 meters thick, has $\alpha = 3.4$ km/sec, $\beta = 1.7$ km/sec, and $\rho = 2.3$ g/cm$^3$. The amplitude at the receiver on the soft site is a factor of 1.45 larger than at the receiver in the same position in the plane-layered model. A simple conservation of energy argument, ignoring the transmission coefficient into the slow layer, would predict an amplification of $(v_2\sqrt{\rho_2})/(v_1\sqrt{\rho_1}) = 1.9$, where $v$ is velocity, $\rho$ the density, 1 refers to the slow media, and 2 refers to the top layer of the plane-layered model. When the loss in transmission into the slow material is considered, the observed amplification agrees with the simple prediction. The particle motion is also more vertical than in the flat-layered case for the receiver at 7 km as the ray is more refracted toward the vertical due to the greater velocity contrast. Reverberations and conversions in the slow media may be seen 1 to 2 sec after the initial pulse.

At the station on the faster material, the amplitude is smaller by a factor of 1.2 compared to the simple prediction of 1.4, before correction for

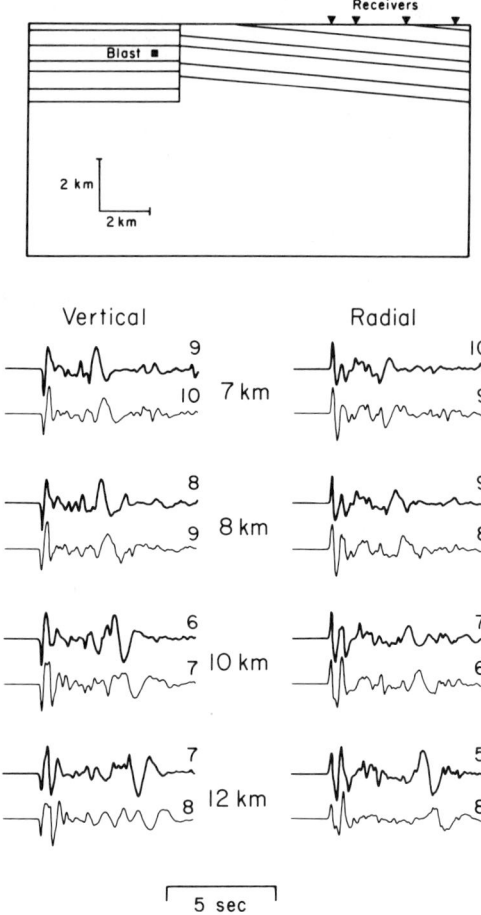

FIG. 15. FD simulation of the effect of a fault with 1-km offset. Synthetic velocity seismograms at ranges of 7, 8, 10, and 12 km (heavy lines) are compared with those from the flat-layered model (light lines). Amplitudes are given in centimeters/second. The model is displayed above the traces and the layering is the same model described in Table I.

transmission, so the estimate does not work as well. Other factors, such as focusing, diffraction, and free-surface interaction may be important. The direct waves and longer period surface waves appear unchanged at the ranges of 10 and 12 km, which are beyond the local station structures.

The best guess for the structure between the blast and station M05 inferred from Orphal *et al.* (1970) is used to generate the comparison shown in Fig. 17. The structure is specified in Table II. The same velocity model is

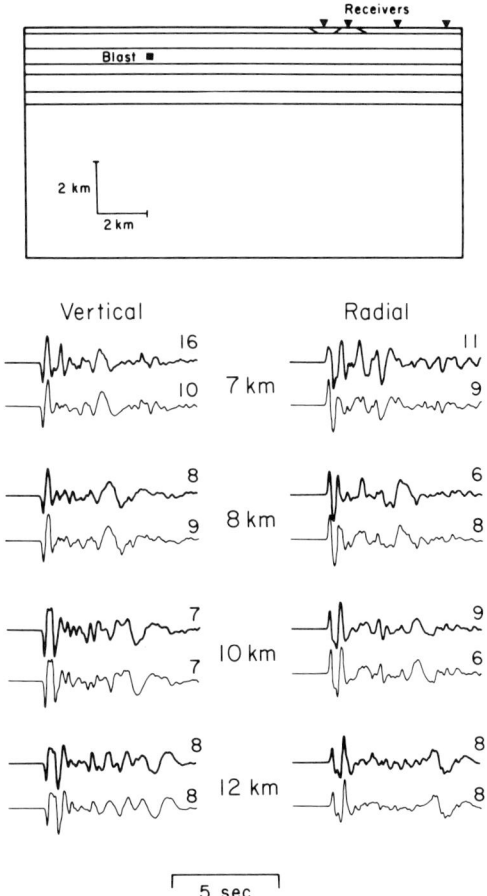

FIG. 16. FD simulation of the effect of a soft rock site next to a hard rock site. Synthetic velocity seismograms at ranges of 7, 8, 10, and 12 km (heavy lines) are compared with those from the flat-layered model (light lines). The soft and hard rock sites are at ranges of 7 and 8 km, respectively. Soft site material has $\alpha$ of 2.0 km/sec, $\beta$ of 1.0 km/sec, and $\rho$ of 1.8 g/cm$^3$, and hard site material has $\alpha$ of 4.6 km/sec, $\beta$ of 2.3 km/sec, and $\rho$ of 2.3 g/cm$^3$. The model is displayed above the traces and the layering is the same model described in Table I.

used below the source as in the layered model, which results in similar waveforms in the body-wave arrival. The waveforms are quite sensitive to the structure where the rays bottom, which agrees with the result of the previous section. The amplitudes fluctuate by 25% on the vertical component and by 50% on the radial component. About 30% more amplitude, which translates to 70% more energy, is converted into the surface wave by

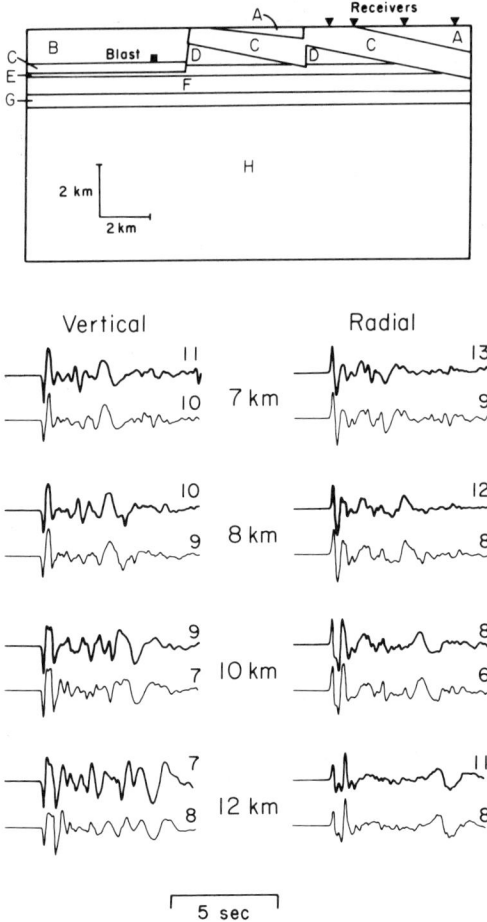

**FIG. 17.** FD simulation of the effect of the cross section derived from Orphal *et al.* (1970) for station M05. Synthetic velocity seismograms at ranges of 7, 8, 10, and 12 km (heavy lines) are compared with those from the flat-layered model (light lines). The model is displayed above the traces, and the velocities and densities that correspond with the letters A–H are given in Table II.

the structures that dip down away from the source. The tendency of dipping layers to convert body waves to surface waves is examined in more detail in Section IV.

Shallow structure is seen to affect the amplitude of body waves as well as surface waves. These effects are difficult to model deterministically because the structures are difficult to determine. Shallow structure may well cause

TABLE II. VELOCITIES IN THE FAULTED AMCHITKA MODEL

| Letter | P-wave velocity | S-wave velocity (km/sec) | Density (km/sec) |
|---|---|---|---|
| A | 3.4 | 1.7 | 2.0 |
| B | 3.4 | 1.7 | 2.1 |
| C | 3.7 | 1.9 | 2.1 |
| D | 4.4 | 2.2 | 2.4 |
| E | 4.6 | 2.3 | 2.5 |
| F | 4.9 | 2.8 | 2.6 |
| G | 5.1 | 2.9 | 2.7 |
| H | 5.9 | 3.3 | 2.7 |

the fluctuation in the ratio of vertical to radial energy seen in the data in Fig. 4 and contribute to the misfit between the synthetic seismograms and the data in Fig. 6. Techniques like the FD scheme proposed in this work are necessary to understand the effects of shallow structure.

## III. Earthquake Sources

We suppose that motions produced by earthquakes can be simulated by assuming a distribution of shear dislocations. A principle objective of Section III is the derivation of line source descriptions to be used in the line-to-point source mapping procedure. In Section III.A, the generalized ray theory (GRT) whole space solution for a point shear dislocation is developed. The generalization of GRT to a layered structure is discussed in Section III.B. The use of a whole space GRT line source solution as a source for a finite-difference is described in Section III.C. The FD method, in combination with the GRT source, allows the effect of complex structures to be investigated. The insertion of a source with a FD grid is discussed in Section III.D, and the line source radiation patterns are shown in Figs. 23–25.

### A. Shear Dislocation Source

Harkrider (1976) has obtained convenient forms of displacement potentials starting with Haskell's representation, which allows a discontinuity in displacement across a fault plane. The results, after Langston and Helmberger (1975), in terms of Laplace-transformed displacements along the vertical, tangential, and radial directions are

$$\hat{W} = \frac{\partial \hat{\phi}}{\partial z} + sp\hat{\Omega},$$

$$\hat{V} = \frac{1}{r} \frac{\partial \hat{\phi}}{\partial \theta} - \frac{1}{spr} \frac{\partial^2 \hat{\Omega}}{\partial z \partial \theta} - \frac{\partial \hat{\chi}}{\partial r}, \qquad (15)$$

$$\hat{Q} = \frac{\partial \hat{\phi}}{\partial r} - \frac{1}{sp} \frac{\partial^2 \hat{\Omega}}{\partial r \partial z} + \frac{1}{r} \frac{\partial \hat{\chi}}{\partial \theta},$$

where $z$, $r$, and $\theta$ are the vertical, radial, and polar angle coordinates, respectively. The P-wave potential ($\hat{\phi}$), the SV-wave potential ($\hat{\Omega}$), and the SH-wave potential ($\hat{\chi}$) are as follows:

(1) P-waves:

$$\hat{\phi} = \frac{M_0}{4\pi\rho} \frac{2}{\pi} \operatorname{Im} \int_c^{i\infty+c} C_1(p) \frac{p}{\eta_\alpha} \exp(-s\eta_\alpha |z - h|) K_2(spr) \, dp$$
$$\cdot A_1(\theta, \lambda, \delta)$$
$$+ \frac{M_0}{4\pi\rho} \frac{2}{\pi} \operatorname{Im} \int_c^{i\infty+c} C_2(p) \frac{p}{\eta_\alpha} \exp(-s\eta_\alpha |z - h|) K_1(spr) \, dp$$
$$\cdot A_2(\theta, \lambda, \delta)$$
$$+ \frac{M_0}{4\pi\rho} \frac{2}{\pi} \operatorname{Im} \int_c^{i\infty+c} C_3(p) \frac{p}{\eta_\alpha} \exp(-s\eta_\alpha |z - h|) K_0(spr) \, dp$$
$$\cdot A_3(\theta, \lambda, \delta), \qquad (16)$$

(2) SV-waves:

$$\hat{\Omega} = \frac{M_0}{4\pi\rho} \frac{2}{\pi} \operatorname{Im} \int_c^{i\infty+c} SV_1(p) \frac{p}{\eta_\beta} \exp(-s\eta_\beta |z - h|) K_2(spr) \, dp$$
$$\cdot A_1(\theta, \lambda, \delta)$$
$$+ \frac{M_0}{4\pi\rho} \frac{2}{\pi} \operatorname{Im} \int_c^{i\infty+c} SV_2(p) \frac{p}{\eta_\beta} \exp(-s\eta_\beta |z - h|) K_1(spr) \, dp$$
$$\cdot A_2(\theta, \lambda, \delta)$$
$$+ \frac{M_0}{4\pi\rho} \frac{2}{\pi} \operatorname{Im} \int_c^{i\infty+c} SV_3(p) \frac{p}{\eta_\beta} \exp(-s\eta_\beta |z - h|) K_0(spr) \, dp$$
$$\cdot A_3(\theta, \lambda, \delta), \qquad (17)$$

(3) SH-waves:

$$\hat{\chi} = \frac{M_0}{4\pi\rho} \frac{2}{\pi} \text{Im} \int_c^{i\infty+c} SH_1(p) \frac{p}{\eta_\beta} \exp(-s\eta_\beta |z-h|) K_2(spr) \, dp$$
$$\cdot A_4(\theta, \lambda, \delta)$$
$$+ \frac{M_0}{4\pi\rho} \frac{2}{\pi} \text{Im} \int_c^{i\infty+c} SH_2(p) \frac{p}{\eta_\beta} \exp(-s\eta_\beta |z-h|) K_1(spr) \, dp$$
$$\cdot A_5(\theta, \lambda, \delta), \tag{18}$$

where $s$ is the Laplace transform variable, $p$ is the ray parameter, $\eta_\nu = (1/\nu^2 - p^2)^{1/2}$, $h$ is the depth of source, $\alpha$ the compressional velocity, $\beta$ the shear velocity, $\rho$ the density, $M_0$ the seismic moment, and $c$ a small constant that offsets the path of integration from the imaginary axis. In the above equations, the $C_i$, $SV_i$ and $SH_i$ are functions of $p$ and are identified with the vertical radiation patterns defined by

$$\begin{aligned} C_1 &= -p^2, & SV_1 &= -\varepsilon p \eta_\beta, & SH_1 &= 1/\beta^2, \\ C_2 &= 2\varepsilon p \eta_\alpha, & SV_2 &= (\eta_\beta^2 - p^2), & SH_2 &= \frac{-\varepsilon \eta_\beta}{\beta^2 p}, \\ C_3 &= (p^2 - 2\eta_\alpha^2), & SV_3 &= 3\varepsilon p \eta_\beta, & & \end{aligned} \tag{19}$$

where

$$\varepsilon = \begin{cases} +1, & z > h, \\ -1, & z < h. \end{cases}$$

The azimuthal radiation patterns or orientation constants $A_i$ are determined by

$$\begin{aligned} A_1 &= \sin 2\theta \cos \lambda \sin \delta + \tfrac{1}{2} \cos 2\theta \sin \lambda \sin 2\delta, \\ A_2 &= \cos \theta \cos \lambda \cos \delta - \sin \theta \sin \lambda \cos 2\delta, \\ A_3 &= \tfrac{1}{2} \sin \lambda \sin 2\delta, \\ A_4 &= \cos 2\theta \cos \lambda \sin \delta - \tfrac{1}{2} \sin 2\theta \sin \lambda \sin 2\delta, \\ A_5 &= -\sin \theta \cos \lambda \cos \delta - \cos \theta \sin \lambda \cos 2\delta, \end{aligned} \tag{20}$$

where $\theta$ is the strike from the end of the fault plane, $\lambda$ the rake angle, and $\delta$ the dip angle.

The geometry displaying the orientation of the fault in the cylindrical coordinate system is given in Fig. 18. Note that a vertical strike-slip event is defined by $\lambda = 0°$ and $\delta = 90°$. In Eqs. (16) through (20), the subscripts 1 and 4 correspond to a pure strike-slip fault (SS), the subscripts 2 and 5 correspond to a pure normal or dip-slip fault (DS), and the subscript 3 corresponds to a 45° dip-slip (DD) fault. Any fault orientation can be obtained by a linear combination of these three (Burridge *et al.*, 1964).

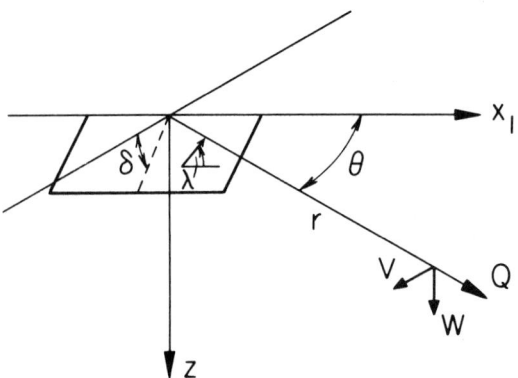

FIG. 18. Description of conventions for mechanism and orientation.

Integrals of the type given in Eqs. (16) through (18) can be transformed back into the time domain by applying the Cagniard–de Hoop technique. For example, suppose we consider the integral

$$\bar{\zeta}(r, z, s) = \frac{2}{\pi} s \, \text{Im} \int_c^{i\infty+c} \frac{p}{\eta_\alpha} K_2(spr) e^{-s\eta_\alpha |z-h|} \, dp \qquad (21)$$

that can be expressed as

$$\zeta(r, z, t) = \frac{2}{\pi} \frac{\partial}{\partial t} \text{Im} \int_0^t \frac{c(t, \tau)}{\sqrt{(t-\tau)(t-\tau+2pr)}} \left(\frac{dp}{d\tau}\right) \frac{p(\tau)}{\eta_\alpha} \, d\tau, \qquad (22)$$

where

$$c[t, \tau(p)] = \cosh\left[2\cosh^{-1}\left(\frac{t-\tau+pr}{pr}\right)\right].$$

The various functions of $p$ are to be evaluated along the Cagniard–de Hoop contour, namely,

$$\tau(p) = pr + \eta_\alpha |z - h|,$$

from the first arrival time $R/\alpha$ to $t$. The integral is computed numerically. This particular integral has a closed form solution

$$\zeta(r, z, t) = \frac{d}{dt}\left[\frac{1}{R} + \frac{2\alpha}{r^2}\left(t - \frac{R}{\alpha}\right)\right] H\left(t - \frac{R}{\alpha}\right) \qquad (23)$$

as discussed by Helmberger and Harkrider (1978).

Thus, one way to proceed is to substitute Eqs. (16) through (18) into Eq. (15) and evaluate the integrals following the above scheme. This is the full Cagnaird solution, and the results for a simple case are shown in the bottom row of seismograms in Fig. 19.

# 6. PATH EFFECTS IN STRONG MOTION SEISMOLOGY

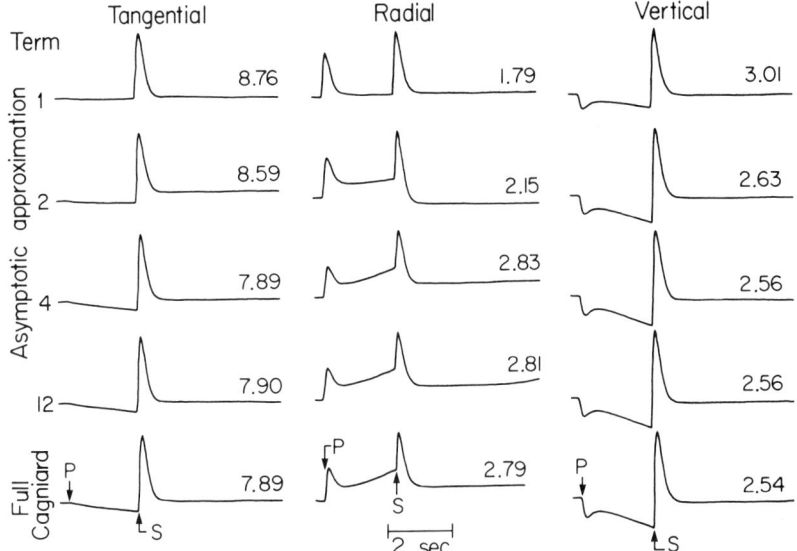

FIG. 19. Comparison of the three components of displacement for a whole space with a strike-slip source. The top four rows contain the asymptotic summation with 1, 2, 4, and 12 terms. The full solution is displayed on the bottom. The source depth is $h = 8$ km, and the range is $r = 16$ km. Model parameters are $\alpha = 6.2$ km/sec, $\beta = 3.5$ km/sec, and $\rho = 2.7$ g/cm$^3$.

A much faster procedure is to expand the modified Bessel functions of the second of order $n$, $K_n$, as

$$K_n(spr) = \sqrt{\frac{\pi}{2spr}} e^{-spr} \left[ 1 + \frac{4n^2 - 1}{8spr} + \cdots \right] \qquad (24)$$

and substitute the potentials in Eqs. (16), (17), and (18) into the displacement Eq. (15) and expand in powers of $s^{-1}$. The first term of such an expansion has the form

$$\mathbf{V}(r, z, t) = \frac{M_0}{4\pi\rho_0} \frac{d}{dt} \left[ \dot{D}(t) * \sum_{j=1}^{2} A_{j+3}(\theta, \lambda, \delta) v_j(r, z, t) \right], \qquad (25)$$

$$\mathbf{W}(r, z, t) = \frac{M_0}{4\pi\rho_0} \frac{d}{dt} \left[ \dot{D}(t) * \sum_{j=1}^{3} A_j(\theta, \lambda, \delta) w_j(r, z, t) \right], \qquad (26)$$

$$\mathbf{Q}(r, z, t) = \frac{M_0}{4\pi\rho_0} \frac{d}{dt} \left[ \dot{D}(t) * \sum_{j=1}^{3} A_j(\theta, \lambda, \delta) q_j(r, z, t) \right], \qquad (27)$$

where

$$v_j = \sqrt{\frac{2}{r}} \frac{1}{\pi} \left[ \frac{1}{\sqrt{t}} * \operatorname{Im}\left( \frac{\sqrt{p}}{\eta_\beta} SH_j p \frac{dp}{dt} \right) \right], \tag{28}$$

$$w_j = \sqrt{\frac{2}{r}} \frac{1}{\pi} \left\{ \frac{1}{\sqrt{t}} * \operatorname{Im}\left[ \left( \frac{\sqrt{p}}{\eta_\alpha} C_j(-\varepsilon\eta_\alpha) \frac{dp}{dt} \right)_\alpha + \left( \frac{\sqrt{p}}{\eta_\beta} SV_j\, p \frac{dp}{dt} \right)_\beta \right] \right\}, \tag{29}$$

$$q_j = \sqrt{\frac{2}{r}} \frac{1}{\pi} \left\{ \frac{1}{\sqrt{t}} * \operatorname{Im}\left[ \left( \frac{\sqrt{p}}{\eta_\alpha} C_j\, p \frac{dp}{dt} \right)_\alpha + \left( \frac{\sqrt{p}}{\eta_\beta} SV_j(-\varepsilon\eta_\beta) \frac{dp}{dt} \right)_\beta \right] \right\}. \tag{30}$$

This is a first term of an asymptotic expansion similar to the expansion used for explosive sources in Section II. The approximation is accurate for $spr \gg 1$, which means it is most accurate for a high-frequency, large-range, and nonvertical takeoff angle. The two arrivals in the $w_j$ and $q_j$ case are the P wave and SV wave. Note that the first term becomes uncoupled in that **V** depends only on $\chi$; and **W** and **Q** only on $\phi$ and $\Omega$, so the SH solution separates from the P–SV solution in this asymptotic form.

## B. Response of a Multilayered Half-Space

Using the concepts of generalized ray theory it is possible to solve the multilayered problem (Helmberger, 1983). For instance, if we were to compute the SH response for a layer containing a source we would modify Eq. (28), to

$$V_i = \sqrt{\frac{2}{r}} \frac{1}{\pi} \left[ \frac{1}{\sqrt{t}} * \sum_{j=1}^{n} \operatorname{Im}\left( \frac{\sqrt{p}}{\eta_\beta} SH_i\, \Pi_j\, p \frac{dp}{dt} \right)_j \right],$$

where the summation is over the $n$ rays, and $\Pi_j$ is the product of transmission and reflection coefficients appropriate for the $j$th ray.

The P–SV elastic case is more complicated than the SH case due to the possible mode changes at each boundary, and the accuracy of the GRT methods must be checked against more complete solutions. The GRT calculations for a simple layer over a half-space case are shown as the dotted lines in Fig. 20 and described in Table III. These results are compared with a more exact wave-number integration solution, shown as the solid lines in Fig. 20 (Apsel and Luco, 1983). A comparison between the GRT results and the computationally slower, but more general FD proce-

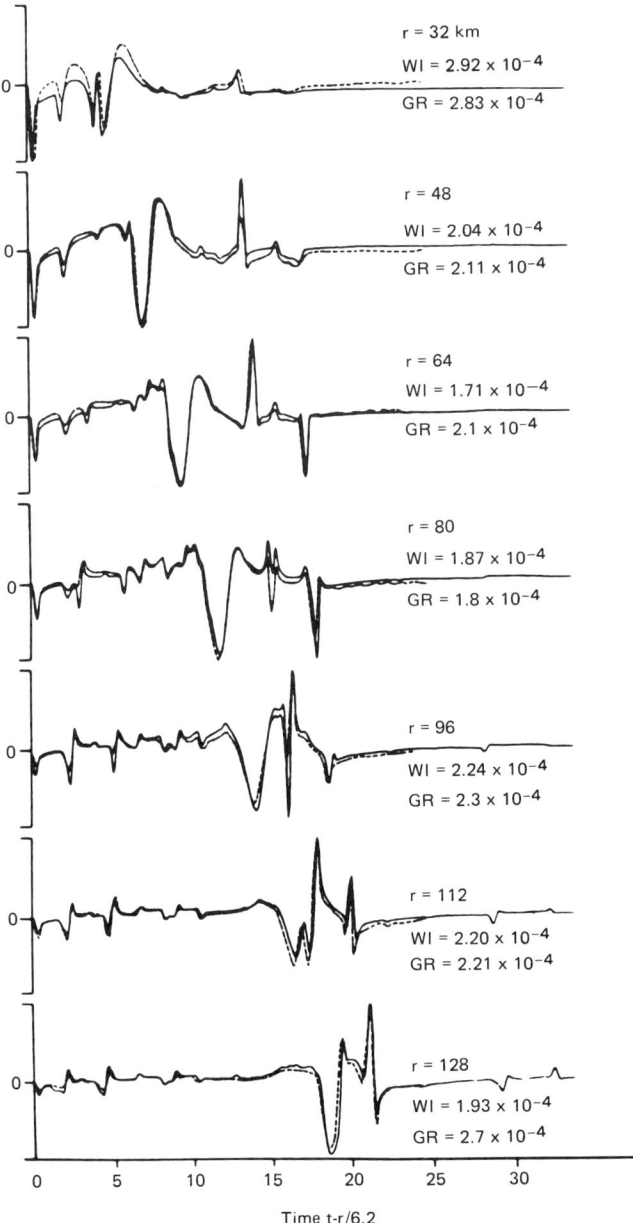

**FIG. 20.** Comparison of the wave number integration solution (———) with the GRT first term asymptotic solution (— — — —) for vertical displacement (45° from strike) on the free surface due to a vertical strike-slip dislocation buried at a depth of 8 km in the model described in Table III. (After Apsel and Luco, 1983.)

TABLE III. LAYER OVER HALF-SPACE MODEL

| P-wave velocity (km/sec) | S-wave velocity (km/sec) | Density (g/cm$^3$) | Layer thickness (km) |
|---|---|---|---|
| 6.2 | 3.5 | 2.7 | 32.0 |
| 8.2 | 4.5 | 3.4 | $\infty$ |

dure described in Sections III.C and III.D is given in Fig. 21, where the results for all three fundamental faults are shown. The methods agree very well.

Adding layers to the model greatly complicates the GR-approach, especially for the latter portion of the record when many ray paths are involved. On the other hand, the FD procedure described below remains unaffected by this complexity and appears well suited for exploring lateral variation effects, including structures that cannot be described by plane layers.

### C. Source Expressions for 2D Numerical Grids

Note that the term

$$V_j = \text{Im}\left(SH_j \frac{p}{\eta_\beta} \frac{dp}{dt}\right) \tag{31}$$

in Eq. (28) and the analogous terms in Eqs. (29) and (30) are solutions to the two-dimensional line source wave equation. This suggests a scheme whereby sources with a radiation pattern may be introduced into a 2D numerical grid. With the source located in a homogeneous whole space we have

$$p = \frac{r}{R^2} t + i\left(t^2 - \frac{R^2}{V^2}\right)^{1/2} \frac{|z-h|}{R^2}, \tag{32}$$

$$\eta_V = \frac{|z-h|}{R^2} t - i\left(t^2 - \frac{R^2}{V^2}\right)^{1/2} \frac{r}{R^2},$$

and

$$\text{Im}\left(SH_j \frac{p}{\eta_\beta} \frac{dp}{dt}\right) = \text{Re}(SH_j p) \frac{H(t - R/\beta)}{(t^2 - R^2/\beta^2)^{1/2}}. \tag{33}$$

Similar expressions may be derived from Eqs. (29) and (30). The effective line source radiation patterns can be obtained by evaluating the various real part operators.

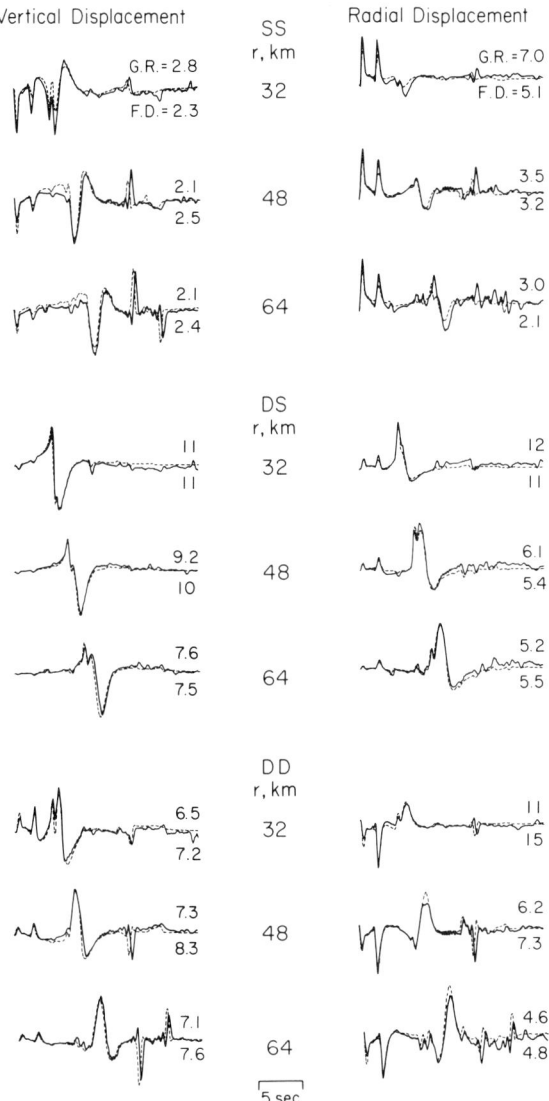

**FIG. 21.** Comparison of FD and GRT seismograms for the ranges 32, 48, and 64 km for the strike-slip, dip-slip, and 45° dip-slip mechanisms. The far-field source time function, $\dot{D}(t)$, is specified by a trapezoidal shape with equal $\delta t_j$'s of 0.2 sec. The parameters for the layers are given in Table III. Amplitudes may be scaled to moment.

It is only a matter of algebra to find the explicit functions that may be used to drive the source box in the FD grid. The source box mechanics are described in Section III.D.

1. SH Case

The following solutions are for a whole space. Let

$$\Gamma_\gamma = \sqrt{\frac{2}{\gamma} \frac{1}{4\pi^2 \rho \, 10^{20}}}, \qquad T_\gamma = \frac{R^2}{t^2 \gamma^2},$$

where $\gamma = \alpha$ or $\beta$. The analytic source expressions at the edges of the source box are

$$V_4 = \frac{\Gamma_\beta}{\beta^2} \frac{\sin i}{R\sqrt{1 - T_\beta}} H\left(t - \frac{R}{\beta}\right) * \frac{dM_0(t)}{dt}, \qquad (34a)$$

$$V_5 = \frac{\Gamma_\beta}{\beta^2} \frac{\cos i}{R\sqrt{1 - T_\beta}} H\left(t - \frac{R}{\beta}\right) * \frac{dM_0(t)}{dt}, \qquad (34b)$$

where $V_4$ and $V_5$ represent the strike-slip and dip-slip cases, respectively; $H$ is the Heaviside step function; $\rho$, $\beta$, and $\alpha$ (used below) are the density, shear wave, and compressional wave velocity at the source; $M_0(t)$ is the moment release as a function of time; $R$ is the source to receiver distance; and $i$ is the angle between the vertical and the line connecting the source and receiver, as shown in Fig. 8a. Convolution with an appropriately smooth time function avoids problems with the singular pulse of energy at the geometrical arrival time.

After the energy propagates across laterally and vertically heterogeneous structure, "line source seismograms" $\tilde{V}_i$ may be extracted from the grid. Note the difference between $V$ and $\tilde{V}$: $V$ are the whole space solutions inserted in the source region of the FD grid; $\tilde{V}$ are the seismograms that are extracted from the FD grid. The point source seismograms are obtained by

$$\mathbf{V}_p = \frac{1}{\sqrt{R}} \frac{d}{dt}\left(\frac{1}{\sqrt{t}} * (A_4 \tilde{V}_4 + A_5 \tilde{V}_5)\right), \qquad (35)$$

where $\mathbf{V}_p$ is the SH displacement in centimeters when moment is in dynes per centimeter, density is in grams per cubic centimeter, $\beta$ is in kilometers per second, and $R$ is in kilometers.

2. P–SV Case

For convenience, we define

$$\Phi_\gamma = \Gamma_\gamma \frac{t^2}{R^6 \sqrt{1 - T_\gamma}} H\left(t - \frac{R}{\gamma}\right) \frac{dM_0(t)}{dt},$$

where $\gamma = \alpha$ or $\beta$. Next, we present the results for the three fundamental faults, the strike-slip, the dip-slip, and the 45° dip-slip cases.

(a) Strike-slip case

$$\begin{aligned} Q_1 &= r\{\Phi_\alpha(r^2 - 3z^2 + 3T_\alpha z^2) \\ &\quad + \Phi_\beta[3z^2 - r^2 + T_\beta(r^2 - 2z^2)]\}, \\ W_1 &= z\{\Phi_\alpha[z^2 - 3r^2 + T_\alpha(2r^2 - z^2)] \\ &\quad + \Phi_\beta[3r^2 - z^2 + T_\beta(z^2 - 2r^2)]\}. \end{aligned} \quad (36)$$

(b) Dip-slip case

$$\begin{aligned} Q_2 &= z\{\Phi_\alpha[6r^2 - 2z^2 + T_\alpha(2z^2 - 4r^2)] \\ &\quad + \Phi_\beta[2z^2 - 6r^2 + T_\beta(5r^2 - z^2)]\}, \\ W_2 &= r\{\Phi_\alpha[2r^2 - 6z^2 + T_\alpha(4z^2 - 2r^2)] \\ &\quad + \Phi_\beta[6z^2 - 2r^2 + T_\beta(r^2 - 5z^2)]\}. \end{aligned} \quad (37)$$

(c) 45° Dip-slip case

$$\begin{aligned} Q_3 &= r\{\Phi_\alpha[9z^2 - 3r^2 + T_\alpha(2r^2 - 7z^2)] \\ &\quad + \Phi_\beta[3r^2 - 9z^2 + T_\beta(6z^2 - 3r^2)]\}, \\ W_3 &= z\{\Phi_\alpha[9r^2 - 3z^2 + T_\alpha(z^2 - 8r^2)] \\ &\quad + \Phi_\beta[3z^2 - 9r^2 + T_\beta(6r^2 - 3z^2)]\}. \end{aligned} \quad (38)$$

Where $r$ is the horizontal component of $R$ and is positive in the direction of the receiver, and $z$ is the vertical component of $R$ and is positive downward.

As in the SH case, the line source seismograms extracted from the FD grid, $\tilde{Q}_i$ and $\tilde{W}_i$, are transformed into point seismograms by

$$\mathbf{Q}_p = \frac{1}{\sqrt{R}} \frac{d}{dt}\left(\frac{1}{\sqrt{t}} * (A_1\tilde{Q}_1 + A_2\tilde{Q}_2 + A_3\tilde{Q}_3)\right), \quad (39)$$

$$\mathbf{W}_p = \frac{1}{\sqrt{R}} \frac{d}{dt}\left(\frac{1}{\sqrt{t}} * (A_1\tilde{W}_1 + A_2\tilde{W}_2 + A_3\tilde{W}_3)\right). \quad (40)$$

Again, $\mathbf{Q}_p$ and $\mathbf{W}_p$ are horizontal and vertical displacements in centimeters.

The extension to higher order terms is simple using the analytical Cagniard–de Hoop expressions since they depend mostly on the temporal integrals of the previous terms. The results of such an expansion are shown in Fig. 19. When the responses are convolved with most instrument responses, the higher order corrections make little difference because they are more long period. Thus, we will limit our discussion to the first term of the asymptotic expansion for the rest of this work.

### D. Insertion of the Source into the FD Grid

1. Source Region

Energy is introduced into the FD grid with the type of source region described by Alterman and Karal (1968). The use of a source region rather than a source point is necessary to avoid singular points such as at the source of an explosion in the displacement field. The fourth-order FD calculations do poorly for propagating energy with wavelengths shorter than 10 grid points per wavelength (Alford *et al.*, 1974), and such energy is abundant near singular points. In each time step, the FD algorithm uses the present and past time steps to compute a future time step, which is written over the past time step. During each time step, energy insertion is a two-stage process. The first stage is to add in the energy that is coming out of the source. The second stage is to remove direct energy from the source and follow the indirect energy that is traveling through the source region. If the displacements computed analytically were simply imposed on the source box, the source area would act as a rigid reflector for energy impinging on the area from the outside. The rigid reflector source box noticeably affects the results of the finite-difference modeling, particularly if the source is near the free surface (Alterman and Karal, 1968).

In this section, the two-stages procedure will be outlined for a second-order algorithm. The FD results in this paper are made with a fourth-order source insertion algorithm, which only differs from the algorithm described in this section in that rings 2 and 3 each contain 2 rings of grid points and the grid is updated by a fourth-order FD scheme. The advantages of the fourth-order FD schemes are discussed in Alford *et al.* (1974).

The FD grid is divided into (1) an exterior, (2) an outer ring of grid points, (3) an inner ring of grid points, and (4) an interior (see Fig. 22). The source is in the center of region 4, the interior, which never sees the direct waves from the source.

(a) Region 4 from the past time step is saved in an array.

(b) The entire grid is updated to the future time step by the FD algorithm.

(c) The outer ring, region 2 in the future time step, is saved in an array for later reinsertion.

(d) Rings 2 and 3 for the past time step are loaded into the source region grid with displacements that have had the source subtracted, and region 4 is loaded with the past time step that was saved in step (a).

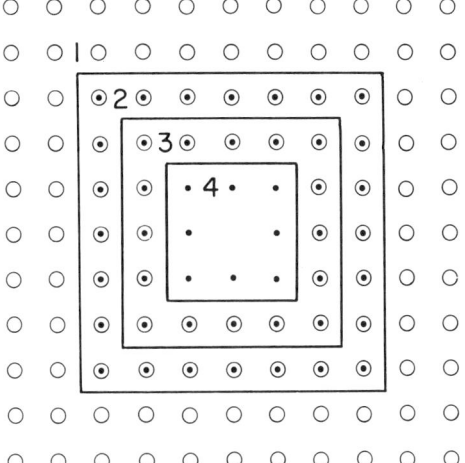

FIG. 22. Source region geometry. Region 1 is outside rings 2 and 3, which are each 1 grid point thick. Region 4 is inside rings 2 and 3. The source is in the center of region 4. Region 1 is only computed with the source included, and region 4 is computed only without the source. Rings 2 and 3 are computed both with and without the source. The geometry for a fourth- rather than second-order FD scheme would use rings 2 and 3 that are 2 rather than 1 grid point thick.

(e) The grid for ring 3 and region 4 is again updated to the future time step by the FD algorithm.

(f) The analytic source values at the grid points in rings 2 and 3 for the present time step are read into arrays.

(g) The source-free ring 3 from the present time step is saved for use in step (d) in the next time step, and the analytic source is added to ring 3. The ring 2, which includes the source, is saved in an array, and the analytic source is subtracted from the array, which will also be used in step (d) in the next time step.

(h) Ring 2 is reloaded with the array stored in step (c), so region 1 and rings 2 and 3 all contain the source.

(i) All the past values at the grid points are now overwritten by the future values. The grid is advanced one time step by making the array of present values into past values and the future values into the present values. The present arrays saved for rings 2 and 3 become the past arrays that are needed for the next time step.

Since this process requires the analytic form of the displacement at many points in space and time, we use a homogeneous source region. This allows the use of whole-space solutions for the analytic functions. The use of even simple half-space solutions, which would allow a source closer to the free surface since the free surface could be within the source region, would lengthen the time required to compute the analytic source functions to more than the time the FD difference code takes to run.

2. Source Vertical Radiation Patterns

We will next show the SH and P–SV sources as a function of space at a fixed point in time to gain some insight as to how they work. The time function used to excite the grid is a Heaviside step function smoothed by convolution with the function $e^{-Kt^2}$, where $t$ is time. The smoothing operator is necessary because the FD method does not treat the highest frequencies correctly. Most FD workers, however, convolve by $te^{-Kt^2}$, or, equivalently, use a spatial function like $e^{-Kr^2}$ to start up the FD grid, where $r$ is the distance to the source. That operation eliminates the low-frequency as well as the high-frequency ends of the source power spectrum. Without the low frequencies in the source, the $1/\sqrt{t}$ amplitude decay that characterizes line sources is not as apparent. We chose to use the $e^{-Kt^2}$ source, which keeps the $1/\sqrt{t}$ tails and the low frequencies in the FD grid, but then we eliminate the tails without losing the low-frequency energy with the line to point source conversion described by Eqs. (35), (39), and (40).

The whole-space source functions for the SH case are shown in time slices in Fig. 23. The explosive case is isotropic, and the strike-slip and dip-slip cases have $\sin\theta$ and $\cos\theta$ vertical radiation patterns respectively, where $\theta$ is the takeoff angle. The explosion is physical only in the acoustic problem, which obeys the same equation as the SH case with a uniform whole space. The traveling energy in Figs. 23, 24, and 25, which has propagated nearly to the edge of the grid from the source by the second time slice, has the character of a line source. The seismograms show an amplitude decay in space of $1/\sqrt{r}$ and have an impulsive arrival followed by a $1/\sqrt{t}$ amplitude decay in time. The "pseudo-near-field," which is the energy seen in Fig. 23 in the region of the source, becomes a static field with time. Note the square devoid of displacement in the center of the source

FIG. 23. SH vertical radiation patterns. The displacement fields due to explosive, strike-slip, and dip-slip sources are shown after 125 and 250 timesteps. The explosive plots have a white background, and all displacements are positive, while the plots for the earthquakes have a grey background, where positive displacements are shown in black and negative in white.

## 6. PATH EFFECTS IN STRONG MOTION SEISMOLOGY

t = 125 time steps    t = 250 time steps

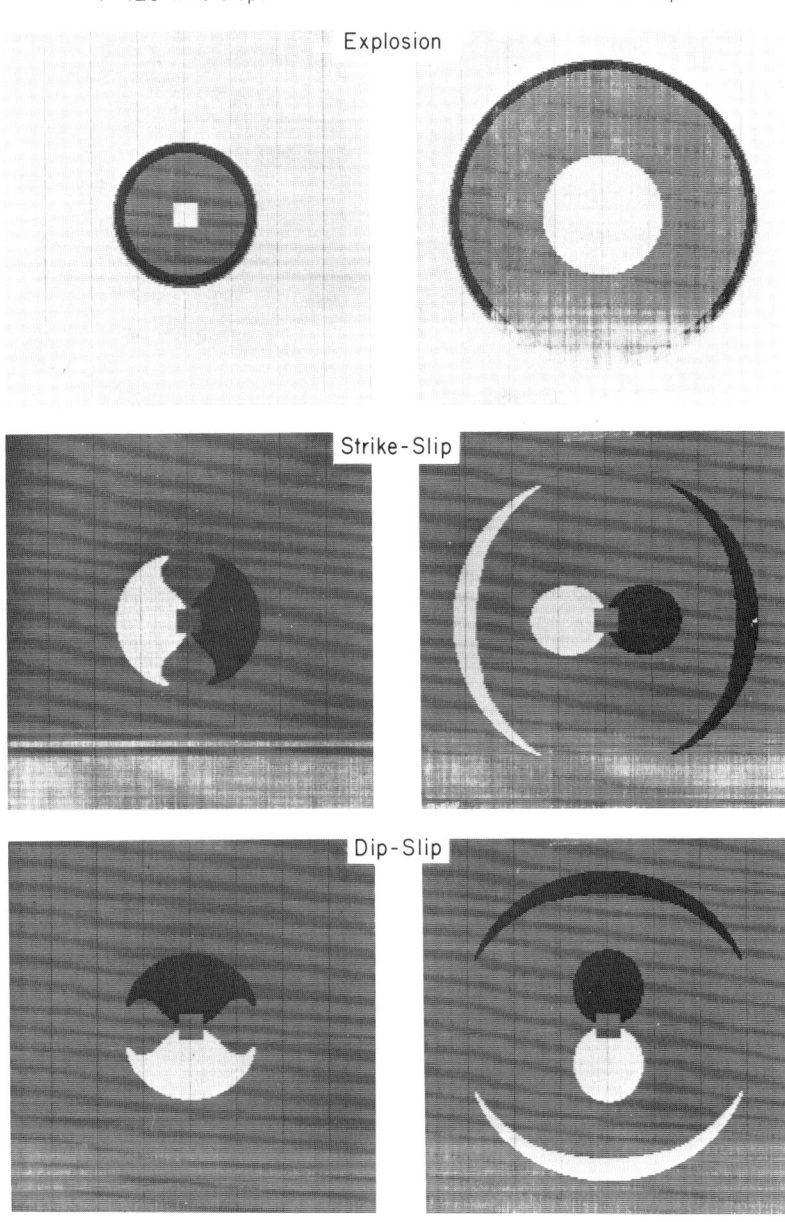

(a)    (b)

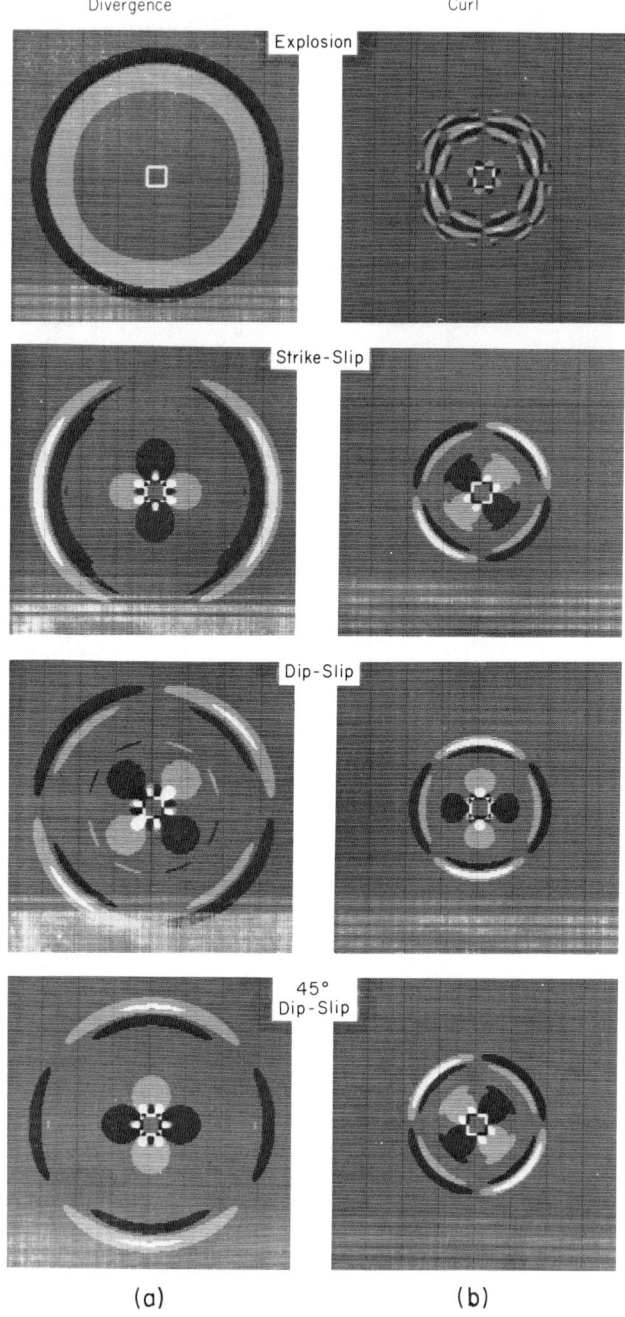

**FIG. 24.** P–SV vertical radiation patterns. The divergence and curl fields due to explosive, strike-slip, dip-slip, and 45° dip-slip sources are shown after 150 time steps. The plots have a grey background, where positive is shown in black and negative in white.

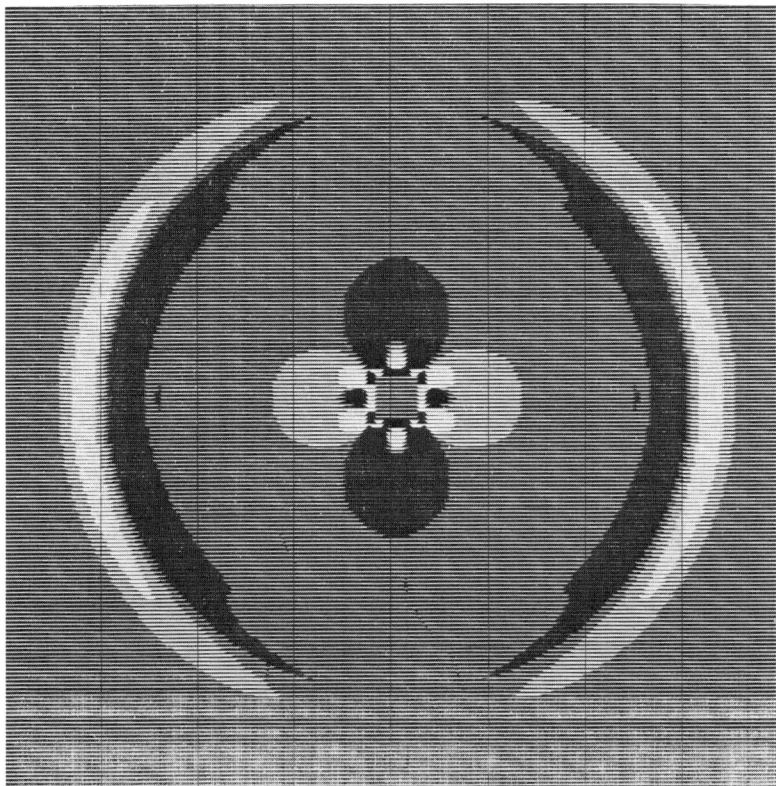

FIG. 25. Strike-slip divergence (P-wave) vertical radiation pattern. The far-field and two "pseudo-near-field" patterns may be seen. The far field has the familiar 2 lobes, the outer near field has 4 lobes and the innermost near field has 12 lobes. The near-field terms are necessary to maintain the correct far-field radiation pattern.

region. This is region (4), the interior of the source, which never sees the direct waves.

The whole-space source functions for the P–SV case are shown in time slices in Fig. 24. The divergence and curl of the displacements that are calculated in the FD grid are shown for the explosive, strike-slip, dip-slip, and 45° dip-slip cases. The divergence is nonzero where there is compressional wave energy and the curl is nonzero for shear wave energy. The divergence and curl are first spatial derivatives and so raise the frequency content slightly over that of the displacements themselves. The elastic explosion has an isotropic, compressional radiation pattern. The time slice of the curl for the explosion shows only scaled-up noise, which is actually much smaller than the energy in the divergence time slice. It is similar to the SH earthquake source in that it has a permanently displaced "pseudo-near-field."

The P–SV far-field radiation patterns are the usual ones for the explosive, strike-slip, dip-slip, and 45° dip-slip cases. For example, in the strike-slip case, the P waves show a 2-lobed pattern in the divergence, and the S waves a 4-lobed pattern in the curl of the displacements. Figure 25 shows the divergence of the strike-slip case on a larger scale. The patterns of energy distribution needed to maintain the correct far-field radiation are complex. The "pseudo-near-field" energy may be seen as the inner cloverleaf patterns. The clover-leaf patterns are followed by more near-field energy in 12-lobed patterns. Again, the inner region of the source area is visible by its lack of disturbance, and the edge of the inner region is visible from the divergences and curls that arise from the truncation of the displacement around the inner region. These "pseudo-near-field" terms grow with time and cause the synthetic seismograms to become inaccurate after some time. These terms have more long-period energy than the earlier energy. For these reasons, the P–SV earthquake sources are most accurate at both earlier time and shorter period, up to the short period limit of the FD grid. The explosive P–SV source and the SH earthquake sources have "pseudo-near-fields," which result in static displacements, that is, the near fields do not grow with time, so they do not become less accurate at longer times. The limitations arise from the asymptotic approximation shown in Eq. (24).

The optimal choice of source box size depends on whether the source is explosive or double couple and whether the calculation is P–SV or SH. The source box introduces energy into the FD grid far enough from the source to ensure that the displacements are sufficiently smooth. Smaller source boxes, however, are better in that they allow structure nearer the source and they require less computation. Elastic earthquake sources are more demanding in the sense that there is more amplitude variation along the source rings than elastic explosive sources, which are in turn about as demanding as SH earthquake sources, which are more demanding than acoustic explosive sources. Elastic earthquakes therefore require the largest source boxes and acoustic explosions can be inserted with the smallest source boxes.

## IV. Earthquake Applications

In this section we will discuss a few applications of these approximate solutions to strong motions produced from earthquakes. First, the effect of upgoing, line source energy incident on a basin is investigated. In this case the FD calculations can be used to check the accuracy of the GRT results. Next, the double-couple source excitation of Love waves in the presence of sloping structure and attenuation is considered. Lastly, we review the strong motions produced by the 1968 Borrego Mountain earthquake as recorded in the Imperial Valley.

A. STRONG MOTIONS IN A BASIN ENVIRONMENT

One of the puzzling features in strong motion seismograms is the noticeably long duration of high-frequency P waves in sedimentary basins, as is seen in Fig. 2. These P waves are seen to rotate onto the vertical component in the strong velocity gradient near the surface (Liu and Helmberger, 1985). The latter portion of these records are generally depleted in low frequency. Thus, one might conclude that there are propagational waveguides that preferentially prolong high-frequency motions. Nonplanar surface layering displays this property; see Fig. 26 where a low-velocity surface layer grows with distance from the source to the receiver. The geometric ray paths are displayed in the upper panel. The broadband results are produced by GRT for a line source. The square-root singularity in Eq. (12) is apparent for the direct arrival. Note that after one bounce the reflection coefficient from the lower interface becomes complex and a head wave and post-critical-angle reflection develops, as described in Hong and Helmberger (1977). The more times a ray bounces, the steeper the angle of incidence when the ray hits the boundary. Once the rays bounce with more than the critical angle of incidence, they lose a significant amount of energy with each bounce and have little amplitude when they reach the receiver. This allows an accurate response to be computed using only a small number of rays from the infinite sum of possible rays that constitute the exact solution.

The middle trace of Fig. 26 displays the broadband results after applying a filter to remove the high frequencies. The bottom trace displays the FD results, and the agreement is excellent.

A comparison of the two methods for the same dipping structure assuming a point source is displayed in Fig. 27. Again the codes yield similar waveforms. The upward drift in the GRT synthetics is caused by including $\sqrt{p(t)}$ in the calculations whereas in FD method only $\sqrt{p_0}$ is used. These plots show the asymptotic step responses, as given in Eq. (28), and performing the derivative in Eq. (25) tends to eliminate this drift problem and enhance the later arrivals.

The S waves that dominate the horizontal motions generally do not show the strong ringing nature associated with the P waves, for example, see Fig. 2. Also, the higher frequency content of the P waves relative to the S waves

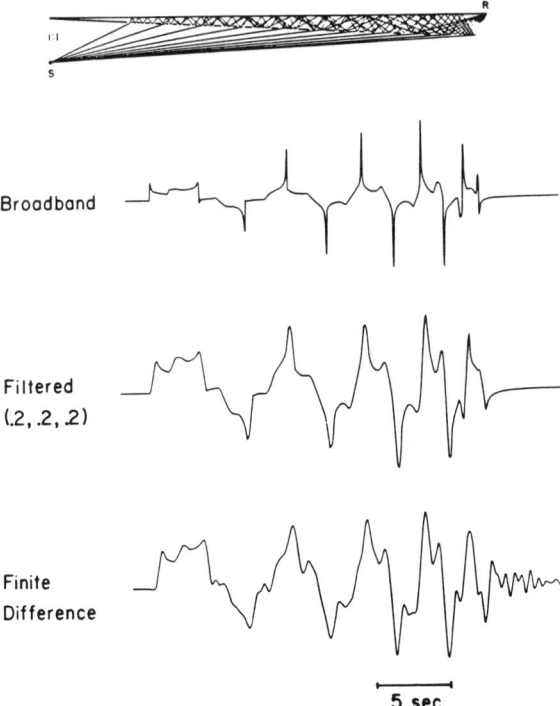

FIG. 26. Comparison of GRT results with finite-difference calculations for a dipping layer over a half-space. (After Vidale et al., 1985.)

is quite apparent. This feature seems to be a property of many basins. Explaining the difference in frequency content is difficult. It may be that this difference is caused by different source histories for P and S waves (Hanks and McGuire, 1981).

Another possible mechanism is a different amount of attenuation for P and S waves. The attenuation of body waves can be parameterized by the quantity $t^*$ as

$$t^* = \int_{\text{raypath}} \frac{dt}{Q} = \frac{T}{Q_{\text{av}}},$$

where $T$ is the total travel time of the ray, and $Q$ is the quality factor; $t_\alpha^*$ refers to compressional wave attenuation, and $t_\beta^*$ refers to shear wave attenuation. Thus, a body wave pulse generated synthetically with the

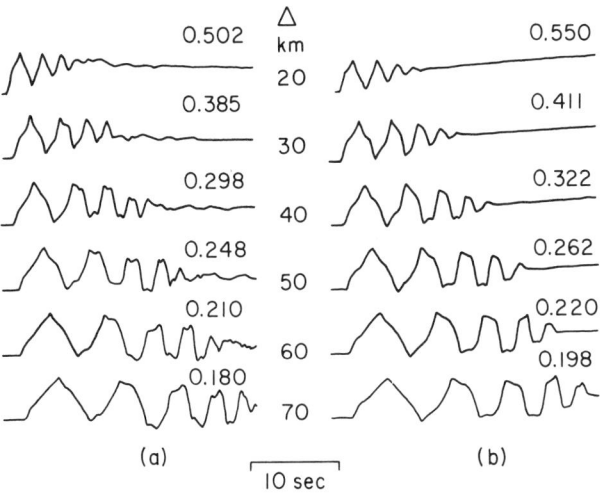

FIG. 27. Profiles for a single layer that dips down 2.8° from the horizontal away from the source. The layer is 0.3-km thick directly above the source. The source is 6 km below the surface. A trapezoidal time function of 0.3, 0.3, 0.3 sec has been convolved into both suites of seismograms. The seismograms on the left (a) are generated by the finite-difference method and those on the right (b) are from the generalized ray method.

assumption of elasticity can be corrected for attenuation by a convolution with the Futterman operator, as described in Carpenter (1967). Estimates of $Q_\alpha$ greater than 250 are obtained from Helmberger *et al.* (1979) who studied the behavior of multibounce P waves trapped between the surface and the bottom of the sedimentary column. $Q_\beta$ can be obtained by supposing that

$$Q_\beta = \tfrac{4}{3}(\beta^2/\alpha^2)Q_\alpha,$$

and since $(\beta/\alpha)$ becomes small at the surface we might expect $t_\beta^*$ to be considerably larger than $t_\alpha^*$. Estimates of $Q$'s for the Imperial Valley based on mapping P-wave shapes into S-wave shapes have been made by Liu and Helmberger (1985). They obtain $Q_\beta$ values of 25 for the uppermost layers. Such $Q_\beta$'s have profound effects of multiple bounces in dipping structures as shown in Fig. 28.

The effect of dipping layers and the effect of different $Q$ factors for P and S waves can combine to qualitatively explain the records shown in Fig. 2. A more definitive explanation of records of the type displayed in Fig. 2 requires detailed knowledge of the crustal structure. Such detailed models are rarely available. Thus, most progress in modeling path effects starts at the longer periods, which we consider next.

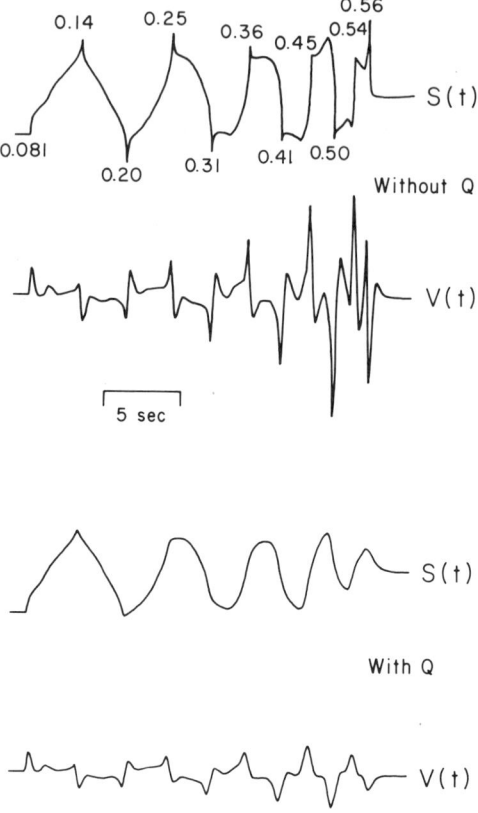

FIG. 28. Effect of absorption for step response and velocity synthetic seismograms for a strike-slip event located within the same geometry described in Figs. 26 and 27 when the range is 70 km. The numbers in the top panel indicate the $t_\beta^*$ encountered by the various multiples when $Q$ in the basin is 70, but $Q$ is only convolved with the traces on the bottom. Both step response synthetics and both velocity synthetics are plotted on the same scale.

## B. Modeling Strong Motions of the Borrego Mountain Earthquake

The Borrego Mountain earthquake ($M_L = 6.4$) occurred on April 9, 1968, just outside the northwestern end of the Imperial Valley, as shown in Fig. 29. Numerous studies of this event have been conducted both with the near-field data (Heaton and Helmberger, 1977; Swanger and Boore, 1978) and far-field data (Burdick and Mellman, 1976; Ebel and Helmberger, 1982) using flat-layered structure. It proves interesting to reinvestigate the

### 6. PATH EFFECTS IN STRONG MOTION SEISMOLOGY

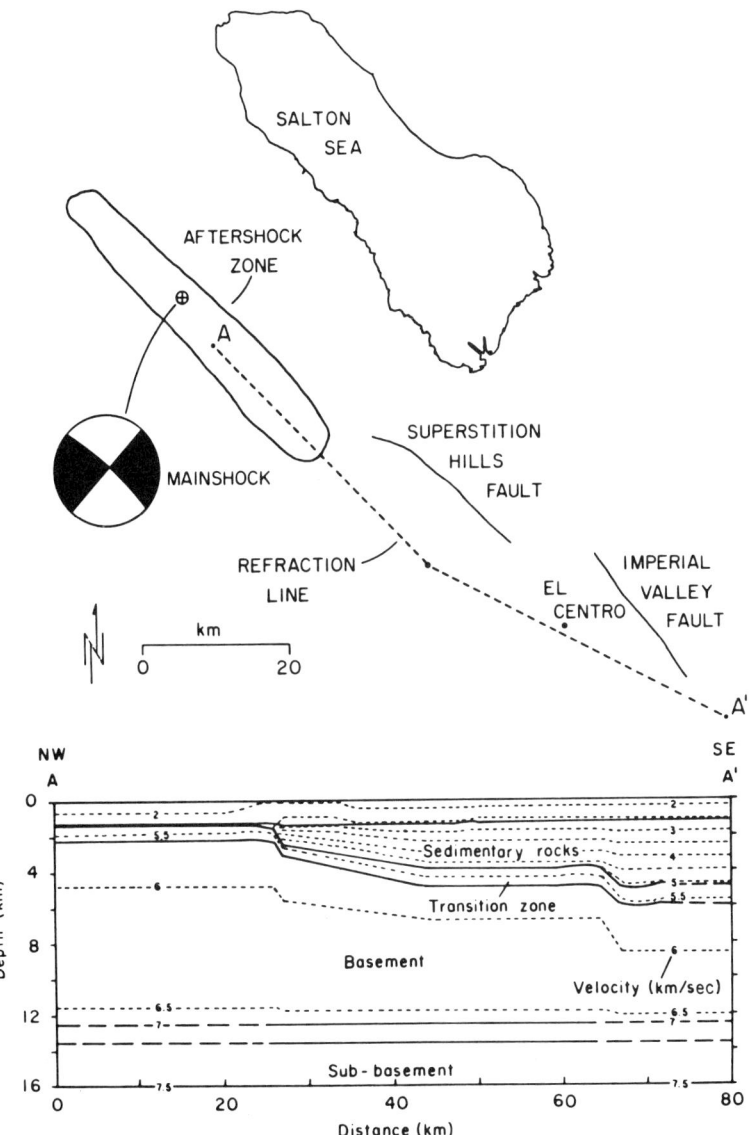

**FIG. 29.** Map indicating the location of the 1968 Borrego Mountain earthquake relative to the Imperial Valley El Centro station and refraction survey conducted by Fuis *et al.* (1983).

motions observed at El Centro in light of the improved knowledge of structure, which is shown in the lower panel of Fig. 29.

El Centro was the nearest site where the event was recorded by Carder displacement meters and a standard accelerograph. A comparison between these two responses that shows the processed records in the displacement domain is displayed in Fig. 30. From the teleseismic long-period results, the fault plane has an orientation of N 45° W, and El Centro is probably less than 5° from the SH maximum and contains predominantly SH energy. Given the geometry, we expect about equal motions on the NS and EW instruments. This is the case for the longer periods. However, note the strong initial pulse on the NS component that is not on the EW component. This feature can be explained by an asperity model with a small portion of the fault striking N 60° W occurring near the start of the rupture process (Ebel and Helmberger, 1982). This type of complex rupturing appears to be compatible with the teleseismic short-period motions. We will be primarily concerned with the longer period, that is, the 1 to 20 sec period motions.

An early attempt at modeling the tangential motion at El Centro was done by Heaton and Helmberger (1977). The strategy followed in that paper was to find a simple layer over a half-space model in combination with a faulting slip history $D(t)$ compatible with this record. Figure 31 illustrates the GRT step function $v_1(t)$ from Eq. (28), the response of a

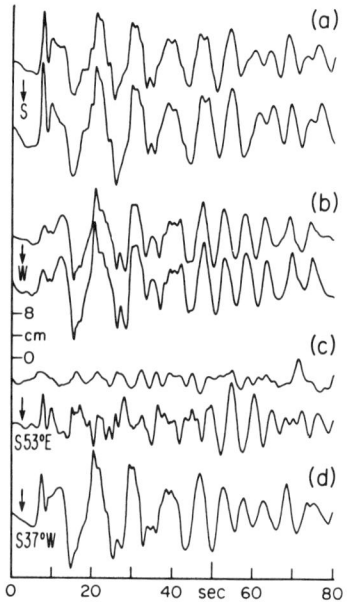

FIG. 30. Summary of observed ground motion at El Centro for the 1968 Borrego Mountain earthquake. (a) Comparison of deconvolved Carder displacement meter record (top) and integrated accelerogram (bottom) for N–S component. (b) Comparison of deconvolved Carder displacement meter record (top) and integrated accelerogram (bottom) for E–W component. (c) Ground motion rotated into vertical (top), radial (middle), and tangential (bottom) components. (From Heaton and Helmberger, 1977.)

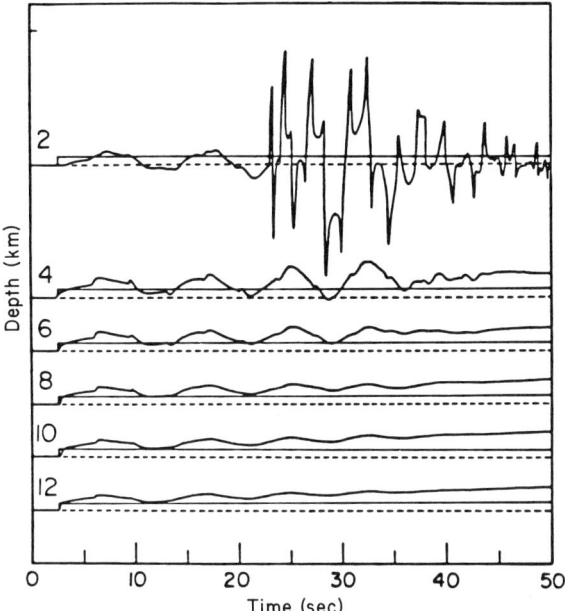

FIG. 31. SH step function response at the surface assuming a point strike-slip dislocation buried at various depths. The top trace is noticeably different because it is within the top layer. The flat-layered geometry is meant to simulate the El Centro recording of the 1968 Borrego Mountain earthquake. The amplitudes are scaled in relation to the top trace. (After Heaton and Helmberger, 1977.)

point source located at various depths. The layer's velocity and thickness were adjusted until synthetics yielded periods similar to those occurring in the observations. For shallower source depths, the direct wave is diffracted and becomes less sharp while the Love waves become strongly developed. When the source is placed in the layer we see the classical head wave development with later arriving high-frequency multireflected arrivals. Since these high-frequency spikes are not apparent in the observation, it was concluded that the source must be below the layer.

It proved difficult to obtain a sharp beginning and simultaneously maintain the strength of the Love waves with a single point source and, thus, a two source model was allowed with the results given in Fig. 32. The lower source provides the sharp beginning while the shallower source excites the Love waves. The moment required for this fit is less than that found in the teleseismic study by about 30%. Several other finite source models, including propagating ruptures, were attempted by Heaton and Helmberger (1977) and found satisfactory. Directivity, the term usually

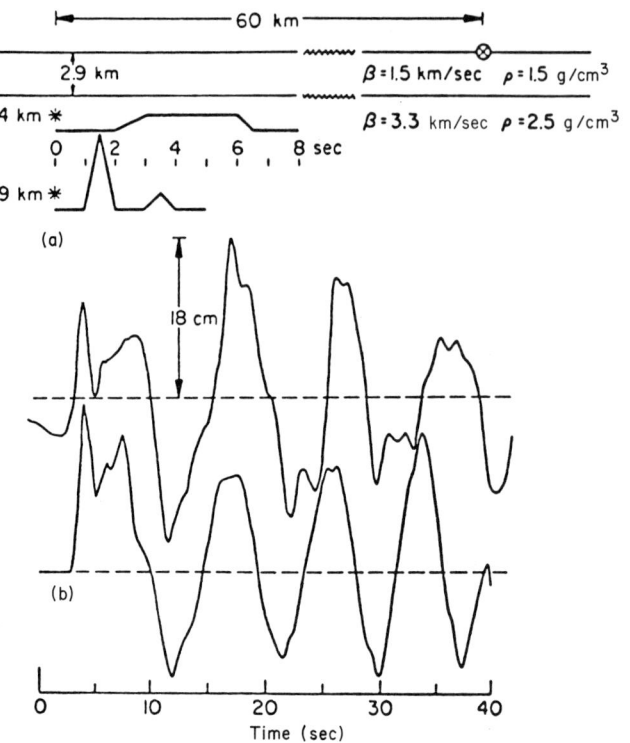

FIG. 32. Comparison of the observed (a) strong ground motion (transverse component) of the 1968 Borrego Mountain earthquake at El Centro, California, with ground motion computed on the basis of the source model (b) schematically illustrated and the flat-layered crustal structure parameterized in the upper right.

used to describe rupture effects, has stronger effects on the body waves than the longer period Love waves. A study by Swanger and Boore (1978) using the normal mode approach found similar results when crustal models more appropriate to that expected for El Centro were assumed. Apparently, the trade-off between the source model and Green's functions computed from the various conceivable crustal models leads to serious ambiguity between source and path effects.

Fortunately, a P-wave velocity cross section along the seismic path is now available (see Fig. 29). The shear velocities are still somewhat uncertain although they are well established near El Centro. The model parameters are given in Table IV with idealized cross sections displayed in Fig. 33. This complicated structure can be easily used in the FD method outlined above to generate synthetic seismograms for El Centro. The FD synthetic seismo-

## 6. PATH EFFECTS IN STRONG MOTION SEISMOLOGY

TABLE IV. VELOCITIES IN THE BORREGO MODEL

| S-wave velocity (km/sec) | Density (g/cm$^3$) | Depth to top of layer (km) | |
|---|---|---|---|
| | | Left side | Right side |
| 1.0 | 1.4 | 0. | 0. |
| 1.55 | 1.9 | 1.6 | 1.6 |
| 1.8 | 2.0 | — | 3.8 |
| 3.0 | 2.3 | 1.8 | 5.0 |
| 3.75 | 2.7 | 2.6 | 6.2 |
| 4.0 | 2.8 | 12.2 | 12.2 |
| 4.125 | 2.8 | 12.6 | 12.6 |
| 4.25 | 2.9 | 13.2 | 13.2 |
| 4.375 | 2.95 | 13.8 | 13.8 |
| 4.5 | 3.0 | 14.2 | 14.2 |
| 4.625 | 3.1 | 14.6 | 14.6 |
| 4.75 | 3.2 | 15.2 | 15.2 |

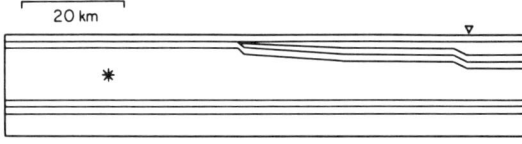

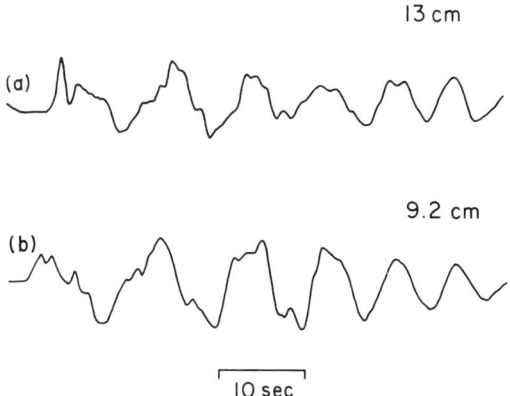

**FIG. 33.** Comparison of observed (a) SH displacement from the Borrego Mountain earthquake and synthetic (b) displacement record from the laterally varying velocity model displayed and the source given by Burdick and Mellman (1976). The velocity model is summarized in Table IV.

gram is also shown in Fig. 33. The source parameters were taken from the teleseismic results and, thus, the FD synthetic seismogram has no free parameters and becomes mainly a prediction.

Comparing the observed and synthetic traces we see that the duration is about right and the overall amplitude slightly small. Thus, to fit the amplitude better would require somewhat more moment or perhaps a slightly shallower source depth. On the other hand, the flat-layered model predicted too large an amplitude. It appears that the main contribution of the more realistic structure was to lengthen the wave train.

## V. Summary

Until recently, most seismic modeling has been done using flat-layered structures. In this work, methods to compute synthetic seismograms for more realistic complex structures are emphasized. In addition, the usefulness of these methods is demonstrated with several applications.

The FD method is the most novel computational scheme developed in this chapter. A whole space, line-source, first-term-asymptotic GRT solution is used as a source in the FD grid. The wave field is allowed to pass through arbitrarily complex structures in the FD grid. The wave field is sampled at receivers to make line source seismograms. These are converted to point source seismograms by a convolution operation.

The results of using flat-layered structures to model nuclear explosion seismograms is reviewed for the case of the explosion MILROW. The relative contribution to the waveform at various ranges of energy initially radiating upward and downward from the source is discussed. This analysis predicts that outside a range of 7 km the waveforms are not affected by spall and suggests that the synthetic seismograms from the flat-layered structure model the data fairly well. This good fit is not surprising as the layered model was constructed to fit the data.

Using the FD method, synthetic seismograms are calculated using more realistic structures based on a geological cross section. The nonplanar structures cause station-to-station variations in peak velocity and apparent incidence angle similar in magnitude to that observed for the explosion MILROW. In addition, the layers that dip down away from the source are seen to convert more direct waves to surface waves than the flat-layered models. The more realistic structure models the observations almost as well as the flat-layered model, which was determined by fitting the observations.

A high-frequency resonance of P waves but not S waves is sometimes observed in basins. This effect is investigated using a dipping layer GRT method. The combination of a gently dipping layer that traps the high-frequency energy and a very low $t_\beta^*$ that attenuates the S waves can reproduce this effect.

Lastly, the FD method is used to predict the long-period SH strong motions observed at El Centro for the 1968 Borrego Mt. earthquake. When the source parameters are taken from the teleseismic studies and the velocity structure is taken from a refraction profile, the resulting displacement record is similar in amplitude and waveform to the El Centro record.

As more accurate geological structures are determined, we will be able to model strong motions better using the techniques discussed in this chapter.

### ACKNOWLEDGMENTS

This work was partly supported by Air Force-Cambridge grant F19628-83-K-0010. J. E. V. was supported by an NSF fellowship. Robert W. Clayton and Arthur Frankel aided in the development of the finite-difference program. We thank Heidi Houston, Gladys Engen, Tom Heaton, and Don L. Anderson for their critical reviews of this manuscript. California Institute of Technology contribution number 4283.

### REFERENCES

Alford, R. M., K. R. Kelley, and D. M. Boore (1974). Accuracy of finite-difference modeling of the acoustic wave equation. *Geophysics* **39**, 834–842.

Alterman, Z., and F. C. Karal (1968). Propagation of elastic waves is layered media by finite difference methods. *Bull. Seismol. Soc. Am.* **58**, 367–398.

Apsel, R. J. (1979). Dynamic Green's functions for layered media and applications to boundary value problems. Ph.D. Thesis, Univ. of California, San Diego.

Apsel, R. J., and J. E. Luco (1983). On the Green's functions for a layered halfspace, Part II. *Bull. Seismol. Soc. Am.* **73**, 931–952.

Bard, P., and M. Bouchon (1985). The two-dimensional resonance of sediment-filled valleys. *Bull. Seismol. Soc. Am.* **75**, 519–541.

Bard, P., and B. E. Tucker (1985). Underground and ridge site effects: A comparison of observation and theory. *Bull. Seismol. Soc. Am.* **75**, 905–922.

Burdick, L. J. (1983). Simultaneous modeling of body waves and surface waves in near field records of nuclear explosions. Woodward-Clyde Consultants Rep. WCCP-R-83-02.

Burdick, L. J., and G. R. Mellman (1976). Inversion of the body waves from the Borrego Mountain earthquake to the source mechanism. *Bull. Seismol. Soc. Am.* **66**, 1485–1499.

Burridge, R., E. R. Lapwood, and L. Knopoff (1964). First motions from seismic sources near a free surface. *Bull. Seismol. Soc. Am.* **54**, 1889–1913.

Carpenter, E. W. (1967). Teleseismic signals calculated for underground, underwater, and atomospheric explosions. *Geophysics* **32**, 17–32.

Cerveny, V., M. M. Popov, and I. Psencik (1982). Computation of seismic wave fields in inhomogeneous media, Gaussian beam approach. *Geophys. J.* **70**, 109–128.

Chapman, C. H. (1978). A new method for computing synthetic seismograms. *Geophys. J.* **54**, 481–518.

Clayton, R. W., and B. Engquist (1980). Absorbing boundary conditions for wave equation migration. *Geophysics* **45**, 895–904.

Cohn, S. N., T. Hong, and D. V. Helmberger (1982). The Oroville earthquakes: a study of source characteristics and site effects. *J. Geophys. Res.* **87**, 4585–4594.

de Hoop, A. T. (1960). A modification of Cagnaird's method for solving seismic pulse problems. *Appl. Sci. Res., Sect. B* **8**, 349–356.

Ebel, J. E., and D. V. Helmberger (1982). P-wave complexity and fault asperities: the Borrego Mountain, California, earthquake of 1968. *Bull. Seismol. Soc. Am.* **72**, 413–437.

Engdahl, E. R. (1972). Seismic effects of the MILROW and CANNIKIN nuclear explosions. *Bull. Seismol. Soc. Am.* **62**, 1411–1423.

Fuis, G. S., W. D. Mooney, J. H. Healy, G. A. McMechan, and W. J. Latter (1983). A seismic refraction survey of the Imperial Valley region, California. *J. Geophys. Res.* **89**, 1165–1189.

Gilbert, F., and L. Knopoff (1961). The directivity problem for a buried line source. *Geophysics* **26**, 626–634.

Hanks, T. C., and R. K. McGuire (1981). The character of high-frequency strong ground motion. *Bull. Seismol. Soc. Am.* **72**, 2071–2096.

Harkrider, D. G. (1976). Potentials and displacements for two theoretical sources. *Geophys. J.* **47**, 97–133.

Heaton, T. H., and D. V. Helmberger (1977). A study of the strong ground motion of the Borrego Mountain, California, earthquake. *Bull. Seismol. Soc. Am.* **67**, 315–330.

Helmberger, D. V. (1983). Theory and application of synthetic seismograms. In "Earthquakes: Observation, Theory and Interpretation" (H. Kanamori and E. Boschi, eds.), Proc. Int. Sch. Phys. "Enrico Fermi," Course LXXXV, pp. 174–221. North-Holland Publ., Amsterdam.

Helmberger, D. V., and D. M. Hadley (1981). Seismic source functions and attenuation from local and teleseismic observations of the NTS events JORUM and HANDLEY. *Bull. Seismol. Soc. Am.* **71**, 51–67.

Helmberger, D. V., and D. G. Harkrider (1978). Modeling earthquakes with generalized ray theory. In "Modern Problems in Elastic Wave Propagation" (J. Miklovitch and J. D. Achenbach, eds.), 499–518. Wiley, New York.

Helmberger, D. V., G. Engen, and P. Scott (1979). A note on velocity, density, and attenuation models for marine sediments determined for multibounce phases. *J. Geophys. Res.* **84**, 667–671.

Helmberger, D. V., G. R. Engen, and S. P. Grand (1985). Notes on wave propagation in laterally varying structure. *J. Geophys.* **58**, 82–91.

Hong, T. L., and D. V. Helmberger (1977). Generalized ray theory for dipping structure. *Bull. Seismol. Soc. Am.* **67**, 995–1008.

Hong, T. L., and D. V. Helmberger (1978). Glorified optics and wave propagation in nonplanar structure. *Bull. Seismol. Soc. Am.* **68**, 1313–1330.

Langston, C. A., and D. V. Helmberger (1975). A procedure for modeling shallow dislocation sources. *Geophys. J. R. Astron. Soc.* **42**, 117–130.

Lui, H. L. (1984). Interpretation of near-source strong ground motion and implications. Ph.D. Thesis, California Inst. Technol., Pasadena.

Lui, H. L., and D. V. Helmberger (1985). The 23:19 aftershock of the 15 October 1979 Imperial Valley earthquake: more evidence for an asperity. *Bull. Seismol. Soc. Am.* **75**, 689–708.

Madariaga, R., and P. Papadimitriou (1985). Gaussian beam modeling of upper mantle phases. *Ann. Geophys.* **3**, 799–812.

Orphal, D. L., C. T. Spiker, L. R. West, and M. D. Wronski (1970). Analysis of seismic data. Milrow Event, Environ. Res. Corp. Rep. NVO-1163-209.

Regan, J., and P. Glover (1985). Modeling surface wave propagation in laterally heterogeneous media. *Ann. DARPA/AFGL Seismic Res. Symp., 7th, USAFA, Colorado Springs, Colo.*, 17–27.

Rodean, H. C. (1971). "Nuclear Explosion Seismology." USAEC, Oak Ridge, Tennessee.

Strick, E. (1959). Propagation of elastic wave motion from an impulsive source along a fluid–solid interface, parts II and III. *Philos. Trans. R. Soc. London, Ser. A* **251**, 465–523.

Swanger, H. J., and D. M. Boore (1978). Simulation of strong-motion displacements using surface-wave modal superposition. *Bull. Seismol. Soc. Am.* **68**, 907–922.

Vidale, J. E., D. V. Helmberger, and R. W. Clayton (1985). Finite-difference seismograms for SH waves. *Bull. Seismol. Soc. Am.* **75**, 1765–1782.

# Index

## A

Acceleration, 41, 48
  amplitude spectra, 1, 65
  components, 30
  digital, 60
  particle, 42
  peak, 24, 183, 186, 197
  radial, 26, 185
  vertical component, 184
Accelerograms, *see* Seismograms
Accuracy, 216
Adaptive integration, 251
Aftershocks, 57, 84, 87, 92, 100, 107, 125, 148, 174, 200, 217, 268
Aki's models, 1, 57, 91, 109, 225, 240, 275
Amchitka nuclear tests, 270, 272, 285
Amchitka structure, 275
Amplification, 179
Amplitude
  decay, 302
  functions, 28
  spectral, 1, 65, 155
Anelastic media, 163, 169, 269
Angular dependence, 11
Antiplane shear crack, 93, 108, 138
  reflections, 31
  slip, 18
  transverse emission, 32
Arbitrary velocity variation with depth, 219
Array
  analysis, 55
  correlation, 63
  differential, 62
  Imperial Valley, 61, 261
  optimum design, 77
  radio telescope, 60
  SMART, 1, 60, 61
  steered, 60
  strong motion, 60
Artificial boundaries, 113

Asperities, 57, 60, 109, 121, 122, 147, 170, 187, 197
Asymptotic
  approximation, 12, 20
  expansion, 294
Attenuation, 2, 179, 215, 279, 308,
  $Q$, 163, 179, 308, 310
Azimuthal variation, 86

## B

Barriers, 57, 109, 128
Baseline vector, 78
Basins, 307, 316
Beamforming, 64
Beamsteering, 66
Bessel transform, 220, 293
Bilateral rupture, 183, 186
BORN structure, 188
BORREGO structure, 315
BORY structure, 190
BRAW structure, 193
Breakout, 146
  phases, 59
Brittle
  fracture, 2, 5, 17
  zone, 141
Brune's theory, 1
Buried point source, 45, 166

## C

Cagniard–de Hoop method, 15, 92, 192, 237, 276, 292, 299
Cannikin, 269, 284
Canonical problems, 12
Capon's method, 67
Caustic surface, 22, 26, 28, 45
Chi-squared distribution, 70, 76

Circular
  coda waves, 60, 190
  crack, 108, 115, 245
  rupture, 48
  slip zone, 254
Coherency, 62, 71, 72, 75, 80, 81, 82
  statistical aspects, 75
Cohesive
  force, 121
  traction, 14
Complex
  ray, 34
  rupture, 312
  structures, 253, 296
Computational effort, 211, 216, 221, 226, 233, 235, 239, 293
Computer, 170, 187
  CRAY-MPX, 188
  IBM 370/168, 187
  time, 170, 187
  VAX11/780, 259
Corner frequency, 1, 155, 216, 274
Correlation
  array, 63
  normalized, 72
Correlator, 63, 77
Courant stability criterion, 259
Covariance matrix, 64
CPU time, 260
Crack
  edge, 9
  growth, 5
  length, 120
  semi-infinite, 34
  shear, 93, 108, 119, 142
  stress-corrosion, 121
  tip, 83, 156
Crack models
  mode II, 2, 3, 48
  mode III, 2, 3, 48
CRAY-MPX, 188
Critical
  intensity factor, 117
  points, 229
  reflection, 4, 27
Cross-correlation method, 62
Cross covariance, 71
Cross-spectral
  estimates, 69
  matrix, 66, 76, 80
Cubic spline, 235

## D

Decibels, 56
Delta function, 10, 247
Delta matrices, 156
Depth dependence, 42, 141
Detection problem, 64
Diffraction, 12, 13, 22, 253, 257
Digital accelerographs, 154
Directivity function, 229, 313
Discrete wave number, 92, 95, 206, 226
Dislocation models, 57, 93, 94, 207, 245
Dispersion, 158, 159, 164, 165
  Rayleigh-wave curve, 171, 173
Displacement
  potential, 275
  stress vector, 220
  time functions, 133
  waveforms, 275
Doppler shift, 57, 62
Double-couple point sources, 91, 237
Ductile zone, 142
Duration, 156, 194, 316
DWFE method (discrete wave number/finite-element method), 95, 100, 227, 257, 258, 260
Dynamic shear crack problems, *see* Crack
Dynamic source models, 208

## E

Earthquake
  1906 San Francisco, 57
  1940 El Centro, 217
  1962 Mexico City, 156
  1966 Parkfield, California, 209
  1968 Borrego Mountain, California, 188, 307, 310, 312, 317
  1969 Central Gifu, 105
  1971 San Fernando, 62
  1975 Oroville, California, 268
  1976 Brawley, California, 192
  1976 Friuli, 171, 200
  1979 Coyote Lake, 253
  1979 Imperial Valley, 61, 87, 214, 215, 233, 257, 268
  1980 Off-Izu Peninsula, 100, 102
  1980 Irpinia, Italy, 196
  1981 January 29, Taiwan, 61, 76
  1984 Western Nagano, 98
  1985 Mexico City, 156

Earthquake sources, *see* Source
Eigenvalues
  accuracy, 161
  automatic computation, 157
Elastic
  rebound theory, 57
  tensor, 208
El Centro, 217, 311, 314, 317
Ellipticity, particle motion, 177, 179
Emission
  coefficient, 18
  spectra, 20
Energy, 55, 83, 275, 300
  flux, 5
  integral, 175, 178
  pulse, 298
  sink, 7
Engineering, 83, 201, 205, 217
  earthquake, 1, 153
  structure, 88
Equations of motion, 111, 243
Explosions, 207, 267, 269, 303, 316
Extended sources, 183, 221

## F

Far-field ground motions, 33, 231, 254
Faster procedures, 211, 293
Fault, 5
  finite, 205, 247
  geometry, 122
  homogeneous, 118
  inhomogeneous (heterogeneous), 123, 128, 199
  patches, 128
  rupture length, 87
  rupture process, 91, 108, 111, 127, 129, 134
  slip, 2, 141, 155, 244
  stochastic model, 104, 147
  surface integral, 212, 216
  vertical, 137
  vertical strike-slip, 127, 134
Fault length factor, 169
FE modeling, *see* Finite-element
Fictitious displacements, 134
Filon's method, 226, 245
Finite-difference (FD), 4, 112, 217, 269, 273, 308
  grid, 300
  modeling, 285, 302
  source insertion, 300

Finite-element (FE), 4, 92, 95, 96, 110, 206, 218, 259, 269, 273
Finite fault calculations, 205, 247
Finite-length sources, 169, 313
Finiteness factor, 169, 183, 263
Fisher's distribution, 71, 76
Fitting problem, 64
Flash point, 52
Focusing, 3
Fourier
  amplitude spectra, 62, 80
  mapping, 79
  transform, 20, 79, 96, 166, 181, 212, 228
Fracture
  criteria, 115, 116
  initiation, 8
  mechanics, 2, 5
  strength, 120
Fraunhofer approximation, 231
Frequency
  domain, 39, 170
  spectrum, 3, 65, 81
Frequency-dependent reflection coefficient, 33
Frequency–wave-number
  analysis, 65
  conventional (CV), 67
  corner, 155, 274
  high resolution (HR), 67
FREUL 7A model, 172, 174, 201
Friction, 10
  laws, 121
  sliding, 117
Frictional
  sliding stress, 120, 123, 142
  strength, 118
  stress, 112, 117
Futterman operator, 309

## G

Geometric ray paths, *see* Ray
GEOSCOPE, 154
Glorified optics, 279
Gouge, 122, 123
Gaussian
  beams, 228, 273
  distribution, 132
  quadrature, 235
  taper, 79
Green's
  displacement, 49

Green's (*Continued*)
  empirical function, 108, 147, 169
  explicit interpolation, 233, 235, 247
  far-field, 252
  function, 4, 35, 58, 92, 94–96, 104, 144, 155, 206, 209, 216, 227, 232
  implicit extrapolation, 238, 247
  reciprocity, 210
  tensor, 38, 47, 49
Griffith's fracture criterion, 108, 116
Ground
  acceleration, 24
  displacements, 144, 240
  peak acceleration, 183
Group velocities, 158, 166, 175
GRT formalism, 278, 294, 307, 312

## H

Haskell's model, 57, 91, 214, 215, 247, 257
  representation, 168, 289
Headwave, 17, 29
Heterogeneous
  faults, 123, 128, 199
  media, 134, 141, 262
  structure, 277
Higher modes, 159
High frequency
  approximations, 20, 224
  behavior, 213
  features, 104
  pulses, 59
  radiation, 201
  seismic waves, 111, 144, 153
  solution, 279
High-resolution (Capon's) method, 67
High-stress concentrations, 125
Hilbert transform, 23
Homogeneous source region, 301
Huyghen's principle, 49

## I

Incoherent rupture propagation, 108
In-plane
  reflections, 24
  shear cracks, 93, 119
  slip, 19
  transverse emission, 27

Instruments, 2
  digital, 60
  response, 96
Integral representation, 34
  Love waves, 35
  Rayleigh waves, 38
Intensity, 78, 79, 88
  seismic, 55
  stress, 5, 6
  stress factor, 128
Interferometer, 77
Interpolation, 233, 234, 236
Inverse transformation, 247
Inversion techniques, 148, 215
IRPI structure, 198
Irwin's fracture criterion, 109, 116, 147
Isochrone, 3, 230, 252
  integration, 218, 256

## J

Japan, 98

## K

Kinematic dislocation
  models, 94, 208, 213, 240
  slip function, 233
  sources, 228
Kirchhoff–Helmholtz theory, 218
Knopoff's method, 156, 157
Kostrov's method, 2, 109, 115, 214
  function, 58, 255
Kummer's series, 38

## L

Lagrange multipliers, 64
Laplace transforms, 12, 15, 275
Large-amplitude S waves, 144
Laterally varying structure, 218, 272, 277
Layered structure, 134, 157, 188, 190, 224, 240
  dipping, 278, 307
Lg phases, 175
Likelihood function, 64
Locked mode approximation, 171
Longitudinal
  emission, 24

Longitudinal (*Continued*)
  particle velocity, 15
  waves, 59
Love wave, 4, 33, 59, 159, 307, 313
  acceleration, 48
  excitation, 47
  frequency domain calculation, 36
  Green's function, 35
  integral representation, 47
  mode, 32, 157, 166
  particle displacement, 36
  transient, 37, 38
Low-velocity
  channel, 162
  layer, 175, 180, 193
  surficial layer, 138, 146
  zone, 171

## M

Maximum likelihood, 64, 69
Microearthquakes, 217
Milrow, 269, 272, 281, 283, 316
  velocity records, 271
Modal summation, 156
Mode
  following, 159
  Love-wave, 4, 157, 166, 180
  Rayleigh-wave, 4, 157, 166, 179, 270
Mode II, 3
Mode III, 3
Mode follower, 160, 165, 175
Modeling
  finite-difference (FD), 217, 285
  finite-element (FE), 218
  forward, 205
  numerical, 91
  three-dimensional, 217, 219
  two-dimensional, 217, 219
  waveform, 62
Moment
  seismic, 1, 155, 168, 181
  tensors, 58, 94, 96
  tensor source, 221
Multilayers, 278, 294
Multimode summation, 159, 171

## N

Near-field
  ground motions, 4, 21, 142, 231
  theory, 134
Near-tip fields, 5
Nondispersive waves, 236
Nonlinear
  absorption, 269, 279
Normal mode method, 92, 270, 314
Nuclear explosion seismograms, 270
Numerical techniques, 273

## O

Optimum design, 77

## P

Particle
  acceleration, 42
  motion, 168, 177
  velocity, 11, 285
Parzen taper, 186
Patches, 128
Path effects, 267
Phase velocities, 157
Point source, 45, 91, 180
  summation, 185, 231, 248
Polarization, 62, 177
Potentials, 224, 243, 275
  P wave, 290
  S wave, 290
  SH wave, 290
Power, 56, 70, 76
pP phase, 281
Precision, 157
Prediction, 58, 91, 142
  ground motion, 4
  models, 257
  synthetic seismograms, *see* Synthetic
Propagator matrix, 220, 240, 241
Pseudo-near-field, 302
P-SV equations, 222
  elastic case, 298
  problem, 246

## Q

$Q$, 163, 179, 190, 308, 310
Quasi-osculations, 174

## R

Radial acceleration component, line source, 185
Radiation, 3, 9, 148, 253
 factor, 24, 27
 patterns, 291, 302, 305
Radio-astronomy, 76
Radio telescope arrays, 60
Ray, 34
 complex, 34
 fields, 4
 methods, 2, 156
 path, 95
Rayleigh waves, 4, 22, 33, 38, 59, 92, 120, 157, 159, 166, 171, 173, 179, 235, 270
 depth dependence, 42
 existence, 42, 45
 frequency domain calculation, 39
 geometrical decay, 45
 Green's tensor, 39, 49
 speeds, 16, 50
 transient, 41
Ray parameter, 276
Rays, 4, 21, 95
Ray summations, 281
Ray theory, 12, 92, 211, 217, 228, 253, 273
 generalized (GRT), 270, 274, 278, 294, 307, 312
Reciprocity relation, 210
Redundancy
 fixed, 77
 variable, 77
Reflection
 antiplane, 31
 coefficient, 32, 33, 35
Reflectivity methods, 92
Representation
 evaluation, 50
 integral, 4, 34, 47, 50
 theorem, 93, 94, 206, 226
Resonance, 316
Rise times, 97, 100, 105, 110, 126, 214, 247
Roughness, 57, 59, 207
 distribution density, 59
Rupture
 bilateral symmetric, 183, 186
 front, 2, 22, 83
 length, 87
 process, 57, 91
 propagation, 129, 239
 super-shear, 87
 time, 119, 229
 velocity, 2, 60, 61, 83, 100, 110, 232, 250, 261

## S

Scattering, 60, 92, 104, 227
Schwab–Knopoff algorithm, 159
Sedimentary
 basins, 307, 316
 layers, 174, 190
Seismic hazard studies, 183
Seismic intensity, see Intensity
Seismic moment, 1, 155, 168, 181
Seismograms, 57, 99, 103, 252, 268
 radial, 181, 193, 252, 277, 288, 297
 vertical, 182, 186, 200, 277, 288, 297
Semblance, 72, 73
Semi-infinite crack, see Crack
Shear
 crack, 58, 93, 108, 119, 138, 142
 dislocation, 268, 289
 strength, 121
SH equations, 222
 displacement, 315
 problem, 246
 radiation patterns, 303
 response, 290, 294, 298
Signal
 detection, 64, 70
 estimation, 64
Similarity conditions, 97
Slant stacking, 237
Sliding, 5
Sliding frictional stress, 117, 120, 123, 142
Slip, 18, 244
 angles, 107
 antiplane, 18
 distribution, 137
 duration, 156, 194, 316
 function, 95, 155, 239, 255
 inplane, 19
 motions, 93
 rate time, 239, 255
 rate vector, 213
Slowness mapping, 62
SMART 1, 60, 61, 76, 80, 88
Soft rock site, 268, 285
Soil deposits, 156

Source
  complicated, 253
  depth, 141
  dip-slip case, 299, 304
  earthquake, 289
  explosion, 269, 316
  parameterization, 205, 296
  strike-slip case, 299, 304
  theory, 2
  time function, 105, 181, 269
  two-point, 191
Spall zone, 270, 272, 280
Spatial coherency, 60
Spectral amplitude, 1, 65
Spontaneous fault rupture, 108
Spreading factor, 44, 252
Stationary phase, 36, 39, 40, 50
Statistical approach, 75, 196, 213
Steepest descents, 36
Stick-slip, 148
Stochastic models, 5, 93, 104, 146, 147
Stopping, 57, 109, 111, 122, 148, 186
  event, 43
  phases, 59, 201
Stress
  corrosion cracking, 121
  drop, 2, 57, 98, 122, 136, 141
  glut, 207, 210
  intensity factor, 5, 6, 128
  tensor, 211
Stress jump factor, 117
Strong motion, 4
  seismology, 80
Strong motion arrays, *see* Arrays
Strong motion records, 99, 154, 193, 257, 262, 268
  interpretation, 59, 270
Structure minimization, 161
Super-shear rupture velocities, 87, 120, 230
Surface
  acceleration, 62, 156, 217, 279, 312
  sources, 209
  waves, 4, 35, 92, 171, 235
S-wave
  coherency, 82
  large amplitude, 144
  spectra, 85
  window, 81, 83
Synthesis, 95
  mapping, 55, 76, 79
  semi-empirical, 96

Synthetic
  accelerograms, 49
  records, 154
  seismograms, 57, 95, 100, 101, 106, 156, 181, 194, 201, 218, 286
  signals, 170
  strong motions, 201
  velocity, 286

## T

$t^*$, 309
Taiwan, 60
Tectonic stress, 110
Temperature, 142, 148
Temporal convolution, 238
Thin-layer approximation, 31
Three-dimensional
  axisymmetric medium, 219
  geometry, 21
  modeling, 108, 123, 217
  problem, 119
Time domain, 180, 292
  convolution, 238
Torsional motion, 13
Transient Love wave, 37, 38
Transmission coefficient, 13, 280
Transverse
  emission, 27, 32
  particle velocity, 15, 17, 21
Trapezoidal-rule quadrature, 234
Two-dimensional
  finite element, 219
  Green's function, 217
  numerical grid, 296
Two-point source, 191

## U

Uniform layers, 224
Unnormalized correlation, 72

## V

Variational techniques, 164
Velocity
  group, 158, 166, 175
  phase, 166
Vertical strike-slip fault model, 127, 134

Visco-elastic stress, 125
Visibility, 77
Volume sources, 207, 210

## W

Wave
 coherence, 71
 dispersion, 158, 159, 164, 165
 intensity, 55
 mixing, 59, 62
 slowness, 228
Waveform modeling, 62
Wave-number
 frequency numerical integration, 271
 method, 92, 206, 226
 spectrum 59, 82
Weakly inhomogeneous layers, 223
Weibull distribution, 123, 125
Whispering gallery mode, 254, 256
Wiener–Hopf method, 12, 15
WKBJ method, 273, 280, 281